BRADY

Fundamentals of
Fire and Emergency
Services

Fundamentals of Fire and Emergency Services

Jason B. Loyd, MA Ed

Texas Engineering Extension Service (TEEX)
Emergency Services Training Institute (ESTI)
Texas A&M University System

James (J.D.) Richardson, MA Ed

San Antonio College Protective Services Department
Assistant Professor of Fire Science
Coordinator of Emergency Management/Homeland
Security Administration

Pearson
Upper Saddle River, New Jersey
Columbus, Ohio

Library of Congress Cataloging-in-Publication Data

Loyd, Jason B.

   Fundamentals of fire and emergency services/Jason B. Loyd, James D. Richardson.
      p. cm.

   Includes bibliographical references and index.

   ISBN-10: 0-13-171835-5 (alk. paper)

   ISBN-13: 978-0-13-232432-8 (alk. paper)   1. Fire protection.   2. Fire
extinction.   I. Richardson, James D.   II. Title.

   TH9146.L69 2009

   628.9'2—dc22   2008043261

Publisher: Julie Levin Alexander
Publisher's Assistant: Regina Bruno
Senior Acquisitions Editor: Stephen Smith
Associate Editor: Monica Moosang
Development Editor: Marion Waldman, iD8 Publishing Services, Inc.
Editorial Assistant: Heather Luciano
Director of Marketing: Karen Allman
Executive Marketing Manager: Katrin Beacom
Marketing Specialist: Michael Sirinides
Marketing Assistant: Judy Noh
Managing Production Editor: Patrick Walsh
Production Liaison: Julie Li
Production Editor: Lisa S. Garboski, bookworks editorial services
Media Product Manager: Amy Peltier
Media Project Manager: Lorena Cerisano
Manufacturing Manager: Ilene Sanford
Manufacturing Buyer: Pat Brown
Senior Designer Coordinator: Christoper Weigand
Cover Designer: Christoper Weigand
Director, Image Research Center: Melinda Reo
IRC Manager, Rights and Permissions: Zina Arabia
Manager, Visual Research: Beth Brenzel
Manager, Cover Visual Research and Permissions: Karen Sanatar
Image Permission Coordinator: Angelique Sharps
Composition: Aptara, Inc.
Printing and Binding: Edwards Brothers Malloy
Cover Printer: Lehigh/Phoenix Color

Pearson Education Ltd.
Pearson Education Singapore Pte. Ltd.
Pearson Education Canada, Inc.
Pearson Education—Japan
Pearson Education Inc., Upper Saddle River,
   New Jersey

Pearson Education Australia Pty. Limited
Pearson Education North Asia Ltd., Hong Kong
Pearson Educación de Mexico, S.A. de C.V.
Pearson Education Malaysia Pte. Ltd.

Prentice Hall
is an imprint of

10 9 8 7 6 5 4 3
ISBN 13: 978-0-13-171835-7
ISBN 10:   0-13-171835-5

Contents

Foreword   *xiii*

Preface   *xv*

About the Authors   *xix*

Acknowledgments   *xxi*

Chapter 1   *History of the Fire Service*   *1*

Introduction   1
The Evolution of Firefighting Services   2
Early Firefighting Apparatus and Fire Vehicles   6
Personal Protective Equipment and Fire Facilities   10
Fire Service Insignias and Traditions   11
   Red Fire Engine   11
   Badges   11
   Maltese Cross   12
   Dalmatian Dog   12
The Fire Dilemma in the United States   12
The Scope of Fire and Emergency Services Today   13

Chapter 2   *Fire and Emergency Services Career Opportunities*   *17*

Introduction   17
Firefighting Career Opportunities   18
   Fire Cadet   18
   Firefighter   19
   Federal Firefighter   20
   Firefighter/Paramedic   20
   Wildland Firefighter   20
   Aircraft Rescue Firefighter (ARFF)   21
   Fire Apparatus Engineer   22
   Search and Rescue   22
   Industrial Firefighter   24
   Hazardous Materials Specialist   24
   Arson Units   24
Non-Firefighting Careers in Fire Protection   25
   Dispatcher/Communications Operator   25
   Occupational Health and Safety Specialist   26
   Fire Equipment Salesperson   26

v

Fire Extinguisher Service Technician 26
Fire Marshal 26
Public Information Officer (PIO) 27
Fire Prevention Specialist 27
Fire Department Training Specialist 28
Insurance Adjuster 28
Fire Inspector 28
Fire Investigator 28
Private Sector Fire Protection Careers 28
Fire Protection Engineer 28
Contract Firefighter Opportunities (Overseas) 29
Inventor 29
Volunteer Opportunities 30
The Future of Fire and Emergency Careers 30

Chapter 3   *The Selection Process*   *34*

Introduction 34
Preparation for a Firefighter Position 35
Prerequisites 35
Written Test 35
Physical Ability/Agility 37
Interviewing Skills 38
First Interview or Chief's Interview 38
Personal History 40
Medical Exam 40
Background History 40
Hiring Conditions 41
Eligibility List 41
Probationary Period 41
Marketing Yourself for the Firefighting job 41
EMT Status 41
Firefighter I and Other Certifications 42
Volunteer Experience 42
Bilingual Ability 42
Clean Driving Record 42
Excellent Physical Fitness 42
Diversity 42

Chapter 4   *Training and Higher Education*   *45*

Introduction 45
Higher Education 46
Training 46
Fire Department Training or Special Operations Bureaus 48
The U.S. Fire Administration and the National Fire
    Academy 48
Advancement and Promotion 51
Certification 51
Specialty Certifications 51
Training Instructor 53
Training Officer 54

Higher Education   54
   The Fire Science Technology Curriculum   56
   Distance and Online Education   57
Setting Standards for Higher Education in the Fire
   Service   58
   Fire and Emergency Services Higher Education (FESHE)
     Guidelines   59
International Fire Service Accreditation
   Congress   60
Pro Board Accreditation   62

Chapter 5   *Fire Department Resources   65*

Introduction   66
Fire Department Facilities   66
   Administration Buildings   66
   Training Facilities   67
   Fire Stations   68
Fire Apparatus   70
   Pumping Apparatus   71
   Wildland Firefighting Apparatus   72
   Mobile Water Supply Apparatus   73
   Aerial Apparatus   73
Specialized Apparatus   76
   Ambulance Apparatus   77
   Rescue Apparatus   79
   Hazardous Materials Apparatus   80
   Aircraft Rescue and Firefighting Apparatus
     (ARFF)   80
   Fireboats   81
   Firefighting Aircraft   81
Personal Protective Equipment (PPE)   82
   Clothing   82
   Self-Contained Breathing Apparatus   87
   Personal Alert Safety System (PASS)   87
Tools and Other Equipment   88
   Thermal Imaging Cameras   88
   Mobile Data Terminal   89

Chapter 6   *Fire Dynamics   93*

Introduction   94
Atoms and Molecules   94
Matter, Energy, and the Combustion Process   95
   Solid Matter   96
   Liquid Matter   96
   Gaseous Matter   97
Heat Energy   99
   Fahrenheit, Rankine Scales, and British Thermal Units
     (BTUs)   99
   Specific Heat and Latent Heat   100
   Conduction, Convection, and Radiation   100

Chemical Energy   101
   Heat of Combustion or Caloric
     Value   102
   Spontaneous Heatings   102
   Heat of Decomposition   103
   Heat of Solution   103
   Heat of Reaction   104
Electrical Energy   105
   Forms of Electrical Heat Energy   106
Mechanical Energy   109
   Friction   109
   Compression   109
   Combustion Process   110
   Burning Process   110
Fire Classification   112
   Class A: Ordinary Combustibles   112
   Class B: Flammable Liquids   113
   Class C: Energized Electrical Fires   113
   Class D: Flammable Metals   113
   Class K: High-Temperature Cooking Oils   113

Chapter 7   *Building Construction*   *118*

Introduction   118
Loads and Forces   119
   Load Imposition   120
   Axial Loads   120
   Eccentric Loads   120
   Torsional Loads   121
Structural Components   122
   Foundations   122
   Walls   123
   Wood Trusses and I-Joists   123
   Ceiling Joists   127
   Roof Configurations   127
Construction Types   129
   Type I Construction   129
   Type II Construction   134
   Type III Construction   134
   Type IV Construction   138
   Type V Construction   140

Chapter 8   *Fire Prevention Codes and Ordinances*   *147*

Introduction   147
Building Codes and Fire Prevention Codes   148
History of Model Codes in the United States   149
   Building Officials and Code Administrators International
     (BOCA)   150
   International Conference of Building Officials (ICBO)   150

Southern Building Code Congress International
(SBCCI) 150
International Code Council (ICC) 150
National Fire Protection Association (NFPA) 150
Historical Basis for the Legal Authority of Code
Enforcement 151
Dillon's Rule, the Cooley Doctrine,
and *Atkins v. Kansas* 152
Warrantless Entry 152
Right of Entry 153
Types of Codes and Code Adoption 154
Minimum Codes versus Mini-Maxi Codes 154
Model Codes 155
Code Administration and Enforcement 155
Documentation 156
Summons and Citations 156

Chapter 9   *Fire Protection Systems*   *160*

Introduction 161
Water Supplies and Distribution Systems 161
Sources of Water Supply 161
Water System Components 162
Pipe Size Requirements 167
Water Distribution Models 168
Fire Hydrants 169
The Base-Valve Dry-Barrel Hydrant 169
The Wet-Barrel (California) Hydrant 170
Dry Hydrants 170
Hydrant Maintenance and Flow Testing 172
Standpipe Systems 172
Class I System 173
Class II System 173
Class III System 174
Types of Standpipe Systems 175
Standpipe Flow Rate Requirements 175
Pressure-Regulating Devices 176
Automatic Sprinkler Systems 176
Automatic Sprinkler System Components 178
Types of Automatic Sprinkler Systems 185
Signaling Systems 190
Types of Signaling Systems 190
Automatic Detection Systems 192
Heat Detectors 193
Smoke Detectors 194
Special Extinguishing Systems 197
Halon Systems 197
Carbon Dioxide Systems 198
Dry Chemical Systems 200
Foam Systems 201

Chapter 10   *Organizational Structure and Emergency Incident Management Systems*   *207*

Introduction   207
Fire Department Basics   207
   Types of Fire Departments   207
   Organization and Rank Structure   209
   Rank Structure   209
Incident Command Systems   210
   FIRESCOPE   211
   Fire Ground Command System (FGC)   211
   FIRESCOPE versus the FGC   212
Incident Management System (IMS)   212
   Functional Sectors of the IMS Command System   212
   Administration and Finance Sector   214
   National Interagency Incident Management System (NIIMS)   214
   National Fire Service Incident Management System Consortium (NFSIMSC)   215
   Incident Command Systems—A Report Card   215
National Incident Management System   216

Chapter 11   *Preincident Planning, Fire Strategy, and Tactics*   *223*

Introduction   223
Strategy   224
   Preincident Plan   224
   Water Supply   226
   Apparatus and Personnel   226
   Occupancy   228
   Weather   228
   Topography   229
   Building Construction   231
   Auxiliary Appliances   231
   Life-Safety Hazards   231
   Time of Day   232
   Street Conditions   232
   Hazardous Materials   233
   Location and Extent   234
   Communications   234
   Command Structure   235
Tactics   236
   Tactical Decision Making   236

Chapter 12   *Support for Fire Emergency Services: Their Vital Functions*   *242*

Introduction   242
National and International Organizations   243
Federal Organizations   245
State Organizations   247
Local Organizations   248
Periodical Publications   248

Support Staff in the Fire and Emergency
    Services   248
    Dispatch   249
    Hazardous Materials (HazMat) Teams   249
    Arson Units   249
    Graphic Arts and Maps   249
    Information Systems   250
    Repair Garages   250
    Radio Shops   250
    Weather Forecasters   250

Appendix A   *On Scene Suggested Answers*   *253*

Appendix B   *Review Question Answers*   *257*

Appendix C   *Stop and Think Suggested Answers*   *265*

Appendix D   *Standard Operating Procedures (SOPs)*   *271*

Glossary   *275*

Index   *285*

Foreword

It has been my pleasure to know author Jason Loyd for nearly five years. I first met Jason, in an academic setting, when he was the Director of Fire Science Technology at Weatherford College. I remember meeting him and thinking that this was a young man full of fresh ideas who was passionate about wanting to build a top-notch Fire Science program. Our friendship started while I was enrolled in a Fire Inspector certification course at the college. Jason managed to convince me to teach in their Regional Fire Academy.

It gives me great pleasure to introduce Jason's textbook because of his integrity and commitment to the fire service. This textbook contains easily understood concepts pertaining to the fundamentals of fire and emergency services. The Fire and Emergency Services Higher Education (FESHE) curriculum is followed, and each chapter outlines specific learning objectives. It also covers the vast responsibilities of the modern firefighter and presents the information in an easy-to-read format. Firefighters not only respond to structure fires but also must handle hazardous materials, emergency medical services (EMS), vehicle fires, and rescue operations, to name a few. As a fire chief, educator, and lifelong learner, I value the importance of higher education. Now more than ever, a good education is the foundation of any career. In this textbook, students new to the field will learn the basics of what it takes to become a firefighter, and career firefighters/fire officers will learn new or updated material.

I believe that this textbook will be a valuable resource for individuals thinking about pursuing a fire service career as well as a reference for experienced firefighters.

Eddie Burns, Sr., Fire Chief
Dallas Fire-Rescue

Preface

Prior to 1970, the scope of fire service was narrow. The main focus was on fire prevention, arson investigation, and fire suppression. After 1970, the nation's fire service began adding duties to increase the level of public protection. Emergency Medical Services (EMS) was integrated into the fire service, followed by Heavy Rescue and Hazardous Materials Response teams. The increased responsibilities required a change in applicants' requirements. Being able to perform the physical aspects of the fire service was no longer adequate. Education became an important prerequisite for employment. Men and women seeking careers in the fire service were now expected to bring a level of critical thinking and problem-solving abilities. The demand for these skills increased greatly after the events of September 11, 2001, when the fire service incorporated emergency management into its list of responsibilities.

The modern fire service is a dynamic profession with ever-increasing challenges. This text is designed to introduce students to the firefighting profession as well as remind career firefighters that fire service is a profession requiring continuous learning.

HOW TO USE THIS TEXT

Throughout the text there are several features that enhance the material presented. Stop and Think questions ask the student to pause, consider what they have just learned, and apply the concept. General safety icons point out areas where extra precaution is required. The icon of the National Fallen Firefighter's Foundation is used appropriately to promote the 16 life safety initiatives that ensure that "Everyone Goes Home." Each chapter concludes with an On Scene bridging the gap between classroom learning and experience. Review questions reinforce the chapter content by providing a final check that the student has mastered the chapter content.

Chapter 1: History of the Fire Service

This chapter introduces students to the history of the fire service,—how it began, where it is now, and what the future of the fire service profession may be. This chapter covers the ever-changing role and needs of the fire service. The instructor may utilize particular historical information pertaining to the audience to personalize the lecture and link the past to the future fire service.

Chapter 2: Fire and Emergency Services Career Opportunities

This chapter will acquaint new and experienced students with the variety of career opportunities that the fire service offers. The basic role of a fire service worker has evolved into many aspects and new careers over the ages. This chapter focuses on the many specialized opportunities available to a firefighter. It also notes that although a specific job may have various titles, depending on the area or locale, the job description remains the same. The text provides basic information that the instructor can expand on to make the information more relevant to the student's environment.

Chapter 3: The Selection Process

This chapter gives an overview of the procedure that is typically followed in the fire service. It is recommended that the instructor give real examples of this process and possibly carry out a mock interview to illustrate the significance of learning this process. This chapter emphasizes that, no matter what a person's background is, the selection process is a structured venue. The more prepared an applicant is for the job, the more likely he or she will get the job.

Chapter 4: Training and Higher Education

This chapter can be used to impress on new students that training and education does not end when they graduate from the fire academy. Experienced students can also benefit from this chapter by learning different avenues that will allow them to further their career in ways that may not have been available before. Continual training and education are paramount for a safe department and for individuals to move through the fire service ranks. To give relevance to the students' environment, instructors can describe firefighters they know who have used training and education to their advantage.

Chapter 5: Fire Department Resources

This chapter can be used either to introduce new students to a variety of apparatus, facilities, and equipment that are used in the fire service to day/today or the instructor's focus on the resources that are available in the instructor's area. The text provides basic information that the instructor can expand on to make the information more relevant to the students' environment.

Chapter 6: Fire Dynamics

This chapter is an overview of the chemistry and physics of fire and the phenomena that occur during the combustion process. The instructor may utilize the entire chapter to give new students an overview of fire dynamics or focus on certain topics to emphasize areas of concern to the more experienced student. The instructor may choose to use demonstrations to enhance the lecture.

Chapter 7: Building Construction

The focus of this chapter is to enlighten the student on the hazards that buildings pose under fire conditions. The emphasis is on the dangers of a building collapse and how to "read" a building during fire operations. It is recommended that the instructor obtain samples of various forms of structural elements for the student to view and manipulate. The information presented is appropriate for new and experienced students.

Chapter 8: Fire Prevention Codes and Ordinances

This chapter is designed to introduce students to the codes and ordinances that regulate the actions of the fire service. Students will learn why the codes and ordinances were created, the historical events that inspired them, and the constitutional challenges that set judiciary precedents. It is recommended that the instructor utilize the entire chapter to maintain a smooth, chronological flow of information.

Chapter 9: Fire Protection Systems

This chapter introduces the various forms of water-based automatic extinguishing systems, public water supplies, special extinguishing systems, and fire detection and alarm systems. The instructor may choose the level of instruction based on the level of experience of the students.

Chapter 10: Organizational Structure and Emergency Incident Management Systems

Using this chapter, the instructor may choose to introduce new students to the fire service organizational and rank structure before detailing the history and functions of incident management systems. For more experienced students, the instructor may want to give a brief history of how the incident management system was developed before moving to the National Incident Management System (NIMS). The dynamic nature of this system should be emphasized.

Chapter 11: Preincident Planning, Fire Strategy, and Tactics

When covering this chapter, the instructor should continually emphasize the importance of preincident planning. This chapter will be best utilized with scenario-based problem-solving activities based on the information from the text. This will challenge new students and maintain the interest of experienced students.

Chapter 12: Support for Fire and Emergency Services: Their Vital Functions

The focus of this chapter is to acknowledge the many roles that members of other organizations play in the fire service, even if they are not firefighters. Support comes in countless forms when it comes to the fire service, and each function is fundamental in maintaining a public service. The instructor may choose to elaborate on certain support systems that will give the student a better grasp of their network capabilities.

About the Authors

Jason B. Loyd started his career in 1996 as a volunteer Firefighter/EMT with Medina Valley in Castroville, Texas, while also attending San Antonio College. In 1998 he completed the San Antonio College Regional Fire Academy and was offered a job overseas as a contract firefighter on a U.S. military base. He was stationed in Hungary for two years and then promoted to Bosnia for two years. In Bosnia, he was responsible for training local national Bosnians to the Department of Defense firefighter standards for Fire Fighter I & II, Hazardous Materials—Awareness and Operations, Airport Fire Fighter, Driver/Operator-Pumper and Driver/Operator-ARFF in addition to being a crew chief.

Mr. Loyd earned his associate of applied science degree from San Antonio College in Fire Science and his baccalaureate degree from Empire State College in Business Management and Economics with a concentration in Fire Service Administration. Mr. Loyd went through the National Fire Academy Distance Degree Program (DDP) to obtain his baccalaureate degree. After completing his BS degree in 2001, he was offered a full time position at Weatherford College over Fire Science Technology. Mr. Loyd and his wife then relocated back to the United States from Bosnia where he was responsible for two associate of applied science degrees, regional fire academy, and all fire service training courses offered at the college. He holds various certifications with the Department of Defense (DoD) and Texas Commission on Fire Protection (TCFP) including Master Fire Education Specialist, Structure Fire Fighter, Aircraft Rescue Fire Fighter, Driver/Operator-Pumper, Fire Instructor, Fire Inspector, Fire Investigator, Fire Officer II, and field examiner. Mr. Loyd was awarded the Texas Association of Fire Educators George Hughes Instructor of the Year Award for 2003 and was recognized in the Who's Who Among America's Teachers. In 2004 he was appointed to the Texas Commission on Fire Protection Curriculum and Testing Committee and in the same year appointed to the International Fire Service Accreditation Congress (IFSAC) Degree Assembly Board of Governors. In 2005 he completed his Master of Arts degree in Education with a concentration in Higher Education. Furthermore, he remains a reserve firefighter with the City of Weatherford Fire Department.

In May 2007, Jason was hired as an instructor for the private sector group the Emergency Services Training Institute (ESTI) Division of the Texas Engineering Extension Service (TEEX), a part of the Texas A&M University System. This is a world-renowned fire training facility and is a new chapter in his fire service career.

He has been married to his wife Melissa since 1993 who also works at Texas A&M University, in the Chemistry Department. They have two daughters, Keeley and Arabella.

James D. (J.D.) Richardson joined the United State Air Force in February of 1968. He was trained as a munitions specialist and served one year in Vietnam. After returning from Vietnam in 1971 he met and married, his beautiful wife, Dianne. In 1972 he was honorably discharged with the rank of sergeant. In January, 1973 he joined the Houston Fire Department. In 1974, J.D. enrolled in the Fire Science Program at Houston Community College. Before completing his Associates Degree, he transferred to the University of Houston and achieved a Bachelor of Science in Technology degree in December of 1977.

J.D. was promoted to chauffeur (engineer/operator) in 1979 and was assigned to an ambulance as an EMT. In 1980 he was re-assigned to the newly created Hazardous Materials/Heavy Rescue Team. When the team was separated into two entities in 1981, J.D. was assigned as the engineer/operator for the Heavy Rescue Team. He remained with heavy rescue until 1989. From 1989 until 1991 J.D. served as an engineer/operator in the Suppression Division. In 1991 he was promoted to the rank of captain.

In 1993 J.D. developed a hydraulic rescue tool training program that was used to train more than 500 Houston firefighters in the use of the Hurst rescue tool. He was temporarily assigned to the Houston Fire Department Training Division to work with the rescue teams in developing rescue training programs. He returned to the Suppression Division in 1995 and was chosen to be a member of a hazardous materials training Team with the International Association of Fire Fighters. As a member of this team, J.D. helped initiate the 24-hour hazardous materials awareness and operations program now being mandated for the fire service.

Captain J.D. Richardson retired from the Houston Fire Department in January, of 1997 and began a new career as an instructor of fire science with San Antonio College. During his tenure with the college, J.D. was instrumental in developing the Associate of Applied Science Degree in Emergency Medical Service in 1998. In 2003 he initiated the first Associate of Applied Science in Emergency Management and Homeland Security Administration degree in the state of Texas. J.D. served as director of the Fire Science Degree Program from 1999 to 2001 and chairperson of the Protective Services Department from 2001 until 2005. He has achieved the rank of assistant professor. J.D. and his wonderful wife, Dianne, have been married for 37 years. They have one son, R.J., a lovely daughter-in-law, Kate, and a beautiful grand-daughter, Cadwyn.

Acknowledgments

The most important person I want to acknowledge is my loving wife, Melissa, for her unwavering support during this endeavor. This text would never have been possible without Melissa's support. I would like to personally thank Katrin Beacom from Brady for giving me the wonderful opportunity to write this text. In addition I would like to thank the entire Brady staff and Marion Waldman and her team who have worked so hard to see this project through to completion. Many thanks are due my coauthor, Professor James D. Richardson from San Antonio College, for his encouragement to never stop learning and to achieve the highest education possible to prepare for the future. I thank the staff at TEEX–Texas A&M System at the Brayton Fire Field for their help in providing photos for the text. Stephen Malley, Department Chair of Public Safety Professions, Weatherford College, never refused to review and give candid feedback on chapters I wrote during this project. Also, Weatherford Fire Department Fire Chief George Teague and, especially, Fire Marshal Kurt Harris consistently lifted me up by encouraging me to finish this text. I am grateful to the entire staff and firefighters at the Weatherford Fire Department for their encouragement as well during this project. Special thanks to Mrs. Bernice Roberts for her support and help in the initial stages of this text with information on the history of the Weatherford Fire Department. This has truly been the most rewarding project with which I have ever been involved.

Finally, I would like to thank the reviewers in helping shape this text into the final version you will read and learn from. I know I probably missed many other people who have given me support and information throughout this project, and to them—thank you.

—Jason B. Loyd

I would like to thank my wife, Dianne, for her patience and understanding while I worked on this project. She has always been my rock and my foundation, and I could not have completed this book without her support. She is, and always will be, the love of my life.

I would also like to thank the following:

Jim Cox, Ansul Incorporated
Arlene Hoffman, Ansul and Tyco Building and Fire Products
Captain Willis Lamm (Retired), Water Supply Officer, Moraga-Orinda, California, Fire Department
Mike Pickett, author of Pickett's Primers and former director of the San Antonio Fire Science Program
Assistant Chief of Training Joe Martinez and the United States Air Force, Randolph Air Force Base

Fire Chief Ross Wallace and the officers and firefighters of the Universal City (TX) Fire Department

Former Fire Chief Scott Lee and the officers and firefighters of the Selma (TX) Fire Department

Lieutenant Richard (Rick) Bachmeier (Retired), San Antonio (TX) Fire Department

Captain John De La Garza (Retired), San Antonio (TX) Fire Department

I would especially like to make a special acknowledgment to Captain Elmer Wayne (Red) Blevins (Retired), who served 52 years and 9 months with the Houston Fire Department. He was an inspiration to three generations of firefighters. God bless you, Red.

—James D. (J. D.) Richardson

REVIEWERS

Jerry A. Asbach, CFEI
Fire Science Technology Chair
Augusta Technical College Augusta, GA

David S. Becker
EMS Program Director
Sanford-Brown College
St. Louis, MO

Richard L. Bennett
Associate Professor of Fire Protection
The University of Akron
Akron, OH

Craig Bryan, Captain, OFE, NREMT-P,
EMS/Fire Instructor
City of Forest Park Fire Department
Division of Training and Education
Forest Park, OH

Greg Burroughs
Program Chair, Fire Protection Technology
Southeast Community College
Lincoln, NE

Michael E. Kavanaugh
Director, Fire Science
Central New Mexico Community College
Albuquerque, NM

Randall Griffin, Lieutenant
DeWitt (NY) Fire District
DeWitt, NY

Richard W. Hally
Professor of Fire Science
University of New Haven

Lieutenant
West Shore Fire Department, West Haven, CT
Haz-Mat Manager of CT Urban Search and Rescue TF-1

Lonnie Inzer, MLS
Fire Science Technology Degree Program Director
Pikes Peak Community College
Colorado Springs, CO

Gary Kistner, Program Coordinator
Fire Service Management
Southern Illinois University
Carbondale, IL

Judith Kuleta
Program Director, Fire Sciences
Bellevue Community College
Bellevue, WA

Jeffrey Lindsey, PhD
Fire Chief (Ret.)
Estero Fire Rescue

Assistant Professor
George Washington University
Washington, DC

Stephen S. Malley
Department Chair, Public Safety Professions
Weatherford College
Weatherford, TX

Matt Marcarelli, Lieutenant
City of New Haven Fire Department
Connecticut Fire Academy
Middlesex County Fire School
Northford, CT

William C. Opsitnik, Captain
Liberty Fire Department
Youngstown, OH

John Rinard
TEEX EMS Program Supervisor
Milano, TX

Thomas Y. Smith, Sr., MS-EFSL, FO-IV
Fire Science Program Director
West Georgia Technical College
LaGrange, GA

Lloyd H. Stanley
Department Chair, Fire Protection
Technology
Guilford Technical Community College
Jamestown, NC

Joshua M. Sunde
Sierra Community College
Rocklin, CA

Tim Swaim
Fire Marshal
Kernersville Fire Department
Kernersville, NC

Gary D. Young
Training Officer/Captain
Copperas Cove Fire Department
Copperas Cove, TX

16 Firefighter Life Safety Initiatives

 In March, 2004, a Firefighter Life Safety Summit took place in order to address the need for reform within the fire service profession. As a result of this meeting, the 16 firefighter initiatives were developed. The National Fallen Firefighter's Foundation promotes programs and resources that facilitate in preventing line–of–duty deaths and injuries through the Everyone Goes Home website. These training programs like Firefighter Life Safety Resources Kit–Volume 1, 2, 3, and Courage to Be Safe to name a few can be downloaded off the website. Examples of the 16 Firefighter Life Safety Initiatives are stressed throughout this textbook in order to show how these initiatives can be applied in practice. Consider what other initiatives you can apply to practice as you read through the chapters.

The following list includes each of the 16 initiatives and how they are applied in the chapters:

1. Define and advocate the need for a cultural change within the fire service relating to safety; incorporating leadership, management, supervision, accountability and personal responsibility.

   Chapter 4 discusses the challenges a firefighter has had to adapt to in order to keep up with today's society while continuing to provide the best service to their community/customer.

2. Enhance the personal and organizational accountability for health and safety throughout the fire service.

   Chapter 3 illustrates the need to be in excellent physical fitness. This aaplie snot only to individuals wanting to become firefighters but to current volunteer and career firefighters as well. For more information on this topic refer to NFPA 1583, *Standard on Health-Related Fitness Programs for Fire Department Members*, 2008 Edition.

3. Focus greater attention on the integration of risk management with incident management at all levels, including strategic, tactical, and planning responsibilities.

   Chapter 11 discusses preincident planning, fire strategy, and tactics which allows the instructor to facilitate talking points to student's concerning greater attention on integration of risk management in the fire service today and in the future.

4. All firefighters must be empowered to stop unsafe practices.

   Chapter 10 discusses not only organizational structure but emergency incident management systems as well. Instructors should emphasize that everyone on the fireground is empowered to stop an unsafe act. For more information on this topic refer to NFPA 1521, *Standard for Fire Department Safety Officer, 2008 Edition*.

5. Develop and implement national standards for training, qualifications, and certification (including regular recertification) that are equally applicable to all firefighters based on the duties they are expected to perform.

   This initiative in Chapter 4 emphasizes the importance of training and higher education in fire and emergency services. Ultimately this initiative is paramount for national standardization for emergency personnel in all ranks and levels alike. Through lessons learned from 9/11, Hurricanes Katrina and Rita we have trained and educated our responders across the nation to be better prepared to mitigate these types of large-scale emergencies. It's important that our future fire and emergency services leaders have the knowledge required to perform to a certain level set by these national standards for the safety of our responders no matter where they're responding from (local or state).

6. Develop and implement national medical and physical fitness standards that are equally applicable to all firefighters, based on the duties they are expected to perform.

   This initiative in Chapter 3 discusses the selection process and the importance of national standards for medical and physical fitness for fire department members. For more information on this topic refer to NFPA 1583, *Standard on Health-Related Fitness Programs for Fire Department Members,* 2008 Edition.

7. Create a national research agenda and data collection system that relates to the initiatives.

   Chapter 12 discusses support functions from different organizations. Instructors should introduce students to research, data collection systems, and the benefits to the department as a whole.

8. Utilize available technology wherever it can produce higher levels of health and safety.

   Chapter 2 discusses the career opportunities in occupational health and safety. An occupational health and safety specialist is an invaluable asset to the fire and emergency services industry. This career position might have a different title like health and safety officer (HSO) or incident safety officer (ISO). Regardless of the title the key point is to promote higher levels of health and safety. For more information on this topic refer to NFPA 1521, *Standard for Fire Department Safety Officer,* 2008 Edition.

9. Thoroughly investigate all firefighter fatalities, injuries, and near misses.

   This initiative in Chapter 2 discusses the career of fire investigators. This role can be different from locale to locale and even state to state. Regardless of the position title it is of the utmost importance that every firefighter fatality, injury, or near miss be documented so others can learn from these events. Although this chapter may touch on basic information the instructor should expand on this subject to make the information more relevant to the students' environment.

10. Grant programs should support the implementation of safe practices and/or mandate safe practices as an eligibility requirement.

    This initiative in Chapter 12 focuses on several support organizations and the vital role they play in the fire and emergency services industry. There are grant programs that assist fire departments with the essential funds to implement safe practices programs, although certain requirements may apply before a department is eligible for this type of assistance.

11. National standards for emergency response policies and procedures should be developed and championed.

This initiative in Chapter 10 discusses the organizational structure and emergency incident management systems concerning policies and procedures for emergency response.

12. National protocols for response to violent incidents should be developed and championed.

    This initiative in Chapter 11 emphasizes the importance of preincident planning. It is vital that instructors stress to students about responding to violent incidents and the need to have national protocols in place before the incident occurs.

13. Firefighters and their families must have access to counseling and psychological support.

    This initiative in Chapter 2 discusses the many dangers firefighters face today. Because of the added responsibilities of the fire and emergency services personnel it is important that firefighters have access to proper counseling and psychological care if needed.

14. Public education must receive more resources and be championed as a critical fire and life safety program.

    Chapter 8 on fire prevention codes and ordinances, discusses the importance of life-safety education programs. It is recommended that instructors utilize the entire chapter so students will have a greater appreciation of fire prevention codes and ordinances, public education, and life-safety programs resources available.

15. Advocacy must be strengthened for the enforcement of codes and the installation of home fire sprinklers.

    This initiative in Chapter 8 discusses fire prevention codes and ordinances and why they were created. Instructors should emphasize to students the importance of becoming an advocate for enforcement of codes and the installation of home fire sprinklers.

16. Safety must be a primary consideration in the design of apparatus and equipment.

    This initiative in Chapter 2 discusses not only career opportunities in the fire and emergency services field but also new technology geared towards safety for responders. For example, new technology in apparatus and equipment has enhanced firefighter safety like Global Positioning Systems (GPS), Automatic Vehicle Location (AVL), and Geographical Information Systems (GIS) to name a few.

The following grid outlines Principles of Emergency Services course requirements and where specific content can be located within this text:

Course Requirements	1	2	3	4	5	6	7	8	9	10	11	12
Describe and discuss the components of the history and philosophy of the modern day fire service.	X											
Analyze the basic components of fire as a chemical reaction, and the major phases of fire; and examine the main factors that influence fire spread and fire behavior.						X	X					
Differentiate between fire service training and education, fire protection certificate program and a fire service degree program; and explain the value of education in the fire service.				X								
List and describe the major organizations that provide emergency response service and illustrate how they interrelate.										X		
Identify fire protection and emergency service careers in both the public and in the private sector.		X	X									
Synthesize the role of national, state, and local support organizations in fire protection and emergency services.										X		X
Discuss and describe the scope, purpose, and organizational structure of fire and emergency services.										X		
Describe the common types of fire and emergency services facilities, equipment, and apparatus.					X							
Compare and contrast effective management concepts for various emergency situations.										X		
Identify and explain the components of fire prevention including code enforcement, public information, and public and private fire protection systems.									X		X	

BRADY

Fundamentals of Fire and Emergency Services

History of the Fire Service

1 CHAPTER

Key Terms

apparatus Aldini, p. 10
conflagration, p. 2
couvre feu, p. 3
EMT, p. 14
familia publica, p. 2
FESHE, p. 14
fire science, p. 2

fire wardens, p. 4
firemark, p. 4
fireplug, p. 5
The Great Fire of
  Boston, p. 8
The Great Fire of
  London, p. 3

hand pumper, p. 6
Homeland Security, p. 13
Maltese cross, p. 11
NIMS, p. 14
NREMT, p. 14
siphona, p. 6
steam engine, p. 7

Objectives

After reading this chapter, you should be able to:

- Describe the evolution of firefighting in the United States and other countries.
- Explain the history of early firefighting organizations.
- Describe early fire vehicles.
- Describe early personal protective equipment.
- Understand fire service traditions.
- Describe the origins of fire facilities.
- Explain the fire dilemma in the United States and its causes.
- Discuss the changing role of the modern firefighter.

◆ INTRODUCTION

Early humans experienced fire as a great and powerful mystery, a force to be feared. Gradually, as their knowledge of fire's properties grew, humans began to appreciate its potential as a tool for safety and protection against cold, darkness, and other dangers posed by harsh environments. The ability to control fire eventually was an important factor in the formation of stable communities of individual family dwellings and businesses. Fear slowly turned into respect for this great force. However, as humans became more and more accustomed to using fire in their everyday lives, their carelessness caused accidents that resulted in death and destruction. As communities grew in size, so did the need for protective measures against out-of-control fires. Unfortunately, most safeguards were not enacted until after a tragic loss of life and property made the need for prevention obvious.

Fire remains something of a mystery to this day, but over the centuries, the study of fire science has disclosed many of its secrets. The work of fighting fires has evolved into a highly respected occupation for many people, both full-time professionals and dedicated volunteers. The history of firefighting is a fascinating, inspiring story. It is the record of how our inventive and courageous forefathers fought to control one of nature's most useful and potentially most destructive forces. Their efforts laid the foundations on which our modern fire service is built and pioneered the field of study that we call **fire science**.

fire science

Study of the behavior, effects, and control of fire.

The term *field of study* is used because fire science is not a single discipline. Topics studied in a fire science curriculum include, but are not limited to, protective systems, building construction, fire codes and ordinances, hazardous materials, chemistry, strategy and tactics, and, of course, fire behavior. Each of these areas has a history that explains its importance to the study of fire science. Each became part of this field of study because lives had been lost in the past, inspiring efforts to prevent similar losses in the future. The fire science student must learn not only the when and where of developments in the field but also the why. We will begin exploring the "why" by looking back at the evolutionary history of firefighting services and firefighting technology.

◆ THE EVOLUTION OF FIREFIGHTING SERVICES

conflagration

A fire that increases in size and spreads beyond human-made and natural barriers.

familia publica

"Servants of the commonwealth." These men were strategically positioned near the city gates to protect the city from fire.

The earliest recorded effort to organize a firefighting service dates to the Roman Republic. In the first century BC, the crowded city of Rome was subject to repeated **conflagrations** that resulted in massive loss of life and property damage. A body of approximately 600 public slaves (the *familia publica*), strategically positioned at the gates of the city, were assigned the task of alerting the public to existing fires and fighting the conflagrations (Dio Cassius, 54.2–3). They were, however, notoriously ineffective and slow to react; perhaps in the interests of preserving their own safety.[1] After a particularly devastating fire in 6 AD, Emperor Augustus organized a corps of 7000 freedmen, the *vigiles* (watchmen), who patrolled the streets at night. They began to combat fires using bucket brigades, as well as poles and hooks to tear down buildings in advance of the flames. It is generally thought that this is where the "hook" in "hook and ladder" originated.

Rome continued to suffer from serious fires, most notably the July 19, 64 AD, conflagration that started near the Circus Maximus and eventually destroyed two-thirds of the city. Emperor Nero was blamed; although it is unlikely that Nero actually ignited the blaze, some historians believe that he did permit it to burn unchecked. Evidence exists that at least one Roman citizen profited richly from this fire, buying properties in advance of the flames and using teams of slaves to defend his recent acquisitions from being consumed.[2] Whatever his faults, Nero was apparently a man of some vision and intelligence who recognized the dangers of the unregulated construction that was the norm in Rome before his reign. Much wealth and many resources had been expended on the construction of enduring public edifices, but almost all other structures were built of wood without care for safety standards. Nero can be credited with rebuilding Rome in accordance with sound principles of construction, sanitation, and utility. From 64 AD until the fall of the Roman Empire, both public and private building projects were closely regulated.

It was not until the 11th century that fire prevention was again addressed. William the Conqueror of England instituted one of the first recorded fire prevention regulations, which required that candles and all fires used for heating and cooking must be extinguished at nightfall. Every day about 8 PM, a bell was rung, which was the signal to "cover the fire" (*couvre feu* in William's native French). ***Couvre feu*** gradually was corrupted to the English word *curfew*. The curfew bell was utilized in parts of England and France for over 800 years, long after its original purpose had been forgotten. However, this simple precaution was not always adequate, and the archives of French towns and churches contain records of many disastrous fires and rebuilding works (see Figure 1.1).

couvre feu

A French name for a metal lid used to cover an open hearth. The English word *curfew* was derived from this name.

In 60 AD, four years before Nero's fire in Rome, a fire ignited by rebellious natives leveled the fledgling Roman town of Londinium. This was the first of many conflagrations that would destroy large sections of London (as the town would be later be named). History preserves a record of major blazes in 675 and 1087, three in the course of the 12th century, and a massive blaze in 1212. These tragedies pale, however, in comparison to the blaze that has come to be known as **The Great Fire of London**. The Great Fire started on September 2, 1666, in the royal baker's shop on Pudding Lane. It burned relentlessly for four days, consuming about 430 acres (80% of the city's center), and destroying 13,000 homes, 89 churches, and 52 Guild Halls. Tens of thousands were left homeless and financially ruined. Much of central London today shows the results of the reconstruction ordered by King Charles II, which called for

The Great Fire of London

A devastating fire in central London that was considered one of the major events in the history of England.

FIGURE 1.1 ◆ The bucket brigade is a part of American history. This drawing captures an early settlement working together as a bucket brigade to keep fire from destroying the settlement.

wider streets and brick buildings to replace the earlier structures built of highly flammable wood and pitch.

One result of the Great Fire was the formation of what may well have been the world's first insurance company. Dubbed "The Insurance Office," it was the first of a number of such companies that took on the responsibility not only of indemnifying the insured against losses but also of actually fighting fires. Each company formed its own private fire brigade that would battle the flames—but only in those buildings insured by the company. Insured buildings were marked with the company's badge, called a **firemark**, and any building without this identifying mark was left to burn. Clearly, this was not the best solution to the problem of protecting the public.

The fire problem was not limited to European countries, of course. In 1631, a major fire broke out in Boston, which prompted the implementation of fire regulations banning thatched roofs and wooden chimneys. Colonial laws also required each house to have a bucket of water on its front stoop for the use of bucket brigades. In 1648, New Amsterdam (later renamed New York) created an organization of volunteer **fire wardens** who patrolled the city, inspected chimneys, and issued fines for noncompliance. Funds collected from fines were used to purchase new fire equipment. New Amsterdam also appointed a so-called Rattle Watch—men dressed in long capes who would patrol the city's streets by night. They carried wooden rattles to sound an alarm, rousing citizens to form bucket brigades whenever a fire broke out (see Figure 1.2).

Public interest in firefighting continued to increase. In 1679, Boston took the first step toward what has become today's professional fire department, establishing the first paid firefighting service. The firefighters were part-time employees of the city who were assisted by volunteer organizations known as mutual fire societies. However, it was not until 1736 that Benjamin Franklin organized what is known today as the first formal volunteer fire company. This volunteer assembly was named the Union Volunteer Fire Company of Philadelphia.[3] The amount of staffing and skill necessary for firefighting eventually prompted Franklin, like the citizens of Boston, to

firemark

Metal marker that used to be produced by insurance companies for identifying their policyholders' properties.

fire wardens

Volunteers who patrolled the cities, inspected chimneys, and issued fines for noncompliance. Funds collected from fines were used to purchase new fire equipment.

FIGURE 1.2 ◆ The fire rattle was used to alert the town of an emergency.

FIGURE 1.3 ◆ American Founding Father Benjamin Franklin wears a fire helmet in a portrait.

institute a paid fire company, but the concept of volunteer companies had caught on and was spreading across America (see Figure 1.3).

Men (this was before women joined the fire service) joined their local volunteer company for social reasons as well as to serve the public and to protect and their homes and towns. Membership was considered prestigious, the mark of a public-spirited citizen, and would earn a man community respect. The result was long waiting lists of eager volunteers. Despite enormous advances in all areas of firefighting, the volunteer firefighter remains a proud American tradition to this day.

Even after the formation of paid fire companies, there were still disagreements and fights over territory, because insurance money was awarded to the company that was first on the scene of the fire.[4] The situation was complicated by the nature of water delivery in the towns. Public water systems consisted of mains constructed of wooden logs hollowed out to carry water down a street. Approximately every half block, wooden plugs were inserted in the logs. These plugs could be removed to access the water supply for firefighting, which has given us the term **fireplug**. Some fire companies would hire saboteurs to race ahead to the scene of the fire and hide the plug from competing fire companies. These miscreants were called "plug uglies." Even after fire hydrants replaced the fireplug, sabotaging a responding fire company's access to the water supply continued. New York City companies were famous for sending runners out to fires with large barrels to cover the hydrant closest to the fire in advance of the engines. Fights would often break out between the runners and the responding fire companies for the right to fight the fire, which continued to rage while they squabbled. Town parades were another occasion for spirited competition as each company marched down Main Street in its own elaborate, colorful

fireplug

A wooden plug inserted in logs that carried the town's water supply approximately every half block. These plugs could be removed to access the water supply for firefighting.

FIGURE 1.4 ◆ Early volunteer fire company with station and equipment in the late 1800s. Courtesy of Andra Franks/Franks Photography and Weatherford Fire Department.

hand pumper

A fire engine that has long bars (also known as brakes or pumping arms) running parallel to the body which operate the pump. Firefighters physically use the pumping arms to make a full up and down motion to build up pressure allowing water to spray out of a hose.

uniforms, helmets, and hand-painted stovepipe-shaped "fire hats." Fancy fire axes and torches, lavishly painted fire buckets, and, if the company had one, a **hand pumper** decorated by a famous artist, added to the finery. Most striking of all, perhaps, were the engraved silver speaking trumpets that each company used to shout insults at its rivals.

The profession of full-time paid firefighter was not a reality until around the middle of the 19th century. There are exceptions, but typically, rural areas were served by volunteer firefighters, while paid organizations dominated urban areas. Today in the United States and around the world, fire and rescue remains a patchwork of both paid and volunteer responders (see Figure 1.4).

◆ EARLY FIREFIGHTING APPARATUS AND FIRE VEHICLES

siphona

A large syringe used to deliver a stream of water.

The use of technology for firefighting dates back to the 2nd century BC. The Greek engineer Ctesibius of Alexandria is credited with inventing a basic hand pump that operated on the principle of a siphon (Latin **siphona**) and could propel a jet of water toward a blaze.[5] For reasons unknown, this technology fell out of use, and we know of it today only because archaeologists have unearthed parts of the machinery.

The only apparatus available to fight The Great Fire of London in 1666 were buckets, hooks, axes, ladders, and hand-operated water syringes called "squirts." This equipment proved woefully ineffective as the fire continued to burn for four days.

Elsewhere in 17th-century Europe and in the American colonies, firefighting equipment was equally rudimentary. The ravages of London's Great Fire were a timely wake-up call, however, and the late 1600s saw a surge of technological advancements in firefighting equipment, especially in Europe. Hand-operated pumps were placed on wagons, greatly enhancing their mobility. A number of different designs were attempted, including one that required 28 men to work its levers and could shoot a stream of water

80 feet into the air. In 1672, the Dutch inventor Jan van der Heiden designed the first fire hose. It was constructed of flexible leather and coupled with brass fittings at intervals of 50 feet—the interval that remains the standard to this day.[6] These inventions changed the way fires were fought, enabling more sophisticated strategies and tactics.

In 1679, Boston imported America's first fire engine. Our nation would continue to rely on European technology until 1743 when Thomas Lote of New York produced the first American-built fire pump

A steam-powered fire pump, capable of pumping 30 to 40 tons of water per hour, was constructed in London in 1829 by John Braithwaite. One of these new marvels was brought to America, but volunteer firefighters saw the invention as a threat to their social organizations because the **steam engine** would eliminate the need for large numbers of men to operate the hand pumps. Acceptance of the new technology was delayed for several years, but by the second half of the century, the steam-driven pump had become the norm (see Figure 1.5).

In 1841, a group of insurance companies in New York contracted the construction of a self-propelled steam fire engine that resembled a train engine. Again, firefighters resisted its use. Existing pump wagons required numbers of men to push and pull them through the streets, and a self-propelled model would put them out of work.

Steam-powered vehicles may have been out of favor, but the early 19th century did see the advent of horse-drawn fire wagons. Volunteers at the New York Mutual Hook and Ladder Company No. 1 voted in 1832 to purchase a horse to pull their engine. A yellow fever epidemic had reduced available manpower in the city, and it is suspected that this led to the purchase of the horse. Another theory may have been that the volunteers had simply tired of pulling heavy steam-driven pumpers through the streets. Whatever the reason, the idea caught on fast, and horses became an integral part of the fire service all over the world. They were often trained to the pitch of

steam engine

A heat engine that performs mechanical work using steam as its working fluid.

Figure 1.5 ◆ Old Silsby steam engine.
Courtesy of Andra Franks/Franks Photography and Weatherford Fire Department.

The Great Fire of Boston
Boston's largest urban fire and one of the largest fires in American history.

the bell that signaled an emergency. This pitch prompted them to get out of their stalls and stand at attention in front of the fire apparatus (see Figure 1.6).

In the United States, the end of the horse-drawn era was no doubt accelerated by **The Great Fire of Boston** in 1872. Tragically, the fire began at a time when the horses used to pull the heavy steam pumpers were suffering from an epidemic of equine influenza. Without horses, firefighters again had to pull heavy steam pumpers through narrow, twisting streets to the scene of the fire. This, of course, greatly reduced their efficiency. The fire burned for 20 hours, consuming over 60 acres of downtown Boston. Thirty people lost their lives, and property losses were estimated at over $70 million. A review of the problems faced during the Boston fire accelerated the search for alternative ways to transport firefighters and equipment to fire scenes, but Boston did not lead the way.

Earlier, about the middle of the century, the citizens of Cincinnati, Ohio, had become disgusted with the delayed responses and inefficient firefighting that resulted from the use of old-fashioned hand-pulled pumps. They insisted the fire department modernize. In 1853, the first steam-powered fire engine in America came into service. Designed by Cincinnati engineers Able Shank and Alexander Latter, it was affectionately named "Uncle Joe Ross." "Uncle Joe" was able to shoot water 225 feet to a fire and required only three men to operate. The advantages of the new-fangled technology could no longer be ignored. By the end of the century, steam fire engines could be found in numerous municipal fire stations, although horses continued to be used well into the 20th century. Chicago retired its last team on February 6, 1923.

Once the popularity of the new engines was established, inventors and firefighters started thinking of ways to further ease the difficulty of their mission. The aerial ladder wagon made an appearance in the fire service by 1870, followed by the hose elevator in 1871.

An Englishman, Captain George Manby, had invented the first portable fire extinguisher in 1816. It consisted of an air-pressurized copper vessel containing three

FIGURE 1.6 ◆ Horses pulling a large hook and ladder truck.
Courtesy of Andra Franks/Franks Photography and Weatherford Fire Department.

FIGURE 1.7 ◆ Early 1913 American LaFrance gasoline fire apparatus.
Courtesy of Andra Franks/Franks Photography and Weatherford Fire Department.

gallons of pearl ash (potassium carbonate). The late 19th century saw the introduction of the soda-acid extinguisher, which consisted of a cylinder containing a solution of sodium bicarbonate in one to two gallons of water.[7] Inside the cylinder was a vial of concentrated sulfuric acid. The device was activated by striking a plunger to break the vial or inverting the extinguisher, which released a lead bung from the vial. When the sulfuric acid met the sodium bicarbonate, it generated carbon dioxide gas, which forced the water out through a nozzle.[8]

Internal combustion fire engines, built in the United States, first saw service in 1907. At first they were used either as pumping engines or as tractors to pull pieces of equipment. By 1910, both functions were combined, and one gasoline-powered engine was used both to propel the truck and drive the pump. On the same day as the city retired its last team of fire horses (February 6, 1923), the Chicago Fire Department, serving 500,000 residents, became the first completely motorized fire department in the United States (see Figures 1.7 and1.8).[9]

FIGURE 1.8 ◆ 1924 gasoline fire apparatus.
Courtesy of Andra Franks/Franks Photography and Weatherford Fire Department.

Firefighting vehicles today have evolved to serve highly specialized functions and to enable firefighters to respond quickly to a wide variety of emergency situations. Usually diesel powered, vehicles include ladder trucks with aerial platform apparatus that can access high-rise buildings, sending heavy steam applications to heights up to 130 feet. Rescue trucks, brush trucks, mobile command vehicles, smoke ejectors, high-pressure spray trucks, and foam trucks add versatility to the firefighter's arsenal. Industry-specific fire stations and equipment make it possible to handle fires of highly specialized origin. For example, airports under the direction of the Transportation Security Administration (TSA) have introduced fire trucks especially designed to handle fires created by airplane crashes. Refineries have also adapted to their own specialized needs by developing various chemical applicators not used by the typical fire service.

The diesel pump on a modern fire engine is capable of delivering up to 2,000 gallons of water per minute through a lightweight hose reinforced with artificial fibers and measuring up to three inches in diameter. A fireboat, not limited to hydrant supply, can easily deliver as much as 10,000 gallons per minute.

Even the fire extinguisher has evolved into a state-of-the-art fire suppression tool. Fire extinguishers now come in various sizes and designs and employ a variety of extinguishing agents, such as water, dry chemicals, dry powder, carbon dioxide, foam, and special halocarbon agents.

These tools have immeasurably increased the effectiveness and ease of firefighting, but the innovations most appreciated by working firefighters are the improvements to their personal protective gear.

◆ PERSONAL PROTECTIVE EQUIPMENT AND FIRE FACILITIES

Early firefighters faced not only fire but also the effects of smoke with little or no personal safety equipment. Try to imagine responding to an out-of-control structure fire without water- and heat-resistant bunker gear, a state-of-the-art fire helmet, and self-contained breathing apparatus.

Early photography shows Victorian firemen using their own long beards as their only defense against smoke inhalation. Before entering a burning structure, they would wet their beards with water, then clench them in their teeth to act as air filters.[10] Firefighting was never an occupation for the faint-hearted!

apparatus Aldini
One of the earliest recorded attempts to improve breathing difficulties in toxic environments was tested in 1825 in France. This was not the most desirable breathing apparatus for firefighters to wear, but scientific testing was conducted under actual fire conditions.

One of the earliest attempts to reduce breathing difficulties in toxic environments was the French **apparatus Aldini**, which was tested under actual fire conditions in 1825. The Aldini cannot have been comfortable to wear; it consisted of an asbestos mask covered by a second mask made of iron wire, but it may have provided some measure of protection. The space between the two masks trapped a small amount of breathing air. In the 1870s, fire departments began buying and using "Neally's Smoke Excluding Mask." The mask was a simple invention, consisting of a snug-fitting face mask with mica or glass eyepieces and a water bag suspended from rubber tubes that led to a mouthpiece. Two sponges kept wet by the water bag filtered clean air to the wearer. It was marketed as "a most perfect apparatus" and sold for an amazing $15.[11] The first successful American self-contained breathing apparatus, known as the Gibbs, came into use in the first decade of the 20th century. The air pack introduced in late 1945, was the product of research aimed at helping World War II pilots breathe at high altitudes. This unit consisted of a pressurized oxygen cylinder made of steel and a pressure-regulating device that allowed firefighters to breathe fresh air in smoke-filled environments.

The modern self-contained breathing apparatus bears little resemblance to these predecessors. The air pack's heavy steel cylinder has been replaced by a lightweight cylinder developed by the National Aeronautics and Space Administration (NASA) and the space industry.[12] Pressure regulators provide positive pressure to the wearer and allow the firefighter to monitor the amount of air remaining in the cylinder through a "heads-up" display. The amount of air in the cylinder has greatly increased from a 15-minute supply to one hour or more.

Firefighting has certainly come a long way since the days of breathing through wet whiskers![13]

Just as changes in personal equipment have kept pace with modern technology, so has the design of the firehouse. In the early volunteer days, firefighters needed only a place to store their fire pumps. A small shed was perfectly suitable, and stables for the horses were added later. The advent of the professional full-time firefighter created a need for sleeping quarters and for personal space where each individual could store equipment and fire apparatus. To meet this need, many communities built a firehouse to serve the needs of firefighters serving 24-hour shifts. A firehouse usually consists of a kitchen, a bedroom, and an apparatus bay where fire equipment is stored. As technology and engineering design improve, so does the state of the firehouse. (The design of the modern fire and emergency services facility is discussed in more depth in Chapter 5.)

◆ FIRE SERVICE INSIGNIAS AND TRADITIONS

Firefighting and tradition seem to go hand-in-hand. Since time immemorial, certain emblems have been virtually synonymous with firefighting. The best-known of these are the red fire engine, fire badges, **Maltese cross**, and Dalmatian fire dog.

RED FIRE ENGINE

The tradition of painting fire engines red dates back to at least the early 1920s. When Henry Ford made motor cars affordable for the average American family, he let it be known that his vehicles could be bought in any possible color—just as long as it was black. Pretty soon the nation's highways and byways were crowded with black vehicles, and quite logically the fire service began painting their vehicles red in an effort to stand out. Today's consumers can buy vehicles in just about any imaginable color, and so can the fire service. In addition to the traditional red fire engine, it is not uncommon to see white, yellow, blue, orange, green, or even black fire trucks. Many fire departments have, however, stuck loyally to the time-honored red engine.

BADGES

Throughout fire service history, the firefighter has been identified by the badge in much the same way as medieval knights were identified by the insignia on their shields. Insignias allowed knights to be recognized on the battlefield despite the heavy armor that concealed their person. Firefighters' badges serve a similar identifying function and provide a means to distinguish rank and departmental affiliation. Badges are a symbol of courage and a commitment to protect the public.

Maltese cross

Symbol representing the traditions and ideals of the fire service. The arms and the tree of the Maltese cross are equal in length. The arms and the tree widen as they extend from a central point. In most cases, the edges are flat or curved slightly outward.

MALTESE CROSS

The Maltese cross was originally the insignia worn by the crusading Knights of St. John. During the battle for Jerusalem, the Saracens atop the city's walls poured oil down on the attackers and dropped lighted torches to ignite the oil. The Knights, witnessing this horror, dropped their weapons and ran into the flames to rescue their comrades. On their way home from the Crusades, the Knights established a hospital on the Island of Malta to care for the injured, and they flew their insignia on a flag to let all know that they would render aid to the needy. Their descendants continued this charitable work for the next 400 years, and the cross of St. John became a symbol of generosity to friend and foe and a mark of protection for the weak. Not knowing the history of the insignia, many referred to it as the Maltese cross, and the name stuck, becoming an international symbol of aid. The fire service adopted this symbol because it proudly upholds the very same standards and traditions as the Knights of St. John.

DALMATIAN DOG

For years, the loyal Dalmatian has been the trusted companion of firefighters. Few realize that this breed was originally chosen because of the strong bonds that the dogs formed with fire horses, protecting them and keeping them company at the station. The dogs were also expected to rouse the horses at the sound of the alarm bell, then run out and bark a warning at anyone who might be obstructing the firehouse exit. The dogs would then chase the fire apparatus all the way to the scene, sometimes barking the whole way. They served the same function, essentially, as the emergency traffic signals located outside many fire stations today and the sirens on fire trucks.

When horses were replaced by steam- or gasoline-driven fire engines, many departments opted to keep their beloved mascots. It is not unusual even today to see a proud Dalmatian riding on a fire engine as it races to the scene of an emergency.

◆ THE FIRE DILEMMA IN THE UNITED STATES

The United States has a severe fire problem that if not addressed, will continue to worsen drastically. Fire statistics show that our nation, one of the richest and most technologically sophisticated countries in the world, lags behind its peer nations in fire security. Nationally, there are millions of fires, thousands of deaths, tens of thousands of injuries, and billions of dollars lost each year—figures which far exceed comparable statistics for other industrialized countries.[14] In 2001, for example, the direct value of property destroyed in fires was $11 billion ($44 billion if the World Trade Center loss is included).[15] More recently in 2004, direct property losses from fires were estimated at over $9.8 billion.

Americans tend to remember the great fires in our history—the Chicago fire of 1871 and the fire set off by the San Francisco earthquake of 1906—but the thousands of smaller tragedies that occur yearly have failed to generate the reforms our nation so badly needs. We get an eye-opening look at how U.S. citizens view fire danger in a 1973 report issued by the United States Fire Administration, *America Burning*—

Revisited. As examples of views and attitudes that the average American holds about fires, the report cites the following:

- "It can't happen to me."
- "Odds are that it won't happen to me."
- "The insurance company will take care of me."
- "It is not a disgrace to have a fire."
- "I can set fires for revenge."
- "They're just children playing; they didn't know any better."[16]

Such casual attitudes contrast starkly with the more responsible approach taken by European and some Asian nations, where advanced training is required of fire professionals, and building codes tend to be more stringent than in the United States. (For a more in-depth discussion of building construction, see Chapter 7.)

The terrorist attacks of September 11, 2001, have greatly affected the fire and emergency services in the United States. The passing of the Patriot Act and the inception of **Homeland Security** have changed the way national fire policy is written and implemented. New laws and newly formed government entities impact fire service administration, and the new threats posed by terrorism have required greater vigilance from all emergency personnel as well as training in new levels of hazardous material handling and transport, new security measures, and new strategies for emergency response.

Homeland Security

The department created after the events of 9/11 that is responsible for assessing the nation's vulnerabilities.

Stop and Think 1.1

List two ways we can change the fire problem in the United States today.

◆ THE SCOPE OF FIRE AND EMERGENCY SERVICES TODAY

In U.S. communities, the fire and emergency services department is the only entity trained and equipped to deal with multiple kinds of disasters. Fire departments respond to all manner of emergencies and confront all manner of risks and hazards.

The fire service safeguards not only our homes and businesses but also our economy and its critical infrastructure—the electrical grid, interstate highways, railroads, pipelines, and petroleum and chemical facilities are all protected by the fire service. In fact, the fire service is itself considered a vital part of the nation's infrastructure. The service also responds to emergencies in federal buildings, provides aid to military bases, and protects interstate commerce. Every passenger airliner that takes off from a runway is protected by a fire department. Every hazardous materials spill results in a call to the fire service to protect lives and property and deal with the dangerous clean-up process.[17]

Today some municipalities, faced with tight budgets and limited personnel, are shifting to a triple-certification requirement for their emergency responders. It is obviously more cost effective to hire one person to act in three roles—firefighter, police officer, and paramedic—at an emergency scene. Like most things that sound too good to be true, however, there is a downside to triple certification. It has resulted in a higher than usual turnover rate, with firefighters tending to burn out and leave the department either for a higher paying position or because they do not want to be police officers.

With or without triple certification, however, there is no doubt that the traditional job description of firefighters has changed radically in recent decades. For instance, they must be prepared to provide emergency medical services such as advanced life support. Many states and organizations now require, as a minimum standard, that all job applicants be certified as Emergency Medical Technicians (**EMT**) or National Registry Emergency Medical Technicians (**NREMT**). Previously, firefighters just needed to be volunteers with no formal medical training to apply for a job. Changes in educational requirements are also impacting the promotional hierarchy. Firefighter seniority was once key to moving up the ranks. Today, higher educational achievement is an equally important critical criterion, often providing a boost up the ranks ahead of candidates who have had less formal schooling. Advanced education can also help firefighters in their public relations role and in educating the public on fire prevention. Today, when we are seeing an increase in the retirement of the veteran firefighters of the baby boomer generation and an influx of new recruits, training and higher education are more important than ever. (This topic will be discussed in more detail in Chapter 4.)

Since 9/11, fire and emergency services have been saddled with yet another responsibility. On February 28, 2003, President Bush issued Homeland Security Presidential Directive 5, which created a National Incident Management System (**NIMS**). NIMS provides a consistent nationwide template to enable all government, private sector, and nongovernmental organizations to work together during incidents that threaten domestic security. As a result, fire professionals must now be trained in emergency strategies designed to mitigate disasters at the national as well as the local level. (Chapter 10 will provide more information about the NIMS and the firefighter's role in Homeland Security.)

Not only have firefighting management and education changed, but also have basic firefighting techniques, gear, and equipment. In the past, the firefighter would leave the firehouse in a knee-length bunker coat and rubber boots. Modern firefighters are more likely to wear EMS gloves than bunker gear. The typical call is still an emergency to which firefighters respond with sirens wailing, but when they get to the scene, they may not go in with axes and firehoses and spray down the building. Instead, they may take out stethoscopes and blood pressure cuffs and begin questioning victims about injuries. It may be time to change the title of firefighter to fire and emergency services technician or emergency preparedness specialist. Some already refer to firefighters as hazard personnel because they respond to just about every type of emergency call you can imagine. In just the last 10 years, the firefighting profession has evolved into the nation's largest prehospital EMS provider as well as a front-line defense in homeland preparedness.

Some of the new standards and regulations that today's fire and emergency services are embracing were established during the annual Fire and Emergency Services and Higher Education (**FESHE**) Conference. The role of FESHE is to provide a new strategic approach to professional development. The organization will help move fire and emergency services from a technical occupation to the status of a full-fledged profession similar to that of physicians, nurses, lawyers, and architects, who, unlike fire service personnel, have standardized course requirements for their respective degree programs. (A thorough discussion of FESHE will be presented in Chapter 4.)

EMT

Emergency Medical Technician. A specified level of medical training that usually consists of around 100 hours of classroom and on-site training.

NREMT

National Registry of Emergency Medical Technicians.

NIMS

National Incident Management System.

FESHE

Fire and Emergency Services Higher Education. Working with coordinators of two- and four-year academic fire and emergency medical services (EMS) degree programs, the U. S. Fire Administration's National Fire Academy (NFA) has established the FESHE network of emergency services–related education and training providers. The FESHE mission is to establish an organization of post-secondary institutions to promote higher education and to enhance the recognition of the fire and emergency services as a profession to reduce loss of life and property from fire and other hazards.

Fire has played a major role in the development of civilization as we know it today. As a human race, we have learned to use fire to our benefit, but we need to further increase our respect for its power and to improve the ways in which we deal with its potential dangers.

The nature of firefighting requires an individual to be available to handle any crisis at a moment's notice and to handle it with professionalism. Never before in the history of our country have the actions of firefighters been held to the level of scrutiny that is applied today. Although recent events across the nation have brought national attention to firefighters' jobs and how they perform their jobs, the modern fire and emergency service is being challenged to take on an increasing number of responsibilities every year. Fire and emergency services must handle the training needed to use the ever-changing technology of new equipment. Current trends suggest that the fire service industry of the future will emphasize fire prevention and education as much as fire suppression. This trend—as well as the requirement that firefighters be proficient in emergency medical procedures and be prepared to respond to homeland security crises—have increased the need for education and training.

The traditions and culture of the fire service are changing. The once white male-dominated profession is now open to women and to all minorities. This diversity has broadened the scope of the fire service by introducing new ideas and cultural concepts.

There is a long, colorful, and distinguished history associated with the fire service, but there is plenty of room for improvement. Although no crystal ball can predict the world of tomorrow, the 21st century will surely carry the fire and emergency services field in directions yet imagined.

On Scene

Tradition and history are deeply entrenched in the fire service, but in today's society, the fire service has also evolved into a new service. Fire and emergency services in the United States deals not only with fire suppression as in the past, but it also deals with emergency medical services, hazardous materials, code enforcement, wildland fires, wildland/urban interface, fire prevention and education, technical rescue, urban search and rescue, aircraft firefighting, arson investigation, explosive response, industrial fire safety, and much more. Some of the same challenges from the past still plague the fire service today. These challenges include staffing, funding, service demands, facilities, health and safety, and training, to name a few.

1. As a firefighter in the 21st century, are you ready and equipped to carry out the duties and responsibilities of the new role a firefighter plays while still carrying on the traditions that the fire service holds dear? What organizations can help the firefighter hold on to the traditions yet move forward and learn from the past?
2. If problems from the past continue to plague the fire service today, where can you find information on what have we learned to help facilitate a resolution to these long-standing problems?

Review Questions

1. The development and use of the first hand-operated fire pump dates back to _____.
2. The first public volunteer fire department in North America was formed in _____ at _____.
3. What is a "rattle watch"?
4. What was the "apparatus Aldini"?
5. Which insignia was worn by the crusading Knights of St. John?
6. What is the role of FESHE?

Notes

1. Fenton, Marc (2002, November 27). *Great fire of Rome.* Thirteen WNET New York, Retrieved January 26, 2007, from http://www.thirteen.org/pressroom/release.php?get=252.
2. Klinoff, Robert W. (2003). *Introduction to fire protection* (2nd ed.). Clifton Park, NY: Thomson Delmar Learning.
3. Ibid.
4. Ibid.
5. Coleman, Ronny J., & Granito, John A. (1988). *Managing fire services* (2nd ed.). Washington, DC: International City Management Association.
6. Answers Corporation. (2005, February 1). *History of fire brigades.* Retrieved January 27, 2007, from Answers.com website: http://www.answers.com/topic/history-of-fire-brigades.
7. Firehouse International, Inc. (2007). *Fire extinguisher questions and answers index.* Retrieved January 27, 2007, from How Stuff Works Related to the Fire Industry website: http://www.firehouseinternational.com/usersite/how_stuff_works/Fire/Fire_extinguishers/Index/extinguisher_Index.aspx.
8. Ibid.
9. Chicago Municipal Reference Library. (1997, August). *Brief history of the Chicago fire department.* Retrieved January 27, 2007, from the Chicago Public Library website: http://www.chipublib.org/004chicago/timeline/firedept.html.
10. Hashagen, Paul. (1998). The American fire service: 1648–1998—The development of breathing apparatus. *Firehouse Magazine,* Retrieved January 28, 2007, from http://www.firehouse.com/magazine/american/breathing.html.
11. Ibid.
12. Ibid.
13. Ibid.
14. United States Fire Administration. (2004). *Fire in the United States 1992–2001* (13th ed., p. 29). Arlington, VA: TriData Corporation.
15. Ibid.
16. Ibid.
17. Bruegman, Randy R. (2003, April 30). *Meeting the needs of the fire service.* Retrieved January 27, 2007, from the International Association of Fire Chiefs website: http://www.iafc.org/home/article.asp?id=/data/gr/GR_043003.

Suggested Reading

Coleman, Ronny J., & Granito, John A. (1979). *Managing fire services.* Washington, DC: International City Management Association.

Cote, Arthur E. (2004). *Fundamentals of fire protection.* Quincy, MA: National Fire Protection Association.

Cote, Arthur E., & Bugbee, Percy. (1988). *Principles of fire protection.* Quincy, MA: National Fire Protection Association.

United States Fire Administration. (2002). *A needs assessment of the U.S. fire service: A cooperative study authorized by U.S. public law 106-398.* Quincy, MA: National Fire Protection Association.

Fire and Emergency Services Career Opportunities

2 CHAPTER

Key Terms

bioterrorism, p. 30
Department of Homeland
   Security, p. 30
ergonomic, p. 26
FAA, p. 21
fire ecology, p. 21

fire lines, p. 21
GIS, p. 29
helitack, p. 21
hotshot, p. 21
pictometry, p. 29
prescribed fire, p. 21

smoke jumpers, p. 21
Smokey Bear, p. 27
Sparky the Fire Dog, p. 27
standard operating
   procedures (SOP) , p. 19

Objectives

After reading this chapter, you should be able to:

- Identify fire protection jobs in the public and private fire service.
- List duties and requirements for the positions of firefighter trainee and firefighter.
- List duties and requirements for the position of firefighter/paramedic.
- Give examples of fire service jobs other than firefighter.
- List duties and requirements of search and rescue teams.
- List duties and requirements of the wildland firefighter.

◆ INTRODUCTION

Firefighting jobs are not easy to obtain. They call for high levels of physical fitness and mental stamina, as well as a strong desire to serve others. Increasingly, these jobs also call for advanced education and specialized training. Candidates who meet these rigorous requirements will, however, find employment prospects to be generally favorable (see Figure 2.1). Job growth estimates through the year 2014 are that growth in fire service jobs will be greater than the average for all occupations. Most of this growth will occur in suburban areas where volunteer firefighting positions are rapidly

FIGURE 2.1 ◆ Firefighter/paramedic sitting at a desk.

turning into paid positions. Replacements will also be needed for firefighters who retire, stop working for other reasons, or transfer to other occupations.[1]

This chapter is designed to describe traditional firefighting jobs and to introduce some of the new positions that have developed over the last decade. In the fire service, as in any other business or industry today, the range of possible career paths, job duties, and skills required is expanding rapidly. This is the result of varied factors, including technological changes, increased concern for the environment, and radical changes in the nature of the dangers confronted by the fire service. In addition, fire service personnel are being asked to take on duties far beyond those they performed 50 or even 10 years ago. Frequently, firefighters double as emergency medical care specialists. They may also lend their expertise in areas as diverse as public education, forestry, architecture, engineering, insurance, and law enforcement.

The fire service offers a great variety of opportunities and challenges, but all professionals in the field share a common goal—bringing order to chaos and creating a safe environment for the public.

◆ FIREFIGHTING CAREER OPPORTUNITIES

FIRE CADET

The fire cadet, an entry-level training position, may be either a private individual interested in pursuing a career in the fire service or a recruit sponsored by a fire department or another organization. In the latter case, the recruit's fire academy education may be financed by the sponsoring group, which also typically guarantees his or her employment after graduation. Students without sponsors will have to finance their own education and conduct their own job search.

Fire academies may be part of a fire department, or they may be college based. In either case, the training is intellectually and physically arduous and demands commitment on the cadet's part. Typically, fire academy training is 12 to 16 weeks but may vary from state to state and even within the same state. A high school diploma or General Educational Development (GED) certificate of completion is generally the minimum educational requirement to apply for cadet training; however, it can vary from state to state. If the academy is sponsored by a fire department, the department's training officers supervise fire recruits until they have mastered basic firefighting skills and completed their examinations. Trainees are then assigned to a duty station where their progress is supervised by a fire company officer. Often, fire recruits remain on probationary status for their first year of service while their performance, both as an individual and as a team member, is assessed. If the academy is based in a college or university, the graduating cadets begin the long road toward finding an entry-level job and establishing their own career paths with a fire department. Regardless of how cadets find employment, the tasks they face in their new job are the same. In addition to actual firefighting, cadets may be assigned rescue work, first aid, ventilation, forcible entry, salvage, overhaul, site inspections, emergency medical care, public fire education, hazardous materials work, dispatching, . . . and the list goes on (see Figure 2.2).

FIREFIGHTER

Hiring requirements for firefighters are comprehensive and the testing competitive, reflecting the critical nature of the job and the need to work under stressful and physically demanding conditions. Applicants typically must pass a written test, an oral interview, a physical ability test, a medical evaluation, and a background investigation. Those who achieve the highest scores are ranked at the top of the candidate list and have the best chances of being hired. New firefighters work under close supervision learning the department's policies and procedures, which are usually referred to as **standard operating procedures** (**SOP**) or guidelines. As with many other jobs, there is usually a probationary period that can last from six months to a year, depending on department policy. If the probationary period is successfully completed according to the probationary agreement, the probation status is lifted, and the firefighter

standard operating procedures (SOP)

Rules by which an organization or fire department operates for its day-to-day functioning. Usually these procedures are documented in a handbook or related source.

FIGURE 2.2 ◆ Cadets attending fire academy.
Courtesy of TEEX-Texas A&M System.

becomes a full-duty fire professional. However, this can vary from town to town or even state to state. Firefighters respond to emergency calls of all types. Although their primary responsibility is to fight fires, they are more often called on to assist with other kinds of emergencies, such as medical calls and vehicle accidents. As a public service, firefighters often visit schools, day-care centers, and senior centers to teach fire safety. Firefighters' schedules vary, but the typical shift is 24 hours on duty, followed by 48 hours off. Of course, firefighters cannot just leave the scene of an emergency and go home when their appointed shift ends. Long and irregular hours are not unusual, and applicants should consider this before entering the profession. Moreover, a firefighter's duty and responsibilities fluctuate, and written job descriptions can vary considerably. Candidates need to gather as much information about a department's requirements before applying for a job. The Internet and job boards are good resources, and it is also helpful to visit agencies in the area.

FEDERAL FIREFIGHTER

The federal government employs both civilian and military firefighters at some of their installations, especially at military bases. These jobs require the firefighter to deal with specialized on-base emergencies, such as aircraft crashes, in addition to performing the same tasks and responsibilities of a municipal firefighter. Goodfellow Air Force Base in Texas is the home of the Louis F. Garland Department of Defense Fire Academy. The academy serves all branches of the Department of Defense (DoD), including the Army, Marine Corps, Navy, Air Force, and Civil Service. It provides training for about 2,000 students a year, including allied forces, and can be a path to a variety of fire service careers in the DoD branches. DoD firefighting opportunities are readily available on military bases for both civilians and military personnel.

FIREFIGHTER/PARAMEDIC

Fire departments in the United States commonly respond to medical emergencies. In fact, about half of them provide ambulance service. This has led to the dual position of firefighter/paramedic, in which the same person serves in two capacities in an emergency situation. The position comes with increased responsibility and requires advanced medical training that can take from at least six months to a year to complete. In addition to performing basic fire-suppressing duties, firefighter/medics must be prepared to assess injuries and acute illnesses, supervise ambulance transports, and take full responsibility for patient care until relieved by proper medical personnel. They provide emergency treatment working under standing orders or under the directions of a supervising physician. Personnel with this type of training are in great demand, with many fire departments offering pay incentives to qualified applicants for the position.

WILDLAND FIREFIGHTER

Fires set by lightning or careless human behavior consume thousands of acres of our precious wildlands annually. For example, in October 2007, more than 500,000 acres in California burned, and this is a small percentage of the total acres burned in 2007. In national forests and parks, non-firefighters are the first line of defense against these fires. Forest rangers patrol assigned areas to ensure that travelers and campers comply with fire regulations. Forest fire inspectors and prevention specialists spot fires from watchtowers and report their findings to headquarters by telephone or radio. When a fire is reported, wildland firefighters are brought in to suppress the blaze. In addition to basic firefighting training, these specialists study **fire ecology**, the science of how fire behaves in natural environments and how it affects both living and nonliving things in the environment.

The work is physically demanding and can be emotionally taxing. Wildland firefighters are frequently required to work for days at a time—sometimes up to 14 days straight, without a day off or home leave. A typical shift is 16 hours or longer, and shifts over 24 hours are not unknown. In addition to the long hours, the crews must brave hazardous environments, enduring extremes of heat as well as smoky, dirty, and dusty conditions.

A wildland firefighter may not always work directly in the front lines of an advancing fire. Other assignments might include creating **fire lines**—cutting down trees and digging out grass and other combustible vegetation in the path of the fire—to deprive it of fuel. This is one of the most efficient means of battling a blaze. Wildland firefighters may also work at setting and monitoring **prescribed fires**. These are fires purposely ignited by fire personnel or agencies under controlled conditions for specific management objectives. For example, setting a prescribed fire is a good, natural way to help put nitrates back into the soil. Furthermore, a prescribed burn can help decrease wildfires because it removes dead and dying trees to prevent heavy fuel loads during dry sessions.

There are a number of challenging specialist careers within the wildland firefighting field. Helicopter attack teams, or **helitack**, dump enormous buckets or belly tanks of water onto the fire. Some helitack crews rappel out of helicopters, gaining access the only way possible to fires in remote areas or in rugged terrain. **Smoke jumpers** parachute out of planes to reach areas not otherwise accessible, and **hotshot** crews hike into the area where the fire is burning to begin extinguishment. Employers for individuals interested in a wildland firefighting career include the U.S. Department of Agriculture Forest Service, California Department of Forestry and Fire Protection, and contract companies, among others.

AIRCRAFT RESCUE FIREFIGHTER (ARFF)

The aircraft rescue firefighter (ARFF) goes through a highly specialized course of training to earn certification and then be assigned to a civilian airport. The Federal Aviation Administration (**FAA**), an agency of the U.S. Department of Transportation, sets minimum ARFF requirements for each airport, based on an index determined by the overall length of the aircraft it handles and the average daily departures of those aircraft.[2] ARFF training guidelines are outlined in the FAA's Federal Aviation Regulations, Part 139 (FAR Part 139). This requires that trainees receive airport and aircraft familiarization instruction, training in emergency aircraft evacuation assistance, and airport emergency communications systems. They also must be experienced in using the specialized equipment needed to fight aircraft fires and to minimize aircraft cargo hazards. In addition to this basic training, aircraft rescue firefighters must typically participate in an annual 24-hour ARFF refresher course as well as annual live flammable-liquids fire-suppression training to maintain their certification. Additionally, they usually take part in annual multiple-casualty exercises at the airport (see Figure 2.3).

fire ecology

The study of the interrelationship of wildland fires and living and nonliving things in the environment.

fire lines

Boundaries around a fire area to prevent access except for emergency vehicles and relevant professionals.

prescribed fire

A fire that is purposely ignited by fire personnel or agencies under controlled conditions for specific management objectives.

helitack

Used during wildfires when the location of the fire(s) is inaccessible to firefighting crews.

smoke jumpers

Firefighters who parachute from airplanes to suppress forest fires in remote locations.

hotshot

Highly trained firefighters used primarily in handline construction.

FAA

An agency in the U.S. Department of Transportation that oversees all aspects of civil aviation.

FIGURE 2.3 ◆ ARFF apparatus.
Courtesy of TEEX-Texas A&M System.

FIRE APPARATUS ENGINEER

The fire apparatus engineer is a promotion from the firefighter position and requires extensive training and experience. Apparatus engineers work in all kinds of firefighting situations (structural, wildlands, and so on) and have sole responsibility for driving and operating fire apparatus such as engines, water tenders, rescue squad vehicles, and aerial trucks. Emergency driving skills, and knowledge of traffic laws, departmental SOPs, and hydraulics are examples of additional training that a person must have to be successful in this position. Special driving certifications/licenses are required, and good driving records are compulsory.

SEARCH AND RESCUE

Search and rescue (SAR) teams consist primarily of volunteers, usually drawn from other emergency services. They are responsible for conducting missions in a wide range of emergency situations and almost any imaginable setting, including aircraft accidents or incidents, swift-water and open-water rescue, dive rescue and recovery, confined space rescue, mine rescue, high- and low-angle rescue (mountain and structure), and avalanche rescue (see Figure 2.4). The broad scope of their activities calls for a high degree of proficiency in a wide range of rescue techniques. Frequent and intensive training is required (see Figure 2.5). Each department or agency may have its own local SAR team. In addition, state-supported or regional SAR teams are deployed to cope with disasters of a larger scale.

One such state-supported team is Texas Task Force 1. Task Force 1, sponsored by the Texas Engineering Extension Service (TEEX), serves as the state's urban search and rescue team under the Governor's Division of Emergency Management and is one of 28 such teams in FEMA's national urban search and rescue system. This task force, with more than 200 members drawn from 60 organizations across Texas, responded to one of the largest natural disasters in U.S. history. When Hurricane Katrina struck the Gulf Coast of Louisiana and Mississippi on August 29, 2005, Texas

FIGURE 2.4 ◆ Water rescue training.
Courtesy of TEEX-Texas A&M System.

Task Force 1 swung into action, performing a variety of functions as the emergency developed:

- An 80-member urban search and rescue team was deployed by the Federal Emergency Management Agency (FEMA) on August 27 in advance of the hurricane's landfall.
- A 41-member water rescue team was deployed by the Texas Governor's Division of Emergency Management on Monday, August 29.
- Another 75-member urban search and rescue team was deployed by FEMA to New Orleans on Tuesday, September 13.

FIGURE 2.5 ◆ Collapsed structures training.
Courtesy of TEEX-Texas A&M System.

The teams assisted with more than 13,000 rescues in downtown New Orleans, where thousands were stranded after the hurricane's storm surge broke levees and left much of the city under water.[3]

INDUSTRIAL FIREFIGHTER

An industrial firefighter is specialized in meeting the needs of the manufacturing, refining, and petrochemical industries. The oil industry began in America in the late 1800s, and with it came fierce and uncontrollable petroleum fires. As recently as 20 years ago, the technology and specialized training needed to extinguish such fires was not available, and they often burned until the fuel ran out. The cost to the environment, business, and human health was immense. The need for industrial fire-fighting was obvious, but there were no facilities capable of handling the large-scale training exercises that would be required. This need is now met nationally and internationally through training facilities such as TEEX at the Brayton Fire Training Field, a part of the Texas A&M University System (see Figure 2.6).

HAZARDOUS MATERIALS SPECIALIST

Hazardous materials specialists are concerned with the discovery, regulation, and recovery of hazardous materials. They must meet the standards for the hazardous materials specialist level, as set by the National Fire Protection Association. The minimum requirement to achieve full status on a hazardous materials response team (HazMat) is approximately 120 hours of training. A 40-hour annual recertification class is also required. In addition to hazardous materials training, personnel must possess basic first aid and rescue skills. As a rule, these specialists are also required to have specialized training in dealing with weapons of mass destruction as well as biological warfare agents and other hazardous materials that might be used in terrorist incidents.

ARSON UNITS

Arson units are charged with combating arson and other fire-related criminal offenses. A large city might have its own arson team, but in smaller towns and in rural districts, arson units tend to be multijurisdictional. The units fulfill their mission by

FIGURE 2.6 ◆ Simulated chemical complex fire training structure.
Courtesy of TEEX-Texas A&M System.

bringing together and utilizing the talents and resources of experienced and dedicated fire, police, and prosecutorial personnel in a manner that maximizes results while minimizing taxpayer expense. Certification in this area often includes peace officer training, which gives arson team members the power of arrest.

◆ NON-FIREFIGHTING CAREERS IN FIRE PROTECTION

Other career paths in fire service are available that do not require active, on-site firefighting. The physical, educational, and certification requirements for these positions differ. All require an understanding of fire and fire safety, and each serves a critical function in protecting the public.

DISPATCHER/COMMUNICATIONS OPERATOR

One of the most challenging civilian jobs in emergency services is that of fire dispatcher. The dispatcher's job is to support fire management operations, primarily by relaying information from those seeking emergency services to the appropriate responders. In most cases, dispatchers are not solely dedicated to fire emergencies but must handle police and medical emergency calls as well. It is their job to receive each incoming 9-1-1 telephone call, obtain essential incident information from the caller, evaluate what type of emergency service is required, determine the priority of the call, and then dispatch the appropriate unit. They must also relay all essential information about the emergency to the responding unit via voice radio or intercom while at the same time monitoring all other emergency frequencies. Dispatchers work at a fast pace and are under constant pressure. They must be expert communicators and proficient multitaskers and be able to handle high-stress situations calmly (see Figure 2.7).

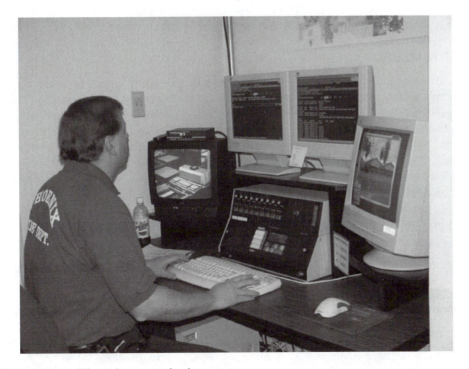

FIGURE 2.7 ◆ Dispatch communications center.

#8

OCCUPATIONAL HEALTH AND SAFETY SPECIALIST

Occupational health and safety specialists are employed in both the public and private sector. Their expertise is analyzing potential hazards in work environments and developing programs and procedures to eliminate or minimize these hazards. They may also serve in an enforcement capacity, conducting inspections and monitoring compliance with health and safety regulations and laws. The safety specialist is able to identify any condition—physical, chemical, biological, or **ergonomic**—that may cause disease or injury. The experience and education required may be different depending on the needs of the public or private sector area. Candidates should always find out what their potential employers may require.

ergonomic

The scientific principle that bases design around human needs, and the profession that applies science and data to this concept.

FIRE EQUIPMENT SALESPERSON

Because firefighters and other emergency personnel can have flexible work schedules, they may be able to work at dual occupations. A firefighter might elect to continue work in the fire service field by working as a part-time fire equipment salesperson. Companies such as Pierce Manufacturing and Tyco Fire & Security, which design and service fire protection equipment, are always looking for experienced personnel to sell their products. Firefighter certification is not required for this position.

FIRE EXTINGUISHER SERVICE TECHNICIAN

Another occupation not directly related to firefighting duties is that of the fire extinguisher service technician who inspects and services firefighting equipment, including all types of hand held fire extinguishers and fire hoses. Fire extinguisher service technicians must have the technical knowledge to disassemble extinguishers and examine them for defects, replace defective parts, clean the fire extinguishers, refill them with fire extinguishing agents, and then test them, using special testing equipment, to ensure that they meet specifications This is another occupation that does not require firefighter certification.

FIRE MARSHAL

The fire marshal typically heads the fire prevention division. However, there are no concrete sets of roles on fire marshal positions. In addition, this type of position could be at a fire department, government entity, or both. Numerous different titles exist, but the basic definitions are the same. Fire marshals conduct inspections of the premises within their jurisdiction to identify unsafe conditions and ensure compliance with fire codes. Additional duties may include fire and arson investigation. Fire marshals also work with developers and city planners to check and approve plans for new buildings. In conjunction with the goals of the fire prevention division, fire marshals often speak before public assemblies and civic organizations to promote fire prevention. This position may require formal education such as a bachelor's degree as well as experience in fire prevention, plan review, inspections, and supervision (see Figure 2.8).

Stop and Think 2.1

Several different avenues are available for a career in fire and emergency services.

Select two different choices in the fire and emergency services field that an individual may pursue besides civil service employment. Explain your answer in detail.

FIGURE 2.8 ◆ FESHE professional model for fire marshal.

PUBLIC INFORMATION OFFICER (PIO)

The job of a fire department's public information officer (PIO) is to communicate accurate information to the community and to members of the fire service in a timely, efficient manner. This information may relate to ongoing emergencies or to other aspects of the department's operations. A public information officer is often interviewed on television or on the radio, responding to questions from reporters. A public information officer must be available at any time of the day or night to cover significant events and serve as liaison between the incident commander and media outlets. The position requires proficient verbal skills and media knowledge. Applicants for the job typically have a bachelor's degree and meet the NFPA's Standard 1035: Standard for Professional Qualifications for Public Fire and Life Safety Educator.

FIRE PREVENTION SPECIALIST

A fire prevention specialist is a unique kind of fire service professional who fights fire by promoting fire prevention training and education. The position involves visiting schools and community organizations to talk about fire safety, and it may also require performing routine checks of fire systems, such as sprinklers and hydrants. Fire prevention specialists use campaigns such as **Smokey Bear**, **Sparky the Fire Dog**®, and Fire Prevention Week to build awareness and educate the public. The Smokey Bear campaign, which was created in 1944, is the most well-known public service prevention program geared toward wildland and forestry fires. Examples of job opportunities as a fire prevention specialist are federal or state forestry services, city or county fire departments, municipalities (such as inspection bureaus), private companies, or through partnership with several of these.

Smokey Bear

Created in 1944. The longest running public service campaign in U.S. history for fire prevention.

Sparky the Fire Dog®

Created in 1951. Used for NFPA's Risk Watch® and Learn Not to Burn® programs and has a special appeal to children's programs.

FIRE DEPARTMENT TRAINING SPECIALIST

A fire department's training specialist or training officer develops and coordinates programs to maintain and enhance skilled performance by the department's firefighters. The position also involves developing standard operating procedures for fire suppression and assuring that these procedures are in compliance with state and federal fire and safety regulations. This is typically a mid management position that reports directly to the fire chief. The specific functions of the training officer are described in more detail in Chapter 4, "Training and Higher Education."

INSURANCE ADJUSTER

Insurance companies frequently hire people with backgrounds in the fire service to work in their loss prevention departments. They especially seek those with knowledge in the fields of fire protection engineering, fire risk management, fire prevention, and arson investigation. Such expertise is invaluable when it comes to inspecting fire-damaged properties to assess whether arson may have been involved. Adjusters also estimate the amount of loss, negotiate with policyholders, and, when necessary, gather evidence to support contested claims. They may be called to testify in court on behalf of their employer.

FIRE INSPECTOR

The major goals of fire inspections are as follows: (1) to raise the public's awareness of fire safety considerations in their immediate surroundings; (2) to identify fire hazards that must be eliminated for a safer environment; (3) to record inspection information for inclusion in the public record; and (4) to verify the proper functioning and maintenance of installed fire protection systems and other building fire protection equipment/features.[4] To achieve these goals, fire inspectors examine the interior as well as the exterior of buildings and structures for fire hazards. Code enforcement plays a major role in their work. Fire inspectors may also collect fees for building licenses and permits. A fire inspector is normally employed by a fire department and may be a member of a team of inspectors.

FIRE INVESTIGATOR

Fire investigators are essentially detectives whose job it is to determine how a fire originated and, in particular, to determine whether it was accidental or the result of negligence or arson. These types of jobs are both private and public. Fire investigators visit fire scenes, photograph and shift through evidence, and interview property owners, building occupants, and witnesses. Once they arrive at a plausible theory, they have the expertise to reconstruct the fire scenario in order to test their theory. They also receive training in interrogating suspects and interviewing witnesses. In some states, fire and arson investigators have the authority to arrest suspects or to subpoena witnesses.[5] In states where they do not have arresting powers, they work closely with law enforcement officials.

◆ PRIVATE SECTOR FIRE PROTECTION CAREERS

FIRE PROTECTION ENGINEER

Fire protection engineers (FPE) often enter the profession after working in some other sector of the fire service. Their work, however, is very different from firefighting.

Their job is to design structures to optimize their fire resistance. The tools they use are mathematical and scientific knowledge. Fire protection engineers may be employed by a large fire department—usually in their fire prevention or code enforcement departments—or they may work as employees or consultants for engineering firms, fire equipment and systems design companies, forensics investigators, hospitals and health care facilities, manufacturing firms, research and testing laboratories, or in the insurance industry.

CONTRACT FIREFIGHTER OPPORTUNITIES (OVERSEAS)

The market for U.S. firefighters overseas is booming. In most cases, they are employees of U.S. companies with operations or contracts abroad. For example, Kellogg Brown & Root (KBR) is under contract to provide fire service for various Department of Defense facilities. The aerospace technology company Lockheed Martin employs firefighters stateside and abroad, as does Boots & Coots International Well Control, Inc., both based in Texas. The Boots & Coots company is known for its role in controlling oil well fires in Kuwait after the Gulf War. These companies provide firefighters exciting career opportunities abroad, including airport rescue firefighting (ARFF), fighting industrial fires, and serving as HazMat technicians, safety specialists, and inspectors. Some companies even seek specialists at the crew chief, fire chief, or fire captain level.

Stop and Think 2.2

As with many specialized occupations, outsourcing is on the rise in firefighting. Why, do you think, is there an increasing need for overseas firefighters?

INVENTOR

As technology advances, firefighting becomes more complex, and the need for innovative equipment and techniques increases. Inventors working to fill this need have found firefighters to be a rich source of ideas. After all, firefighters know best what would make the job safer and more efficient. One new innovation is the use of the Global Positioning System (GPS) in fire apparatus. Several large cities, including Chicago, have invested in state-of-the-art emergency response systems built with GPS-based Automatic Vehicle Location (AVL). A city the size of Chicago may have to respond to over 14,000 emergency calls a day, so a system that speeds response to those calls is well worth the investment; the use of GPS pays off in both public safety and the efficient use of firefighting equipment and labor. A related innovation is geographic information systems **(GIS)**, which provide access not only to street maps but also to geographic data such as water sources, hazardous materials storage sites, residences with disabled persons, and municipal hydrants. Another new tool is **pictometry**, an enhancement of GIS that generates a 3-D perspective of a given area. Understandably, pictometry is an enormous help to firefighters, allowing them to assess a fire scene even before arriving at the site. The size of the building, access to the building on all sides, and the characteristics of the surrounding area are revealed. These inventions all require a person with vision to understand the need and create the product that answers that need.

#13

GIS

Provide access not only to street maps but also to geographical data such as water sources, hazardous materials storage sites, residences with disabled persons, and municipal hydrants.

pictometry

An enhancement of GIS that generates a 3-D perspective of a given area.

◆ VOLUNTEER OPPORTUNITIES

The Community Emergency Response Team (CERT) is a FEMA outreach program designed to train citizens in elementary emergency response procedures. CERT team members are volunteers who receive education in disaster preparedness for the kinds of hazards that may impact their area. They learn basic disaster response skills, such as fire safety, light search and rescue, team organization, and disaster medical operations. Using this training, which is taught both in the classroom and during exercises, CERT members can assist others when an emergency arises in their neighborhood or workplace and professional responders are not immediately available to help.

◆ THE FUTURE OF FIRE AND EMERGENCY CAREERS

#16

Department of Homeland Security

Created in 2002 primarily from a conglomeration of existing federal agencies in response to the terrorist attacks of September 11, 2001. Its purpose is to protect the nation against threats to the homeland. Its efforts affect firefighters and emergency personnel.

bioterrorism

Relating to or involving the use of toxic biological or biochemical substances as weapons of war.

The future of careers in the fire service is bright. As baby boomers retire, new job opportunities and opportunities for promotions will become available. The fire service will undoubtedly continue to take on new tasks and revise old ones, and this evolution will offer further opportunities as the need for technical specialists grows. The increase in educational requirements will lead to career opportunities in the educational field as new degree programs and specialty training courses are created at fire academies, colleges, and universities.

After the disaster of 9/11, the **Department of Homeland Security** has come to the forefront of national issues and is the center of a growing source of new career opportunities. New jobs have already been created, not only in the fire and emergency fields but also in military and law enforcement, scientific research, cybertechnology, biomedical technology, disaster assistance and relief, air marshal operations, and intelligence operations, to name but a few. For example, as recently as 10 years ago, the threat of **bioterrorism** was not of much concern to our nation's firefighters. Today, all emergency personnel must be familiar with its deadly potential and trained in ways to contend with that danger—as well as with the other threats posed by hostile governments and extremist groups. These dangers, already grave, are likely to escalate. Fire and emergency service professionals will need to be prepared to meet these new challenges. A reality check shows that firefighters are first in, last to know, and required to do the most with the least, but it is their responsibility to anticipate the worst and plan the best possible response.

◆ SUMMARY

Although fire and emergency service jobs may be hard to obtain, the profession will continue to be a stable source of livelihood. Firefighting is forecasted to remain a very competitive field for the next several years and to provide a considerable amount of job security while also providing an honorable sense of public service. The nation's economic status plays a role in the fire and emergency service profession. A healthy economy results in the growth and development of communities, and this growth increases the need for additional fire services and individuals to fill those new positions.

Over the past years, the firefighter has taken on new roles and responsibilities and begun the evolution from tradesperson to professional. Titles in this profession and the nature of the job can vary greatly depending on circumstances. Personnel may be employed in either the public or the private sector, but all fire and emergency professionals protect communities against the loss of life, injury, and destruction of property by fire. These professionals have the vital goal of bringing order to chaos. The challenges are real—and so are the rewards.

Stay Safe

On Scene

This chapter covers the professional opportunities available in the fire and emergency service. Most firefighters started out as volunteers in their local community—and many remain volunteers. For those who do decide to take up firefighting as a paid profession, laws and policies are in place to protect their jobs in the event that they are called to help in national disasters. At present, however, there is no similar protection against termination or demotion for the volunteer firefighter or EMS worker.

On March 9, 2006, Senate Bill S. 2399, the Volunteer Firefighter and EMS Personnel Job Protection Act, was introduced to protect the jobs of volunteer emergency services personnel responding to a presidentially declared national disaster for up to 14 days per calendar year. This bill would protect volunteers from having to choose between their careers and serving their country. As of October 25, 2007 the Volunteer Firefighter and EMS Personnel Job Protection Act had not passed Congress.

1. Give concrete examples of how volunteer firefighters can be deployed away from their homes to respond to national disasters and why these volunteer firefighters should not have their livelihood jeopardized.

Review Questions

1. What is the difference between the firefighter recruit and firefighter positions?
2. In the fire and emergency professions, employees can work with various entities or departments on a single incident. Name a few examples that firefighters and emergency personnel might encounter or be governed by on an emergency call
3. What additional requirements are placed on a firefighter paramedic over that of a regular firefighter?
4. Due to recent terrorist attacks in the United States and abroad, new career opportunities have been formed. Can you name a few?
5. What are the differences between a fire prevention specialist and a fire department training specialist?
6. What is the longest running public service campaign in history used for teaching forest fire prevention?
7. What jobs are available in the fire service that do not require actual firefighting?
8. What are some of the private sector jobs that would help you to prepare for the position of firefighter?

Life and Career of Eddie Burns Who Rose Through the Ranks to Become the First Black Dallas Fire Chief

Eddie W. Burns, Sr., became chief of the Dallas Fire-Rescue Department on April 19, 2006. He is 46 years old with 27 years of firefighting experience with the Fort Worth Fire Department. He started his career in the fire service in July 1979. After serving nine years in the Operations Division, where he rose to the rank of lieutenant, he accepted the demanding position as a member of the department's Fire Investigation/Bomb Squad. The assignment required him to become a certified peace officer, and he obtained certification as an explosive ordnance technician through training from the FBI and U.S. Army. On his promotion to captain in 1994, he returned to the Operations Division and worked at one of the busiest stations in the city. In 1997, he obtained the rank of battalion chief and was assigned as the shift commander for the cities west-side battalion. He was selected in 1999 to assume the challenging role of chief training officer. In August 2000, he was promoted to deputy chief of Educational Services. He was reassigned to the Executive Services Division in September 2003, where he served as the executive deputy fire chief until coming to Dallas. He has been active in the fire service as an adjunct instructor with Weatherford and Tarrant County Colleges. He obtained an associate degree in Fire Administration from Weatherford College in August of 2004. He graduated cum laude from Dallas Baptist University and is also a graduate of the Executive Fire Officer Program from the National Fire Academy in Emmitsburg, MD, and John F. Kennedy School of Government from Harvard University.

He is married to Jackie Burns, and they have three grown children: Talia, Eddie, Jr. (firefighter in the Houston fire department), and Cory. He volunteers with the Sickle Cell Foundation as a camp counselor and serves on the advisory committee for the Boy Scouts. Chief Burns has his own foundation called "All That I Am I Owe." This foundation was formed to provide training and certification to underprivileged young men and women.

Notes

1. *Firefighting occupations*. (2005, December 20). Retrieved January 29, 2007, from U.S. Bureau of Labor Statistics website: http://www.bls.gov/oco/ocos158.htm.

2. Nilo, James R. (2000, February 1). Have trainer, will travel. *Fire Chief Magazine*, Retrieved January 27, 2007, from http://firechief.com/mag/firefighting_trainer_travel/index.html.

3. *Hurricane Katrina*. (2005). Retrieved January 29, 2007, from Texas Engineering Extension Service (TEEX) website: http://www.teex.com/teex.cfm?pageid=Agency&area=teex&templateid=1547

4. Blackmon, Kelley (2002). *"Sworn" versus "civilian" career development paths in the fire service in Nevada*. Greenspun College of Urban Affairs: Professional Papers. University of Nevada, Las Vegas Library Special Collection Section.

5. *Career: Fire investigators*. (2005). Retrieved January 29, 2007, from Minnesota Internet System for Education and Employment Knowledge website: http://www.iseek.org/sv/13000.jsp?id=100320

Suggested Reading

Coleman, Ronny J. & Granito, John A. (1979). *Managing fire services*. Washington, DC: International City Management Association.

Cote, Arthur E. (2004). *Fundamentals of fire protection*. Quincy, MA: National Fire Protection Association.

Cote, Arthur E. & Bugbee, Percy. (1988). *Principles of fire protection*. Quincy, MA: National Fire Protection Association.

CHAPTER 3

The Selection Process

Key Terms

ability/agility test, p. 37
background check, p. 40
ceiling breach
   and pull, p. 37

criminal conviction
   record check, p. 41
diversity, p. 42
NFPA 1582, p. 40

probationary period, p. 41
stair climb, p. 37

Objectives

After reading this chapter, you should be able to:

- List the standard phases of the hiring process.
- Describe the importance of preparation in the selection process.
- Give examples of body gestures that can be interpreted in an interview.
- Explain the purpose and importance of the probationary period.

◆ INTRODUCTION

Firefighting is a demanding career that requires very special qualities. Therefore, fire departments devote great care in choosing their entry-level employees. They look for candidates who like to work with people and with machinery, enjoy teamwork, are physically fit, and have an interest in public safety. Potential firefighters should also be adept at following instructions and making quick decisions and should be willing to take risks while remaining calm in stressful situations. The hiring process is extremely competitive. It is not unheard of for hundreds, if not thousands, of applicants to vie for a small handful of jobs. Success depends on much hard work and thorough preparation.

In this chapter, we will consider the various stages of the selection process. It is important to remember that the hiring process, rigorous though it is, is only the first step toward a firefighting career. On the average, becoming a full-time paid firefighter takes anywhere from three to seven years.[1] Nonetheless, getting hired is the first step. We will explore the do's and don'ts of each phase of the process and suggest strategies to optimize success. We will recommend how to build a résumé filled with the right kinds of education and experience and suggest tips for succeeding at test-taking and in face-to-face interviews.

PREREQUISITES

The job search begins with finding where job openings are posted in your area. Most fire departments and fire service organizations post employment opportunities on their city's website and in local newspapers. Some even hold job fairs at local colleges or community events. For a broader search, it is extremely useful to subscribe to job opportunity mailing lists or firefighter examination notification services, such as www.firecareers.com or www.firerecruit.com. Both of these services provide websites with notifications of when departments nationwide are accepting applications, and both are worth every penny of their subscription price.[2] They will save you the time and effort of calling each fire department and asking when they are next hiring.

Once you have identified a job for which you want to apply, it is wise to do a considerable amount of research on the position and the organization. The more you know, the better prepared you will be for the selection process. Talk to people who know the organization, look up its website if it has one, and familiarize yourself with its requirements. Entry-level firefighting jobs usually require that you are at least 18 years of age, possess a valid driver's license, and have, at the minimum, a high school diploma or GED. Be sure to check local and state requirements, however, as they do vary.

Once you have found a job opening for which you want to apply, obtain the necessary paperwork and take note of the application deadline. Most applications require high school transcripts to be attached. However, a candidate must read carefully the necessary paperwork because it may also require references or a résumé. Supplying the proper paperwork is the first impression that an employer has of you. Be certain that you get your application to the correct address by the deadline. The application should be typed or neatly handwritten in ink. Make additional copies of the application to use as trial drafts while you are gathering your information. That way, any mistakes or changes that you make will not appear on the final copy. Remember that all the information you provide will be considered fact, and facts should always be accurate and truthful. Once you have completed the application, make a copy and keep it in your files. This will help you fill out future applications accurately and more quickly.

If you meet the minimum qualifications and have submitted a complete and timely application packet, you will be invited to participate in the next phase of the selection process, a written test.

WRITTEN TEST

Two basic kinds of written tests may be given: (1) tests of general knowledge and (2) skills and tests of material related to firefighting. Most fire service organizations use standardized tests prepared by professional testing services. These tests are backed by extensive research and are inexpensive and easy to administer. They may be scored by the department or by a testing company. These tests can be civil tests or standardized tests that are bought from testing companies. Again, it is the candidate's responsibility to research the type of test that will be given for that particular position.

The typical exam consists of approximately 150 to 200 multiple-choice questions and is timed. Most fire departments require that you score at least 70% to continue in the selection process. (Keep in mind that preference points may be given to military veterans, so it is theoretically possible for them to score over 100%). Scoring well on a written exam does not necessarily mean that you will be an excellent firefighter, of course, but it does mean that you are eligible to go on to the next phase of the weeding-out process. Moreover, some departments use the written exam score as a part of an overall candidate ranking score, so a high score gets you off to a good start.

Some areas that may be assessed on the written exam are as follows:

- Reading comprehension
- Mechanical reasoning and spatial scanning
- Reasoning and problem solving
- Mathematical calculations and comparisons
- Basic emergency medical care
- Interpersonal skills/teamwork

Reading comprehension items usually relate to fire service terminology and problems commonly encountered by firefighters; they evaluate your ability to understand fire department data and training manuals. Mechanical reasoning items test your knowledge of the principles of mechanics and physics commonly encountered by firefighters. Reasoning and problem-solving skills test your ability to react and respond in emergency situations. The mathematical computation items determine your ability to perform the kinds of calculations that measure friction loss, engine pressure, nozzle pressure, and so on. Interpersonal skill items test your ability to understand orders and instructions issued on the fire ground and the ability to communicate with other firefighters. If the department handles medical emergencies, a test of basic medical knowledge may be included as well.

Candidates for a job opening will be ranked according to their score on the exam, from the highest to the lowest grade, so a good score is vital to going on to the next phase of the selection process (see Figure 3.1). As with any test, the key to success is preparation. It is the rare candidate who can walk in and get a high score without first doing some serious studying. Bookstores carry a wide selection of test preparation materials, including study materials and sample tests. It is also possible to borrow these from a public library.

FIGURE 3.1 ◆ Potential candidates taking a written exam.

PHYSICAL ABILITY/AGILITY

Fire departments usually administer an **ability/agility test** to determine the candidate's physical agility and general ability to meet the physical requirements of the job. A widespread test is the Candidate Physical Ability Test (CPAT). This standardized test is accepted by most employers and is being used increasingly in the selection process. Typically, this involves completing an obstacle course consisting of tasks similar to those a firefighter may encounter in the line of duty. These can include the dry hose drag, charged hose drag, halyard raise, roof walk, attic crawl, a ventilation exercise, a victim removal exercise, ladder removal and carry, a **stair climb** with hose, a **ceiling breach and pull**, a crawling search, and a stair climb with air bottles and hose hoist. Whatever the details of the course, this is a grueling test of strength and stamina. It is normally done wearing full firefighting gear, although some departments have changed to using a weighted vest to simulate the weight of the bunker gear. Scoring methods vary, as the test can be pass/fail, based on the candidate's ability to complete all the tasks within a specified time limit, or candidates may be assigned a score based on how long it took them to complete the course, with higher points given for faster times.

To do well, you must get in shape and stay in shape. You must develop a rigorous workout routine that emphasizes muscle groups used in lifting, pushing, pulling, carrying, and other actions related to firefighting work. Your program may include distance running, sprints, rowing, stair climbing with a weight pack, and weightlifting to develop both the upper and lower body. You may be allowed to visit the testing site to familiarize yourself with the testing layout and even to practice with the equipment that will be used in the actual test. Practice is bound to result in a better time.

On the day of the test, arrive early, well nourished, and well rested. Take appropriate clothing such as sweats, good athletic shoes, and gloves, if required. If the testing process will be long, make sure you have water and high-energy snacks on hand. If the day is hot, make sure you drink plenty of water before you start the test. Warm up and stretch just as you would before any strenuous workout (see Figure 3.2).

ability/agility test

A demanding physical test that can include dry hose drag, charged hose drag, halyard raise, roof walk, attic crawl, ventilation exercise, victim removal, ladder removal and carry, stair climb with hose, ceiling breach and pull, crawling search, or stair climb with air bottles and hose hoist— each measuring physical strength and stamina.

stair climb

Applicant must lift a prepared hose bundle from the floor onto the shoulder and climb a specific number of the stairs in a specified time frame. Applicant must be wearing an air tank and harness (without valves, hose, or mask) during this evolution.

ceiling breach and pull

This event is designed to simulate the critical task of breaching and pulling down a ceiling to check for fire extension. It uses a mechanized device that measures overhead push-and-pull forces and a pike pole. The pike pole is a commonly used piece of equipment that consists of a six-foot-long pole with a hook and point attached to one end.

FIGURE 3.2 ◆ Physical ability/agility test.
Courtesy of TEEX-Texas A&M System.

Stop and Think 3.1

The fire chief has selected you to develop and implement a written exam and physical ability test for potential new hires. Based on the readings or your personal experience, answer the following question:

> Why is it important for fire departments to develop and implement medical and physical fitness standards that apply equally to all firefighters, based on the duties they are expected to perform? Explain your answer in detail.

◆ INTERVIEWING SKILLS

Many individuals who want to go into the fire and emergency services discover that they do not have the skills necessary to go through a job interview with confidence. Although curriculum requirements vary from state to state, rarely is time devoted to interviewing techniques, so it is hardly surprising that many cadets find themselves uncomfortable when facing the interview situation for the first time. Some become so frustrated that they actually give up on a fire/emergency career path completely. Homework and practice can make all the difference between success and failure in the interview arena.

FIRST INTERVIEW OR CHIEF'S INTERVIEW

Never underestimate the importance of first impressions. The candidate's initial interview, which can be held with the chief alone or with a committee of department representatives, is crucial. Typically, if the first interview is held with a panel of committee members, they will narrow the selection down to a select few. Then, a second interview may occur, depending on the organization or fire department, with the candidate and the fire chief alone. This interview can actually alter an applicant's position in the rankings. A candidate should learn as much as possible about the department and the chief before the interview. For example, what are some of the core values of the organization? What are some of the key issues or projects with which the department is involved? Although you will be answering many questions, don't be afraid to ask questions as well. Make it clear that you know something about the department and are eager to learn more. The fire service is looking for good communicators. Also, don't be overly modest. Talk about yourself in terms that demonstrate your ability to work as part of a team and your knowledge of firefighting terminology and firehouse etiquette. Show that you care about honesty and integrity.

Another way you can prepare for the interview is to have experienced firefighters give you a mock oral interview. The mock interview should be with people who have been there and who know what to expect. Their questions can give you insights about issues that may be raised during the interview and prepare you to give thoughtful answers. Sometimes writing out a practice answer or rehearsing your delivery in front of a mirror or using an audio or video recorder can be helpful. The more practice you have and the more feedback you get from others, the more confidence you will have during the actual interview.

Sample Interview Questions

1. Describe your work history and experience.
2. Describe credentials you hold that qualify you for this job (education, certifications, and so on.).
3. What are your major strengths?
4. What do you consider development areas?
5. What characteristics do you feel mark a good supervisor?
6. Describe a conflict you experienced with a previous supervisor and how it was resolved.
7. What do you think is a satisfactory attendance record?
8. How do you define "pressure"? How well do you perform in that type of situation?
9. In what type of work environment do you excel?
10. Describe when you went above and beyond the call of duty in order to get a job done.

The interview is typically casual in tone, but be sure you dress appropriately—for example, wear a suit and tie (seeFigure 3.3). Don't overlook the importance of body language. Sometimes candidates can be so concerned with conveying their knowledge and skills that they forget the nonverbal message their appearance is sending to the interviewer. Seeming at ease and confident can be just as important as giving the right answers. Interviewing tips are listed below:

Posture. Sit in a comfortable posture, relaxed but alert. Avoid stiffness or nervous, fidgeting movements such as foot tapping. Breathing should be calm and even. These are signals that you are ready to communicate.

Arms. Sit with your arms uncrossed and your hands relaxed and open. This suggests that you have nothing to hide.

FIGURE 3.3 ◆ Potential candidate dressed in a suit for a first interview.
Courtesy of TEEX-Texas A&M System.

Eyes. Look into the other person's eyes, particularly when they are speaking. Glancing away indicates that you are nervous or not interested. And don't wrinkle your brow!

Nodding. Using nods to punctuate key things the other person has said signals agreement, interest, and understanding.

Smile. A smile signals that you're warm and personable.

Leaning closer. Reducing the distance between yourself and the interviewer, particularly when the interviewer is speaking, indicates that your interest is up and your barriers to communication are down. Leaning back or slouching in your chair can signal boredom or a lack of respect.

Gestures. Talking with your hands, particularly with your palms open, indicates involvement in the conversation and openness to the other person.

Don't interrupt. Be sure the interviewer is finished speaking before you answer or make a comment.[3]

◆ PERSONAL HISTORY

MEDICAL EXAM

NFPA 1582

Standard on Comprehensive Occupational Medical Program for Fire Departments. Standard covers minimum medical requirements for firefighters, including full-time or part-time employees, and paid or unpaid volunteers.

background check

Verification that a candidate's experience or credentials are what they represent.

A preemployment medical exam is required for fire and emergency service employment. The National Fire Protection Association (NFPA) has established standards for the exam to ensure that it is an accurate test of a candidate's physical ability to perform firefighting tasks. **NFPA 1582**, *Standard on Comprehensive Occupational Medical Program for Fire Departments*. This standard covers minimum medical requirements for firefighters, including full-time or part-time employees, and paid or unpaid volunteers. This physical is similar to any other comprehensive medical examination. It typically begins with a patient history that includes, but is not necessarily limited to, the following:

- Medical history
- Surgical history
- Hospitalization history
- Social activities including smoking, alcohol use, exercise, stress, and diet
- Drug screening
- Employment history
- Health conditions including head/nose/mouth/throat, ears/hearing, eyes/vision, respiratory system, cardiovascular system, gastrointestinal system, liver/spleen/gallbladder, reproductive system, neurological system, psychological/mood, musculoskeletal system, endocrine/metabolic system, blood/lymph system, cancer, skin, allergies, and infectious/childhood diseases

You will be required to give blood and urine samples. The exam may also include a body fat composition analysis, hearing and vision tests, a treadmill stress test, a physical condition assessment, and any other medical tests that the department may request. The purpose of the exam is to assure that candidates are healthy and to establish a health baseline against which future test results can be compared.

BACKGROUND HISTORY

A **background check** is used to verify identity and to examine records for information about an applicant's character and behavioral history. References will be checked, and the information provided on the candidate's application will be confirmed: education, past employment history, and any certifications or licenses. An address history and social security number trace may be run, and a **criminal conviction record check** will be made at the county, state, and federal levels, including the sexual offender registry, county and federal civil records, drivers' history, credit reports, workers' compensation records, and results of previous drug testing.[4] Some departments or organizations require that potential employees submit to a polygraph exam.

criminal conviction record check

A search of an individual's history of any convictions under the law for felony, misdemeanor, and motor vehicle convictions.

◆ HIRING CONDITIONS

ELIGIBILITY LIST

After the final candidate list is compiled, it will be certified for a specified time frame, such as one year. If a position becomes available during that period, the top candidates including any with tied scores, will be called for interviews. As more positions become available, the department will move down the list. A candidate should call periodically to check the status of his or her name on the list. The eligibility list is typically valid for six months to one year; however, a candidate should verify the terms of the list. If time expires for the eligibility list, the department will normally schedule testing procedures again, and the whole process starts over.

PROBATIONARY PERIOD

New hires are on probationary status for a period of time that can run between three months and one year. Probation gives supervisors and fellow firefighters a chance to assess the candidate's abilities and character in real-life situations. This trial period should be considered a part of the selection process, just like the interview or the medical exam. Probation is also an opportunity for the new employee to become familiar with standard operating procedures and department guidelines and to take part in ongoing training programs. During the **probationary period**, an individual may be terminated for any reason. The probationary employee has the burden of proof to show that he or she can do the job. Supervisors should, however, keep track of all probationary employees and document their progress.[5]

probationary period

Introductory period of employment that allows the employee and agency to determine if the employee is suited for the job.

◆ MARKETING YOURSELF FOR THE FIREFIGHTING JOB

There is more than one path to a career in the fire and emergency services. Some firefighters start out in a civilian job, such as dispatch, administration, or fire prevention, before making the transition to firefighting. Working as a paramedic, nurse, medical assistant, or in any direct service capacity in an emergency environment are other possible routes to firefighting. A candidate who comes to the fire service already experienced in these fields has a definite edge over a raw recruit. Other avenues of valuable training and experience are listed below.

EMT STATUS

Many fire departments now require a minimum of EMT Basic status as a prerequisite for application. Depending on a department's needs, paramedic status or police officer certification can further improve your chances of getting a job.

FIREFIGHTER I AND OTHER CERTIFICATIONS

Some departments that do not conduct their own cadet training may require that the candidate be certified to the Firefighter I level before they will even accept his or her application. Other certifications that improve the chances of employment include hazardous materials technician, investigator, and prevention specialist.

VOLUNTEER EXPERIENCE

Volunteering is a great way to expand your horizons. In addition to the experience you gain, you will be demonstrating community spirit and a readiness to help others. Joining a volunteer fire department is an obvious move in the right direction, but you may also contact your local fire department and ask if you could volunteer in other areas, such as in administration or in a fire prevention or HazMat program. In fact, almost any kind of volunteer experience will enhance your résumé, showing that you are well rounded and service oriented.

BILINGUAL ABILITY

In our increasingly diverse society, language skills are becoming valuable in many occupations, and the fire service is no exception. Some departments give bilingual candidates preference in the hiring process.

CLEAN DRIVING RECORD

A clean driving record means no accidents and no moving violations. This is important for safety reasons, because at some point you may be driving fire apparatus. Also, if you have too many accidents and moving violations on your record, you may not be eligible for coverage by the department's insurance. This can eliminate you from the hiring process.

EXCELLENT PHYSICAL FITNESS

If you don't have a physical fitness routine now, you need to get one! Pair up with a buddy or two and develop a regular training schedule. Obtain guidance from a professional on what to do and how to do it to obtain the maximum results. Weightlifting alone won't be enough. You need a combination of aerobic, cardiovascular, and strength-building activities. Also, a well-balanced diet will enhance and speed up the results of your fitness training. Physical training is critical preparation for the academy.[6]

DIVERSITY

diversity
Ethnic, gender, racial, and socioeconomic variety in a situation, institution, or group.

Many fire departments are increasing the **diversity** of their applicant pool through targeted recruitment of minorities and women. To be sensitive to a community's needs, the department needs a workforce reflective of that community. This means that the ideal fire department should include both women and men from the various racial, ethnic, and socioeconomic groups that the department serves.[7] Women and minority applicants should be proud of what they have to offer and should not hesitate to point out that their cultural or ethnic background will be an asset to the department, breaking down barriers that would otherwise isolate the department from sectors of the community.

Stop and Think 3.2

The fire chief has received a complaint that the fire department's hiring practices are not fair toward women. You have been appointed to a committee made up of city employees that will examine the department's hiring practices and decide if its workforce reflects the makeup of the community. Based on the readings and your personal experience, answer the following questions:

1. What is your definition of diversity? How does it affect your everyday life?
2. Why is it important to have diversity in the fire service?
3. What are the benefits of having a diversified workforce in the fire and emergency services field?

◆ **SUMMARY**

The selection process for fire and emergency personnel is not easy, and it takes time. Given the critical nature of the job, the responsibility to the public, the dangers and rigors involved, the need for specialized knowledge, the willingness to be ready to face any emergency, to work irregular and long hours as part of a team, and to be an articulate communicator, it is not surprising that the process is so demanding. The department needs to be certain that each new hire has the necessary skills and qualifications to do the job and, very importantly, that he or she has the character traits that make for a good community role model.

Although the process is challenging, the various stages of the process are fairly predictable. This chapter explained what a candidate should expect for the application process and suggested ways to improve performance at each stage. In addition, various ways were listed in which the prospective firefighter can strengthen his or her résumé, particularly through related training, certifications, and volunteer experience.[8]

According to the *U.S Department of Labor Statistics Handbook for 2006–2007*, employment of firefighters is expected to grow faster than the average for all occupations through 2014. Firefighting is a highly respected, rewarding profession with a secure future. One of the keys to success, however, is to understand the challenges and to prepare for them. The road to that full-time paid job may take you months or even years. It will involve hard work—research, study, and physical exercise—and personal sacrifice.

■ ■

On Scene

Although there will be an increase in hiring of personnel in fire and emergency services over the next decade, competition will be keen. The fire service attracts all types of applicants from candidates with a GED to individuals in the white-collar professions such as accountants and attorneys with graduate degrees.

1. Why, do you think, does the fire service attract so many different types of individuals?
2. Since the fire service attracts various individuals, what can you do to make yourself more marketable?

Review Questions

1. What two pieces of documentation do most departments require of applicants before they may apply for the firefighter exam?
2. What is generally the first step in the selection process?
3. List two ways to find out when application periods are open for firefighter exams.
4. List a resource for study material used to prepare for the written firefighter exam.
5. List two ways to prepare for the physical agility and ability test.
6. List two ways to prepare for the oral exam.
7. What is the purpose of the probationary period?

Notes

1. Prziborowski, Steve (2006, August 15). *So, you want to become a firefighter? 5 guidelines to assist you in becoming a firefighter.* Chabot College, Hayward, CA.
2. Ibid.
3. *How to become a firefighter.* (2006, May 16). Retrieved January 30, 2007, from the FireEmployment.com website: http://www.fireemployment.com/enav. asp?nav=nav2.
4. *Types of background check: You can find out the truth about anyone!* (2006, May 16). Retrieved January 29, 2007, from the Net Detective website: http://www.softsite.org/background-check-online.
5. Graham, Gordon J. (2004, January 15). *Why things go right–why things go wrong: Risk management considerations in the fire service operations.* Texas Association of Fire Educators Conference.
6. *How to become a firefighter.* (2006, May 16). Retrieved January 30, 2007, from the FireEmployment.com website: http://www.fireemployment.com/enav. asp?nav=nav2.
7. Crawford, Brian (2004) *Patchwork Force.* Retrieved May 12, 2008, from http://firechief.com/leadership/management-dministration/firefighting_patchwork_force.
8. *Fire fighting occupations.* (2005, December 20). Retrieved January 29, 2007, from the U.S. Bureau of Labor Statistics website: http://stats.bls.gov/oco/ocos158.htm

Suggested Reading

Bennett, Lawrence T. (2008). *Fire service law.* Upper Saddle River, NJ: Pearson Education, Inc.

Bruegman, Randy R. (2009). *Fire Administration I.* Upper Saddle River, NJ: Pearson Education, Inc.

Chikerotis, Steve (2006). *Firefighters from the heart: True stories and lessons learned.* Clifton Park, NY: Thomson Delmar Learning.

Lasky, Rick (2006). *Pride and ownership: A firefighter's love of the job.* Tulsa, OK: PennWell Corporation.

Mahoney, Gene (2007). *Fire department interview tactics.* Clifton Park, NY: Thomson Delmar Learning.

Training and Higher Education

4 CHAPTER

Key Terms

andragogy, p. 57
Associate of Applied
   Science, p. 56
baccalaureate programs,
   p. 57
burn building, p. 47
certification programs, p. 56
drafting pit, p. 47

drill tower, p. 47
higher education, p. 46
International Fire Service
   Accreditation Congress
   (IFSAC), p. 57
manipulative training, p. 47
National Fire Academy
   (NFA), p. 51

NFPA 1041, p. 53
NFPA 1403, p. 47
Pro Board, p. 62
tactical training, p. 46
training, p. 46
training bureau, p. 54

Objectives

After reading this chapter, you should be able to:

- Discuss National Fire Protection Association standards geared toward training.
- Explain the difference between training and higher education.
- Describe the various types of certification and degree programs available.
- Explain FESHE's professional development model.

◆ INTRODUCTION

In this chapter, we will explore firefighter education, including both training and higher education. We will explore the training options that are available to prospective firefighters, as well as the opportunities for higher education that are available to fire professionals. We will describe certification programs, two-year associate's degree programs, and higher degree programs. We will describe the roles of the National Fire Academy, International Fire Service Accreditation Congress, and Pro Board in improving and standardizing fire service education in the United States.

It is critical to understand that a firefighter's education is open-ended, a life-long, or at least career-long, process. Skills once learned must be refreshed, and then with the advent of new technology, new skills need to be acquired. The knowledge that was sufficient to a firefighter 10 years ago is not sufficient for today's challenges.

training
Obtaining knowledge for the maintenance of skills required for actual tactical tasks performed during an emergency response.

Firefighting is physically and mentally challenging work and, of course, highly dangerous. By **training**, we refer to the knowledge and skills that enable the firefighter to do this vital work with efficiency and safety. Some of the knowledge modern-day firefighters need may surprise you. For example, an understanding of building construction is essential because how a building is designed and the materials used in its construction influence how a fire will behave. The more firefighters know about these factors, the better they will be able to anticipate a fire's location and its movement.

As a firefighter's career advances, he or she will want to take advantage of opportunities to earn specialty certifications in areas such as hazardous materials and fire investigation. We will look at the requirements of some of these certifications programs.

◆ HIGHER EDUCATION

higher education
Obtaining knowledge for certifications, professional development, or college credit. The act of accumulating new facts to augment your personal and mental database.

By **higher education**, we refer to advanced learning at the college, university, or equivalent level. The terrorist attacks of September 11, 2001, taught us how important it is to have a well-educated fire service workforce—one that is always prepared. This is true at every level of firefighting, but it is particularly critical within the ranks of management. We will consider the growing value of associate's, bachelor's, and even advanced degrees for firefighters seeking promotion to higher or more specialized positions within their organizations.

Despite the importance of training and education, agencies and fire departments vary in their support of education and training for their members. Often the amount and kind of training they encourage are based on the department's mission and financial capabilities. Some departments protect only airports, whereas others do structural firefighting and provide emergency medical services. Understandably, training and education generally must conform to the goals and mission of the organization.

As a firefighter, it will be your personal responsibility to identify your career goals and to research opportunities for advanced training and education. These opportunities may be available from your employer, or they may involve on-site or online learning offered by an educational institution or by the National Fire Academy. The more you learn, the better prepared you will be to succeed and further your career in the fire science field.

◆ TRAINING

Firefighter training is the single most important factor in determining if firefighters are prepared to meet the challenges of the environments in which they work. Training means the difference between success and failure.

We can distinguish two basic types of training: tactical and manipulative. **Tactical training** is another word for know-how. The goal of tactical training is to teach what works and what doesn't in responding to different emergency situations. This is the mental part of firefighting. **Manipulative training** is hands-on practice. In manipulative training, skills are learned by actually operating firefighting equipment and tools and carrying out rescue procedures. The more often a firefighter practices a procedure, the more effective and confident he or she will be in a real emergency situation. Both types of training are designed to teach the firefighter to know the capabilities and limitations of their equipment. Just as important training teaches them to recognize their own capabilities and limitations.

Formal training takes place in either a traditional, instructor-led classroom or in field exercises. Most fire departments have classrooms or at least a designated space within the department for training. Field exercises require a specially designed setting with equipment and props that simulate fire conditions. These could include a **drill tower** for ladder training or multistory operations, a **burn building** for actual live fire training (see Figure 4.1), and a **drafting pit** for maintenance of pump-operation and pumper-testing skills. A drafting pit is usually a large square concrete hole in the ground that may hold 50,000 to 70,000 gallons of water from which a fire apparatus can draft water. These structures can be expensive to build and maintain, so a small fire department may partner with larger departments to gain access to their facilities. It is paramount that fire instructors know, understand, and not deviate from the **NFPA 1403**, *Standard on Live Fire Training Evolutions*. This standard provides the framework to improve safety during live fire evolutions and exercises. NFPA 1403 covers conditions such as safety, site preparation, water supply, training plan, fuel, and ventilation. For further

tactical training

Comprehensive training, hands-on learning, and analysis for emergency incidents.

manipulative training

Hands-on learning of operations, equipment, and tools. In the fire service, this type of learning exhibits safety and proper techniques.

drill tower

A tower and training facility for firefighters.

burn building

Training tower used so that new recruits can practice their skills.

drafting pit

A poured-in-place concrete pit used to simulate drafting operations.

NFPA 1403

A standard that provides the framework to improve safety during live fire evolutions and covers conditions such as safety, site preparation, water supply, training plan, fuel, and ventilation.

FIGURE 4.1 ◆ There are a variety of burn buildings, including the burn building and tower at the Tennessee Fire Training Center.

information referring to this standard, please refer to the National Fire Codes under NFPA 1403.

As in any field of study, informal training also has a role to play in preparing firefighters for their jobs. Mentoring and peer networking are invaluable. Beginning firefighters should take every opportunity to discuss their career goals and concerns with senior professionals. Training may even be obtained before entering the fire service. Preservice learning can begin by joining the Fire Department Explorers, a scouting program designed for teens interested in firefighting, or by signing up for reserve or cadet programs open to young adults. Joining a volunteer fire department can also be the first step to a professional career.

For most new firefighters, however, the initiation to bona fide training is the fire academy or rookie school. This is only the first stage, however. After a firefighter completes academy training and enters the fire service, the need for continuing training becomes obvious. New technology generates new hazards, such as chemicals and other hazardous materials. Technology also provides new and improved methods and equipment for dealing with these hazards. Firefighters must constantly update their knowledge and skills. They will spend a good proportion of their time attending in-service programs run on-site at their place of employment or at state or national training facilities. Training is not just an accessory to the firefighter's job; it is an integral part of that job.

FIRE DEPARTMENT TRAINING OR SPECIAL OPERATIONS BUREAUS

Most larger fire departments have training divisions or special operations bureaus whose mission is to provide the highest quality training to their firefighters. These divisions are in charge of developing and teaching the department's educational initiatives, utilizing both on-site classroom training and any needed field exercises. The bureaus also provide training that is periodically mandated by the state.

A secondary but very important responsibility of the Training or Special Operations Bureau is to archive the department's training records. These provide proof of the education given to each firefighter. Training records are essential for compliance with the law in most states and must be as complete as possible. Complete records are important when the department is being evaluated, audited by an outside agency, or investigated as a result of a workplace accident, harassment issue, or patient care situation. In today's litigious culture, training records become a vital safeguard for the firefighter and the department (see Figure 4.2).

THE U.S. FIRE ADMINISTRATION AND THE NATIONAL FIRE ACADEMY

In the early 1970s, the federal government recognized that the nation's fire problem was changing and appointed a commission to study the situation. The commission published a document titled *America Burning*: *The Report of the National Commission on Fire Prevention and Control* (1973), which recommended the formation of a federal fire agency that would provide support, including training and educational support, to state and local governments and to private fire organizations. A year later, Congress passed Public Law 93-498, The Federal Fire Prevention and Control Act of 1974. This law created the U.S. Fire Administration (USFA), which today operates under the direction of the Federal Emergency Management Agency (FEMA). The branch of the USFA that is concerned with education—designing fire control courses

LOYDVILLE FIRE & EMERGENCY SERVICES TRAINING REPORT

2. NAME AND TITLE OF INSTRUCTOR	3. DATE	28 March 2007	
	STARTED	FINISHED	TOTAL
Jason Loyd, Battalion Chief	1230	1355	85 minutes

4. SUBJECT	5. NO. ATTENDING CLASS
Sprinkler Systems	5

6. LOCATION OF TRAINING	7. EQUIPMENT AND TRAINING AIDS USED
Fire Station 5	TV/VCR
	Training Room
6A. TYPE OF TRAINING	IFSTA Essentials of Fire Fighting, 4th Edition
	IFSTA Sprinkler Systems Video
Shift Training	IFSTA Private Fire Protection, 2nd Edition

8. GALLONS / POUNDS AGENT / FUEL USED			
FUEL	DRY CHEMICAL	HALON	FOAM
N	O	N	E

9. TRAINING OBJECTIVE

Upon completion of the class fire personnel will be familiarized with sprinkler systems and components that are being used in the fire service today in but not limited to:

- ∞ Water Supply Sources for Sprinkler Systems
- ∞ Water Flow Through Fire Department Connection
- ∞ Sprinkler Control Valves Location and Appearance
- ∞ Main Drain Valve
- ∞ Operating Main Drain Valve
- ∞ Complete Coverage vs. Partial Coverage
- ∞ Automatic Sprinkler Systems

10. DESCRIPTION OF TRAINING CONDUCTED

Fire personnel were briefed on water distribution systems and components. Personnel also watched an IFSTA Essentials Sprinkler Systems video that touched on the techniques and safety procedures of: sources of water supply, types of sprinkler systems, and how sprinklers work. After the video a reading from IFSTA Essentials of Fire Fighting, 4th edition was discussed about sprinkler systems. In addition reading from IFSTA Private Fire Protection, 2nd edition was conducted in a classroom setting. Questions and answers were continual during the training class and safety was stressed at all times.

11. BADGE NUMBER	12. PERSONNEL ATTENDING TRAINING *(CONTINUED ON REVERSE)*	
	PRINT NAME	SIGNATURE
088	Joe Brown	
090	Steve Johnson	
091	Kevin Michaels	
092	Mike Smith	
093	Melissa Rocha	

FIGURE 4.2 ◆ Training forms.
Courtesy of Jason Loyd.

12. PERSONNEL ATTENDING TRAINING (CONT.)		
11. BADGE NUMBER (CONT.)	PRINT NAME	SIGNATURE

13. MAKE UP TRAINING:		14. DATE: 28 March 2007
BADGE NUMBER	PRINT NAME	SIGNATURE
089	Kurt Harris	

APPROVAL	
INSTRUCTOR *(SIGNATURE)*	ASST. CHIEF *(SIGNATURE)*

FIGURE 4.2 ◆ (*Continued*)

and programs and making them available to firefighters across the nation—is the **National Fire Academy (NFA)**.

The NFA maintains a campus in Emmitsburg, MD, where it offers education and training courses aimed mainly at management-level fire personnel and instructors. It also provides classroom courses for volunteer and career firefighters that are held at locations across the country, usually in conjunction with state or local fire services. On-line courses and a variety of audio and video training materials are available as well. It is estimated that the NFA has helped to train over 1,400,000 students since 1975. This organization has benefited not only the students but also the American public. Countless lives have been saved and property losses prevented as a direct result of this training and education.[1]

National Fire Academy (NFA)

Concerned with education—designing fire control courses and programs and making them available to firefighters across the nation.

◆ ADVANCEMENT AND PROMOTION

Advancement in most fire agencies depends on successfully completing competitive examinations or by earning advanced degrees. A firefighter seeking a promotion is often required to participate in continuing education programs that lead to **certification** or in higher education programs that lead to college degrees.

CERTIFICATION

Fire personnel may obtain specific certifications by successfully passing a written and practical state certification test in the specialty discipline. Certification tests for competencies are provided by the bureau or by a professional association or institution that acts as the host entity. Candidates must meet minimum scoring requirements on the written exam, and practical skills are typically graded on a pass/fail basis.

Although it is true that a college degree does not put out a fire, the increase in skills and training required to meet today's operational needs in a fire department successfully has made formal education necessary. Entry-level firefighters who are equipped with a higher education degree may receive faster promotions to management positions and to specialized roles such as fire inspector, fire investigator, or even fire chief.

Whatever their educational background, firefighters must be articulate, educated, skilled, and credentialed in order to earn the credibility required by the public safety and public service communities. These requirements are essential for administrators but are also important for shift commanders, company officers, and firefighters. Individuals holding these positions are the most visible in the fire service because they develop relationships with other agencies and with the public.[2]

SPECIALTY CERTIFICATIONS

Tables 4.1 and 4.2 list NFPA certifications and the requirements for each certification. The certifications in Table 4.1 reflect levels of general professional preparedness and preparedness for special job designations, such as instructor or investigator. The certifications in Table 4.2 refer to high-level technical specializations.

Note: For the most current and up-to-date information on the standards for each of these certifications, please refer to the National Fire Protection Association (NFPA).

TABLE 4.1 ◆ Specialty Certifications

Level of Certification	Basic Requirements
Firefighter I	Obtain a general knowledge of fire department organization, operations, tools and equipment, the role of the firefighter, hazardous materials awareness, and the mission of the fire service.
Firefighter II	Meet the requirements of Firefighter I. Obtain a more advanced knowledge of fireground operations and fire suppression, hazardous materials operations, and rescue techniques.
Fire Instructor I	Learn basic techniques and methods of teaching, use of teaching aids, and proper presentation skills.
Fire Instructor II	Meet all requirements for Fire Instructor I. Emphasis is on curriculum development and design.
Fire Investigator	Learn basic techniques and skills to determine origin, cause, and development of a fire.
Public Fire and Life Safety Educator I	Obtain basic knowledge and skills in a realm of different areas such as fire prevention education, babysitter training, and CPR.
Marine Firefighting for Land-Based Firefighters—Awareness Level	Meet the requirements of Firefighter I and II. Obtain knowledge and skills that structural firefighters may encounter during marine fire suppression in a maritime environment.
Fire Officer I	Obtain basic knowledge and skills that demonstrate the ability to perform tasks at the supervisory level.
Fire Officer II	Meet all requirements for Fire Officer I. Obtain advanced knowledge and skills that demonstrate the ability to perform tasks at the supervisory/managerial level.
Inspector I	Obtain basic knowledge and skills in fire prevention, fire safety, plans review, hazardous materials storage and handling, fire extinguishing systems, building construction, building codes, and code compliance.
Hazardous Materials Technician	Meet all requirements for Hazardous Materials Operations Level. Become trained in stopping the release or potential release of a hazardous material.
Airport Firefighter and Other Firefighting Levels	Meet the requirements of Firefighter I and II. Obtain a more advanced knowledge of fireground operations and fire suppression, hazardous materials operations, and rescue techniques. In addition, meet the basic knowledge and skills of airport firefighter.
Fire Department Incident Safety Officer	Obtain basic knowledge and skills that demonstrate the ability to recognize and handle safety responsibilities and duties.
Driver/Operator-Pumper	Meet the requirements of Firefighter I. Learn general principles of pump operations and obtain the knowledge and skills needed to be a safe fire apparatus driver/operator-pumper.

TABLE 4.2   Other Firefighting Levels

Level of Certification	Basic Requirements
Rope Rescue Technician	Meet the requirements of fire department medical first responder. Learn general knowledge and skills pertaining to technical rescue and rope rescue.
Surface Water Rescue Technician	Meet the requirements of rope rescue technician. Learn general knowledge and skills pertaining to swiftwater/flood rescue.
Vehicle and Machinery Rescue Technician	Meet the requirements of basic technical rescue. Learn general knowledge and skills pertaining to vehicle rescue.
Confined Space Rescue Technician	Meet the requirements of rope rescue technician and hazardous materials operations. Learn general knowledge and skills pertaining to confined space rescue.
Structural Collapse Rescue Technician	Meet the requirements of basic technical rescue, rope rescue technician, confined space rescue technician, and trench rescue technician. Learn general knowledge and skills pertaining to structural collapse rescue.
Trench Rescue Technician	Meet the requirements of basic rope rescue and trench rescue operations. Learn general knowledge and skills pertaining to trench rescue.
Dive Rescue Technician	Meet the requirement of open-water certification and basic technical rescue. Learn general knowledge and skills pertaining to rapid deployment search and rescue/recovery.
Wilderness Rescue Technician	Learn general knowledge and skills pertaining to wilderness rescue.

It is a good idea to subscribe to NFPA (website: www.nfpa.org) to have the most current information.

Stop and Think 4.1

Let us assume that you have been selected by the fire chief to review the different types of specialty certifications available to fire and emergency services personnel. Based on the readings and your experience in the firefighting field, answer the following questions:

1. List two types of specialty certifications and explain how they could help you advance your career.
2. List the NFPA standards for the two specialty certifications you have chosen.
3. Are any continuing education requirements needed to maintain the certifications?

NFPA 1041

A standard that requires instructors to be cognizant of the safety of their students to ensure that classroom and practical evolutions are conducted in a safe, controlled manner. Prerequisites and requirements are mandated.

TRAINING INSTRUCTOR

The NFPA has also established codes or standards that outline the responsibilities of the training officer or instructor. **NFPA 1041**, *Standard for Fire Service Instructor Professional Qualifications*, identifies three distinct levels of responsibility for instructors.

- Instructor Level I. Must assemble necessary course equipment and supplies, review prepared instructional materials, and deliver instructional sessions based on the prepared materials.
- Instructor Level II. Expected to be able to develop course materials independently and deliver those materials to students.

♦ Instructor Level III. Applies primarily to administrators and requires personnel to develop agency policy for training programs.

At all levels, NFPA 1041 requires that instructors are cognizant of the safety of their students. Safety is stressed to ensure that classroom and practical evolutions are conducted in a safe, controlled manner (see NFPA 1403).[3]

TRAINING OFFICER

The training officer occupies a mid management staff position, normally reporting directly to the fire chief. He or she is responsible for developing and coordinating all training for the fire department, including training in fire suppression, fire prevention, hazardous materials, rescue, and other associated areas. This position may require performing various command functions as well as all the regular duties of a firefighter. It is the responsibility of the **training bureau** or training officer to determine adequate levels of training based on several criteria—for example, whether the task is mandated by law, how often the task is performed, the time needed to complete the task, whether any incidents occurred while performing this task, and safety to others. These tasks would be considered job performance requirements (JPRs).

training bureau

A division of an organization or department whose mission is to provide training to firefighters so that they have the knowledge, skills, and abilities to mitigate emergency incidents safely while minimizing the risks to themselves, civilians, and the environment.

♦ HIGHER EDUCATION

As stated previously, the world changed on September 11, 2001. The terrorist attacks on that day affected all American citizens, but in no profession was the impact greater than in the U.S. fire service. New responsibilities were placed on the fire service, and along with these came the requirement for better-educated professionals to meet these responsibilities.

The increasing importance of advanced tactical and manipulative training for the career firefighter has been emphasized. In the past, vocational training, continuing education courses, and in-service training sessions fulfilled .most of these educational needs. Today there is an ever-increasing need for higher education, a need that goes hand in hand with the maturing of the fire service industry from the status of a skilled technical job to a full-fledged profession. The rewards of higher education come in many forms. For the public, the reward is improved safety. For the firefighter, the rewards include career advancement, promotions, pay incentives, and personal satisfaction.

Undergraduate college degrees at the associate's or bachelor's level are recommended for entry-level firefighters and for career firefighters interested in promotion. Advanced degree programs, such as master's and even doctoral degrees, will assist fire professionals interested in further advancement

A recent survey of metropolitan fire departments shows that approximately 75% of fire service professionals currently possess some type of two-year degree. Fewer than 25% of the same group have a four-year degree. This percentage is certain to rise, and it may rise quite rapidly as the need for higher education becomes increasingly clear and as fire departments translate that need into policy. The Charlotte (NC) Fire Department, for instance, stipulated that beginning in 2005 a four-year degree would be mandatory for advancement to a chief officer position.[4] This kind of proactive strategy will ensure that emerging fire service leaders are well equipped to handle the multitude of ever-changing and increasingly challenging problems that the future will bring (see Figure 4.3).

U. S. FIRE ADMINISTRATION
NATIONAL FIRE ACADEMY

DEGREES AT A DISTANCE PROGRAM

Certificate of Completion

This Certificate of Completion is to acknowledge that

Jason B. Loyd

has successfully completed six Degrees at a Distance
Program courses offered by:

SUNY Empire State College

The courses completed are:

Managerial Issues in Hazardous Materials
Advanced Fire Administration
Political and Legal Foundation of fire Protection
Incendiary Fire Analysis and Investigation
Personnel Management of the Fire Service
Analytic Approach to Public Fire Protection

Superintendent, National Fire Academy   Institution Representative

December 11, 2001   December 11, 2001

Date   Date

16-52 (6-98)

FIGURE 4.3 ◆ U.S. Fire Administration National Fire Academy Degrees at a Distances Program Certificate of Completion.
Courtesy of Jason Loyd.

certification programs

Shorter educational programs that provide a certificate verifying expertise in a specific trade or discipline.

Associate of Applied Science

Degree program usually two years in length.

THE FIRE SCIENCE TECHNOLOGY CURRICULUM

A firefighter can pursue numerous avenues when it comes to obtaining a degree related to fire service. Some colleges offer undergraduate **certification programs** in addition to credit-based degree programs. Certification programs generally focus on expanding students' understanding of one area of fire service work, such as firefighting, fire inspection, or fire investigation. Courses universal to these certification programs include fire prevention, fire chemistry, and firefighting strategies and tactics.

Most colleges allow students to transfer credits earned in a certification program to a degree program from state to state. Students may earn a certificate without completing all of the requirements of the degree program, but if they later decide to work toward an associate's degree, their accrued credits will count toward that degree. For example, a college may offer a Basic Fire Fighter certificate program in which a student completes fire academy and emergency medical technician training. If he or she decides to pursue an associate's degree in fire science, the college credit hours of the certificate program may be transferable toward that degree. This flexible approach encourages firefighters to expand their knowledge base and to obtain the higher education that will help them to advance their careers.

Courses required for a two-year **Associate of Applied Science** degree in fire science focus on providing the student with a theoretical understanding of fire science—how fire behaves in different environments—as well as technical and practical training in such areas as rescue procedures and the basics of fire extinguishment. Hands-on instruction and internships are normally part of the required curriculum.

The following is a list of standard fire science courses that may be offered for an associate's program. This list of courses is not all-inclusive and may be different from state to state.

- ◆ Fundamentals of Fire Protection. An introductory course on the philosophy, history, and fundamentals of public and private fire protection.
- ◆ Public Education Programs. Covers firefighter and fire officer awareness as well as how to implement fire and public safety programs in an effort to reduce the loss of life.
- ◆ Fire Administration I and II. Provide an introduction to the organization and management of fire departments and then cover fire service management in greater depth, considering such issues as budgetary requirements, the organization of divisions within the fire service, and relationships among the fire service and outside agencies.
- ◆ Firefighting Strategies and Tactics I and II. Analyze the nature of fire problems and the strategies and tactics that are effective in dealing with them. Included is an in-depth study of the efficient, effective use of staffing, equipment, and incident command systems to mitigate emergency situations.
- ◆ Legal Aspects of Fire Protection. Covers the legal rights and responsibilities of firefighters and fire protection agencies and addresses liability concerns.
- ◆ Fire and Arson Investigation I and II. An in-depth study of basic fire and arson investigation practices. Include instruction on preparing reports, courtroom demeanor, and the role of expert witnesses.
- ◆ Fire Prevention Codes and Inspections. A study of national, state, and local building and fire prevention codes with an emphasis on fire prevention inspections, practices, and procedures.
- ◆ Firefighter Health and Safety. Covers occupational safety and health in emergency and nonemergency situations, as well as the role firefighters play in risk management and in preventing accidents.

- Building Codes and Construction. Examines national, state, and local building codes and requirements, construction types, and building materials.
- Fire Chemistry I and II. Explores the chemical nature and properties of inorganic compounds and carbon compounds, particularly as they affect firefighting. Primary emphasis is on the hydrocarbons and their application in various industrial processes.
- Hazardous Materials I and II. Studies the chemical characteristics and behavior of various materials, with in-depth coverage of mitigation practices and techniques to control hazardous materials spills and leaks effectively.
- Wildland Fire Control. Covers wildland firefighting strategies and tactics including the different types of wildland firefighting.
- Internships. Real-world working experiences for advanced students in a specialized field. Based on a written agreement between the educational institution and a business or industry. Mentored and supervised by a workplace employee, the student achieves objectives that are defined and documented by the college and that are directly related to specific occupational outcomes.

Several four-year **baccalaureate programs** and a handful of master's programs in the United States offer degrees related to fire and emergency management. Accreditation of these programs by the **International Fire Service Accreditation Congress (IFSAC)** aids in promoting sound curriculum and degrees that are widely respected. One of the best-known, fully accredited institutions is Oklahoma State University, which offers a bachelor of science degree in fire protection and safety technology and a master's degree in political science in fire and emergency services administration. The university also recently implemented the nation's first doctoral degree in fire and emergency management. The National Fire Academy Higher Education website has a comprehensive list of institutions offering associate's and higher degrees.[5]

DISTANCE AND ONLINE EDUCATION

The fire service needs to be proactive in utilizing new adult teaching methodology, called **andragogy**. Most adults are interested in study that has immediate relevance to their job or personal life. They prefer courses that are problem centered rather than content oriented. Distance education, particularly, is suited for andragogical methodology.[6]

An excellent source for higher education in fire and emergency services is the National Fire Academy's Degrees at a Distance Program (DDP), which is an independent-study degree program. The NFA has agreements with seven accredited colleges and universities throughout the country to offer bachelor's degrees with concentrations in fire administration/management and fire prevention technology. The DDP provides a way for fire service personnel to earn a bachelor's degree or to pursue college-level learning without having to attend on-campus classes. In recent years, independent study and distance learning have become popular with working adults nationwide, and the DDP is particularly practical for fire department personnel whose work shifts normally make classroom attendance almost impossible. DDP institutions emphasize faculty–student interaction through written, electronic means (e-mail), and telephone contact. Students receive detailed guidance and feedback on the required assignments and take proctored final exams at hometown locations.[7] Distance education courses for college credit are becoming the norm for firefighters who want to advance through the ranks and move up at a faster rate. Online education (Internet) courses are the wave of the future. Their validity was at first disputed,

baccalaureate programs

Degree programs usually four years in length. Students receive a BS or BA degree.

IFSAC

Focuses on ensuring that training and certification within its member jurisdictions meet strict National Fire Protection Association Professional Qualification standards. IFSAC is divided into two Assemblies: the Certificate Assembly and the Degree Assembly.

andragogy

The term relevant to lifelong learning for adult education methodology. Distance education is particularly suited for andragogical concepts.

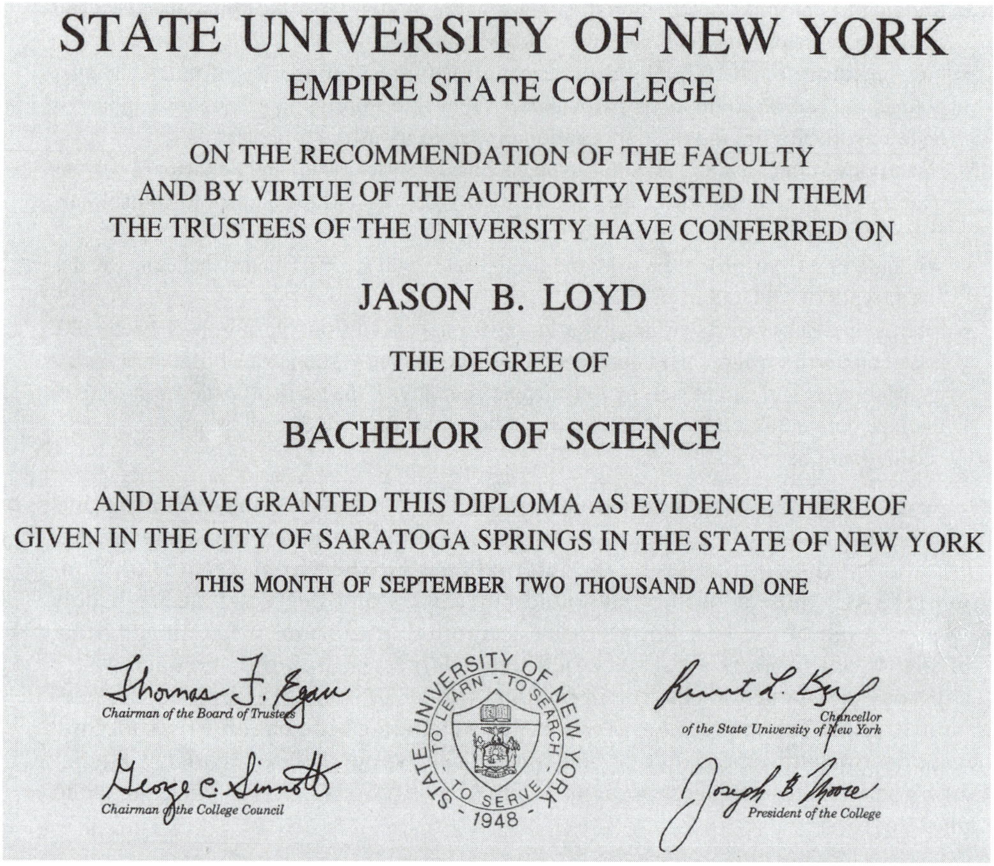
STATE UNIVERSITY OF NEW YORK

EMPIRE STATE COLLEGE

ON THE RECOMMENDATION OF THE FACULTY
AND BY VIRTUE OF THE AUTHORITY VESTED IN THEM
THE TRUSTEES OF THE UNIVERSITY HAVE CONFERRED ON

JASON B. LOYD

THE DEGREE OF

BACHELOR OF SCIENCE

AND HAVE GRANTED THIS DIPLOMA AS EVIDENCE THEREOF
GIVEN IN THE CITY OF SARATOGA SPRINGS IN THE STATE OF NEW YORK
THIS MONTH OF SEPTEMBER TWO THOUSAND AND ONE

Chairman of the Board of Trustees

Chancellor of the State University of New York

STATE UNIVERSITY OF NEW YORK · TO LEARN · TO SEARCH · TO SERVE · 1948

Chairman of the College Council

President of the College

FIGURE 4.4 ◆ Empire State College Bachelor of Science diploma awarded through the Degrees at a Distance Program (DDP).
Courtesy of Jason Loyd.

but in today's technological society, they are fast becoming accepted as a practical and effective teaching tool (see Figure 4.4).

◆ SETTING STANDARDS FOR HIGHER EDUCATION IN THE FIRE SERVICE

The National Fire Academy's mission is to support professional development for the fire and emergency response community and its allied professionals. In April 1999, the Academy sponsored its first conference on Fire and Emergency Services Higher Education (FESHE). Throughout the conference, many participants voiced the same concern—they did not have adequate information on the current status of fire service degree programs in the United States. What types and levels of degree programs were being offered? How was distance education being used in these programs? Should a model curriculum be developed?[8] Since then, an annual conference has been held to answer these questions and react to the growing needs of higher education in the fire service.

This annual conference (FESHE) and the International Fire Service Accreditation Congress (IFSAC) have taken on the task of establishing a national professional development model for fire service professionals and of maintaining uniform high standards for their training and education.

Stop and Think 4.2

Why is it important for the fire service to have a national professional development model? Based on the readings and your experience, answer the following questions:

1. Which fire science technology courses listed could enhance or benefit your career?
2. How has the ability to obtain online fire education changed the face of education for firefighters?

FIRE AND EMERGENCY SERVICES HIGHER EDUCATION (FESHE) GUIDELINES

A result of the 2000 FESHE conference was the development of a model curriculum for an associate's degree in fire science. FESHE attendees identified six core associate-level courses in the model curriculum, including the following:

- Building Construction for Fire Protection. Explores the components of building construction that relate to fire safety. The focus is on firefighter safety and on learning which elements of construction and design are key factors to consider when inspecting buildings, preplanning fire operations, and operating at emergencies.
- Fire Behavior and Combustion. Teaches the fundamental principles behind fire behavior: how and why fires start and spread, and how they can best be controlled.
- Fire Prevention Codes and Inspections. Teaches the fundamentals of history and philosophy of fire prevention, organization, and operation of a fire prevention bureau, use of fire codes, identification and correction of fire hazards, and the relationships of fire prevention with built-in fire protection systems and fire investigation.
- Fire Protection Hydraulics and Water Supply. Provides a foundation in the principles of hydraulics, as well as practical instruction on the use of water in fire protection. Students learn how to apply hydraulic principles to analyze and to solve water supply problems.
- Fire Protection Systems. Covers the design and operational features of fire alarm systems, water-based fire suppression systems, special hazard fire suppression systems, water supplies for fire protection, and portable fire extinguishers.
- Principles of Emergency Services. Provides an overview of a broad range of topics relating to a emergency service work, including career opportunities in fire protection and related fields; the philosophy and history of the fire protection/service; fire loss analysis; the organization and function of public and private fire protection services; fire departments as a branch of local government; laws and regulations affecting the fire service; fire service nomenclature; specific fire protection functions; basic fire chemistry and physics; an introduction to fire protection systems; and an introduction to fire strategy and tactics.

In 2002, FESHE conference attendees approved these model courses and outlines. The major publishers of fire-related textbooks are committed to writing texts for some, or all, of them. Fire science associate's degree programs are encouraged to

FIGURE 4.5 ◆ The National Professional Development Model pyramid illustrates the progression of education and training in the fire and emergency services as noted by the USFA/NFA for Higher Education.

require them as the theoretical core curriculum on which their major is based. Many schools already offer these courses in their programs, while others are in the process of adopting them. Once adopted, these model courses will answer the need for a high national standard of fire service education. They will also enable problem-free student transfers between schools and promote crosswalks for those who want to apply their academic coursework toward meeting the national qualification standards necessary for firefighter certifications and degrees (see Figure 4.5).[9,10]

The professional development model was also finalized at the 2002 FESHE conference. It is not a promotion model addressing credentials, but rather an experience-based model that suggests an efficient path for fire service professional development. This path is supported by collaboration among fire-related training, higher education, and certification providers. The model recommends what these providers' respective roles should be and how they should coordinate their programs.[11]

◆ INTERNATIONAL FIRE SERVICE ACCREDITATION CONGRESS

The International Fire Service Accreditation Congress (IFSAC) focuses on ensuring that training and certification within its member jurisdictions meet strict National Fire Protection Association Professional Qualification standards. The IFSAC, based at Oklahoma State University, is divided into two Assemblies: the Certificate Assembly and the Degree Assembly. As of 2008, 49 state and provincial fire-training entities were accredited by the Certificate Assembly, which is the area of greatest interest to the fire service. There are 35 entities that are members of the Degree Assembly, which provides accreditation to programs that are within the realm of higher education (see Figure 4.6). Many of these entities are community colleges that conduct accredited courses and programs under the auspices of the larger fire service entities in their jurisdiction (see Figure 4.7).

FIGURE 4.6 ◆ IFSAC Degree Accredited Degree Program Certificate.

FIGURE 4.7 ◆ IFSAC Degree Accredited Degree Program.

◆ PRO BOARD ACCREDITATION

Pro Board

National Board on Fire Service Professional Qualifications Accrediting Fire Service Training Organizations.

The **Pro Board** is very similar to IFSAC. Its primary focus is to serve the needs of accreditation of organizations that certify uniform members of public fire departments, both career and volunteer. Other organizations with fire protection interests may also be considered for participation. Pro Board works with the individual states or provinces to issue certifications under that jurisdiction. As with IFSAC, a common goal is to bring professionalism to the fire service.

◆ SUMMARY

As the fire service moves from a skilled trade to professional status, a college or university education will become essential. Greater administrative demands, advances in technology, and the changing social and political environments impacting our profession necessitate advanced training and higher educational opportunities, especially for those firefighters who will be tomorrow's leaders of the fire service.

The fire service professional's access to traditional higher education is limited by geographic, financial, and time constraints. The establishment of national standards, as well as increased acceptance of distance and Internet learning, should reduce some of the difficulties. Agencies or departments that work together on a normal basis would best serve their own interests and those of their employees by arranging for joint training and educational opportunities. Such cooperation would give firefighters access to increased, more effective learning experiences, would lead to greater safety, and would minimize the loss of life and property.

Ultimately, however, it will be the responsibility of individual firefighters—whether new recruits or experienced professionals—to take the steps necessary to get the education they need to meet their own goals and to serve the public to the best of their personal ability. Education opens the door to promotions, certainly, but its real value is in teaching the leadership skills that the motivated firefighter will need to take on increased responsibility.

On Scene

Training and higher education are tools for professional development in the fire service. These are not only a requirement for advancement but also are sometimes preequisites for initial hiring. Eddie Burns, Dallas Fire-Rescue Chief states, "Our goal is to create a national standard for fire science, similar to law, architecture and criminal justice." Chief Burns is passionate about training in the firefighting profession. After earning bachelor's and master's degrees at Dallas Baptist University, he opted to finish his associate's degree in fire services administration at Weatherford College. Chief Burns was a key player in the development of a fire management program at Texas Wesleyan University and was involved with the National Fire Academy in an effort to standardize curricula for firefighting professionals.

Another example is Fire Chief Martin D. Davila, who has 23 years' experience in military and civilian fire protection. He states, "Training and higher education are the cornerstones of success. Alone they don't make you smarter . . . but with self-motivation they open many doors for you. I believe that you cannot succeed in our business without all three of them." (1) Training teaches you to do the job. (2) Higher education (college) opens you to new ideas and theories, enhancing your past experiences and influencing your future decisions. (3) Motivation (the desire to make it better) is the glue that brings it all together. Training has taught us what to do. Higher education will make us think of the "what ifs" and help us formulate new approaches. Motivation will push us to implement our new ideas. Training + Higher Education + Motivation = Success. Great leaders aren't born, but through training, experience, and higher education, a great leader is formed.

1. What model does the National Fire Academy have in place to promote training and education?
2. Why, do you think, are more fire departments today setting training and education standards for initial hiring?

Review Questions

1. The fire science technology curriculum is aimed at providing the student with what types of skills?
2. Training in the fire service can lead to certifications. Name a few certifications in the field that can be beneficial to career advancement.
3. What is the basic college degree in fire science technology?
4. Training programs are offered at various levels, such as federal, state, and local. List two others.
5. List two preservice opportunities for gaining firefighting experience.

Notes

1. NFA Higher Education, USFA Training & Education (April 19, 2006). U.S. Fire Administration. Retrieved November 13, 2006, from the Higher Education website: http://www.usfa.dhs.gov/training/nfa/higher_ed/.
2. Lindsey, Jeffery (2006). *Fire service instructor.* Upper Saddle River, NJ: Pearson Education, Inc.
3. Murphy, D. L., & Stanton. L. M. (2004). A hybrid approach to distance education technology: Tailor-made for the United States fire service. In J. Mathews (Ed.), *Session T1C: Distance learning: Methods and technologies 1* (pp. T1C1–T1C4). Savannah, GA: Frontiers in Education Clearing House.
4. Ibid.
5. NFA Higher Education, USFA Training & Education (September 20, 2006). U.S. Fire Administration. Retrieved November 13, 2006, from the Higher Education website: http://www.usfa.dhs.gov/training/nfa/higher_ed/model/.
6. Murphy, D. L., & Stanton, L. M. (2004). A hybrid approach to distance education technology: Tailor-made for the United States fire service. In J. Mathews (Ed.), *Session T1C: Distance learning: Methods and technologies 1* (pp. T1C1–T1C4). Savannah, GA: Frontiers in Education Clearing House.
7. NFA Higher Education, USFA Training & Education (January 23, 2006). U.S. Fire Administration . Retrieved November 14, 2006, from the Higher Education website: http://www.usfa.dhs.gov/training/nfa/higher_ed/degree_programs/distance/.

8. Sturtevant, Thomas. B. (2001). *A study of undergraduate fire service degree programs in the United States.* Retrieved November 14, 2006, from www.dissertation.com, pp. 1–152.
9. NFA Higher Education, USFA Training & Education (September 20, 2006). U.S. Fire Administration. Retrieved November 13, 2006, from the Higher Education website: http://www.usfa.dhs.gov/training/nfa/higher_ed/model/.
10. Ibid.
11. Ibid.

Suggested Reading

Chikerotis, Steve (2006). *Firefighters from the heart: True stories and lessons learned.* Clifton Park, NY: Thomson Delmar Learning.

Golway, Terry (2002). *So others might live.* New York, NY: Basic Books.

Lasky, Rick (2006). *Pride and ownership: A firefighter's love of the job.* Tulsa, OK: PennWell Corporation.

Lindsey, Jeffery (2006). *Fire service instructor.* Upper Saddle River, NJ: Pearson Education, Inc.

National Commission on Fire Prevention and Control (2003). *America Burning.* Final report. Washington, DC: GPO.

Willis, Clint (2002). *Fire fighters: Stories of survival from the front lines of firefighting.* New York, NY: Thunder's Mouth Press.

Fire Department Resources

5 CHAPTER

Key Terms

advanced life support
 units (ALS), p. 77
aerial ladder
 platform, p. 74
aerial ladders, p. 68
aerial platform
 apparatus, p. 75
articulating boom, p. 75
Bambi Bucket, p. 82
basic life support units
 (BLS), p. 77
compressed air foam
 systems (CAFS), p. 72
heavy rescue squad, p. 76
hydraulically
 operated, p. 74

infrared thermal imaging
 camera (TIC), p. 88
initial attack pumper, p. 72
Level A Ensemble, p. 85
Level B Ensemble, p. 85
Level C Ensemble, p. 85
Level D Ensemble, p. 86
mobile data terminal
 (MDT), p. 89
NFPA, p. 66
NFPA 1001, p. 66
non-fire suppression
 activities, p. 67
OSHA, p. 80

OSHA Code of Federal
 Regulations 29 CFR
 1910.120, p. 80
personal alert safety
 system (PASS), p. 82
personal protective
 equipment (PPE), p. 82
pump-and-roll
 operation, p. 72
regulations, p. 81
self-contained breathing
 apparatus (SCBA), p. 82
standards, p. 66
telescoping boom, p. 75
wildland firefighting
 apparatus, p. 72

Objectives

After reading this chapter, you should be able to:

- Describe the various types of fire department facilities.
- Identify and describe the uses of pumping apparatus.
- Explain the functions and operations of various aerial apparatus.
- Explain how specialized fire apparatus are utilized.
- Describe the different levels of protection for the work uniform, proximity, structural, wildland, and Level A, B, C, and D ensembles.
- Explain why SCBA and PASS devices are important to firefighter safety.

◆ INTRODUCTION

The modern fire department is a multifaceted organization that is expected to perform an astonishingly wide range of critical tasks. To do its work effectively, the department must be equipped with an equally wide range of resources. In this chapter, we will introduce the student to fire service resources, by which we mean all of the material aids that firefighters use in their work. We will look first at fire service buildings, including their functions and design. Next, we will consider the specialized vehicles that answer the specific needs of firefighting, both standard fire vehicles and those designed for very particular tasks, such as aircraft and marine firefighting. We will look at protective gear and breathing apparatus designed to protect firefighters from the diverse dangers that they encounter in different environments. Finally, we will consider traditional handheld firefighting equipment, as well as some of the high-tech inventions, such as GPS equipment and infrared imaging devices, that have helped to bring the science of firefighting into the 21st century.

The National Fire Protection Association (NFPA) establishes manufacturing **standards** for all fire service apparatus. The codes that pertain to apparatus discussed in this chapter are listed at the end of each section. Printed or digital copies of the **NFPA 1001** standards can be requested by contacting the National Fire Protection Association at One Batterymarch Park, Quincy, MA 02269-9101.

standards

A recommended course of action that is not mandated by law.

NFPA 1001

Standard for firefighter professional qualifications.

◆ FIRE DEPARTMENT FACILITIES

Fire department facilities vary in configuration and design. Smaller fire departments normally consolidate their facilities into one or two buildings, whereas larger departments find it more practical to have separate facilities for each functional area. Regardless of the department's size and finances, its facilities must cover three basic functions: administration, training, and fire suppression.

ADMINISTRATION BUILDINGS

The amount of space required for administration purposes depends on the size of the department's staff and the complexity of its operations. Smaller fire departments may need only office space in a fire station. Larger departments may maintain a separate headquarters building for administrative staff. Each arrangement has its advantages and its disadvantages.

When the headquarters office is located in the fire station, administrative personnel are physically close to the day-to-day line operations of the department, and this may result in better communication. The administrators are better informed about firefighters' concerns, and the firefighters in turn have a better understanding of the reasoning behind administrative decisions. This promotes a greater sense of teamwork and can lead to better compliance with policy and procedural changes. The trade-off is that administrators who work in the fire station may have to put up with distractions. Fire stations tend to be noisy, especially during the day when vehicles and other motorized equipment are being tested and personnel are cleaning the working and living areas. Holding conferences can be difficult under these circumstances. There are disadvantages for the firefighters as well. Department headquarters

FIGURE 5.1 ◆ Fire department headquarters offices may be incorporated into a fire station or in their own administration building.
Courtesy of James D. Richardson.

are often visited by local politicians and news media, whose presence may make the firefighters feel that they are constantly on show and under scrutiny.

Other fire service departments that may be located in a separate building include dispatch, fire and arson investigation squad, inspection and codes departments, and training facilities. The purpose of having a separate location may also be to consolidate these **non–fire suppression activities** (see Figure 5.1).[1]

TRAINING FACILITIES

The size of a fire department's training facility varies according to the size of the department or the community being served. There is no universal template for the design or construction of the training center, but it must provide a learning environment and equipment adequate to train firefighters to the NFPA's standards. A classroom with adequate lighting, acoustics, and audio visual equipment should be available, as well as a building or props to use in training nonburn skills. Nonburn skills are those that do not expose trainees to actual fire scenarios, such as ladder carries and raises, hose lays and fire streams, and fire protection systems. At a minimum, the training center should feature a burn building that permits firefighters to practice extinguishment of Class A structural fires. The facility should also have an adequate water supply, from either a pressurized or static source, capable of providing a minimum of 500 gallons per minute (GPM) for 30 minutes (see Figure 5.2).[2]

NFPA 1402, *Guide to Building Fire Service Training Centers; NFPA 1403, Standard on Live Fire Training Evolutions*

non–fire suppression activities
Any fire department activity that is not directly involved in fire suppression.

FIGURE 5.2 ◆ The size and complexity of a fire training facility will vary, depending on the requirements of the state certifying agency.
Courtesy of James D. Richardson.

FIRE STATIONS

Fire stations are the most complex facilities in a fire department. Because of the different functions they serve, they must include elements of a residential house, an office building, a heavy-duty garage, and a maintenance shop. There are many variations in layout and design, but all stations must meet the same safety standards:

- The building must be structurally sound to house heavy vehicles and machinery.
- It must provide protection for the occupants and equipment from changing weather conditions.
- It must provide safe working conditions for the occupants.
- It is vital that it has good security.

In addition to these safety requirements, the National Fire Protection Association recommends the following space accommodations for a typical fire station.[3]

Apparatus Bay

An apparatus bay is a garage area inside the fire station that is used to house fire apparatus and other emergency vehicles. The bay floor must be constructed to support safely apparatus that may weigh 20 tons or more when fully outfitted. It should be large enough to allow for minor apparatus maintenance and include a ventilation system to remove exhaust gases when apparatus are operated indoors.

The bay doors must be oversized to allow clearance for wide apparatus, and the ceiling of the bay must be tall enough to allow for apparatus that has high head clearance requirements, such as **aerial ladders** or platforms. The area should be well lighted and have a heating system that keeps the bay above freezing during winter (see Figure 5.3).

aerial ladder

A rotating, power-operated ladder mounted on a self-propelled automotive fire apparatus.

FIGURE 5.3 ◆ The apparatus bay must be large enough to allow for free movement around the apparatus.
Courtesy of James D. Richardson.

Sleeping Accommodations, Bathroom Facilities, Kitchen, and Living Areas

In most departments, firefighters work 24-hour shifts and then are off duty for a designated period of time. It is estimated that during their careers, they spend approximately one-third of their lives at the fire station. Therefore, the station must have adequate facilities to support many of the activities of ordinary daily living. The more comfortable and home like these facilities are, the better for company morale.

Sleeping accommodations should be separate for male and female personnel and allow firefighters to rest between emergency calls during their long shifts. Some stations are designed with private sleeping quarters or cubicles, but dormitories are more common. To prevent injuries that can occur when descending stairs or sliding a pole, many newer fire stations position the dormitory or sleeping area on the same level as the apparatus bay. Separate bathing and toilet facilities also must be provided for male and female personnel, and these should be constructed of materials that are easily maintained and have a long service life (see Figure 5.4).

The kitchen area of a fire station is often where firefighters congregate during breaks or for fraternizing after "normal" station work is completed. Many fire departments encourage shift personnel to have common meals, as this helps create a sense of camaraderie. Some companies prepare two or three meals every work shift. The kitchen area should be segregated from the apparatus bay to prevent possible contamination from fuels and oils.

The living area is similar to a residential family room. Lounge furniture allows personnel to relax between responses. A TV and video player are often found in this area and can be used for entertainment or training purposes.

FIGURE 5.4 ◆ Dormitories in newer fire stations provide more privacy by separating individual sleeping areas into rooms or cubicles.
Courtesy of James D. Richardson.

Offices

The station officers should have a room in which to conduct station business. Ideally, this room should be located in a private area of the station to allow for confidential counseling of personnel. All personnel files should be located in this office and secured in a manner that prevents unauthorized access.

Other Areas

There are other areas and equipment that the ideal modern fire stations might provide to enhance the safety and comfort of its personnel:

- Personnel decontamination areas for biological and chemical agents.
- Medical supply storage.
- A specially designed laundry room for cleaning turn-out gear and work uniforms. During firefighting operations, turn-out gear and the work uniform worn underneath absorb contaminants from the smoke and runoff water. These contaminants may be toxic or carcinogenic and pose a risk to those exposed. Therefore, the laundry room must be equipped with a water retrieval system that prevents the water used for cleaning from entering the public wastewater system. The laundry water must be properly stored and disposed of according to Environmental Protection Agency (EPA) guidelines.
- A training room for continuing education.
- An exercise area where personnel can work out to maintain physical fitness.

◆ FIRE APPARATUS

Fire apparatus come in many makes and models, each designed for a certain type of firefighting operation. We will look first at pumpers, and then we will consider vehicles designed to transport water, to allow access to the upper stories of structures, and to serve in other specific situations.

This standard establishes the basic requirements that must be met by all motorized fire apparatus.

NFPA 1901, *Standard for Automotive Fire Apparatus*[4]

PUMPING APPARATUS

The fire department's Type I pumper is its most basic fire apparatus. It is usually staffed by four personnel: an officer, an operator, and two firefighters. Pumpers are designed to fight structural fires or to support any operation that requires sustained pumping activity. The standard pumper may be equipped with pump capacities ranging from 750 to 2,000 GPM.

Type I Requirements

- Minimum pump capacity of 750 GPM
- Minimum water supply tank of 500 gallons
- 1,200 feet of supply hose larger than 2½ inches in diameter
- 400 feet of 1½ -, 1¾-, or 2-inch diameter hose
- Assorted tools and equipment (ladders, forcible entry tools, hose adapters and appliances, self-contained breathing apparatus, pike poles, portable fire extinguishers, salvage covers, and so on)

Some fire departments expand the versatility of their pumping apparatus by adding an elevating water tower. The towers range in height from 50 to 75 feet and can supply up to 1,000 GPM through a master stream appliance, a nozzle designed to flow 1,000 GPM or more, mounted on the top of the tower (see Figure 5.5).[5]

FIGURE 5.5 ◆ Elevating water towers enable the pumping apparatus to operate in both offensive and defensive modes.
Courtesy of James D. Richardson.

initial attack pumper

A fire pumping apparatus that is utilized for fast response and rapid attack on a fire.

wildland firefighting apparatus

A fire pumping apparatus with a high ground clearance and off-road capabilities.

Other types of pumping apparatus are the Type II **initial attack pumper**, also referred to as a first or fast attack pumper, and Type III **wildland firefighting apparatus**. The Type II initial attack pumper consists of a custom pumping module and a water storage tank mounted on a pickup truck chassis. It is primarily used to initiate a rapid attack on a structure fire to try to hold the fire in check until the Type I fire apparatus arrives. It is not adequate for sustained operations. Initial attack pumpers can also be used in combating small grass or vehicle fires. The NFPA requires that the tank have a capacity of at least 200 gallons and that the pump has a minimum capability of delivering at least 250 GPM.[6]

> NFPA 1902, *Standard for Initial Attack Fire Apparatus*

WILDLAND FIREFIGHTING APPARATUS

Wildland firefighting apparatus are generally Type III light- to-medium-duty trucks equipped with a pump, a water tank, and a small complement of hose and equipment. These apparatus must have off-road (all-wheel-drive) capability and be able to support **pump-and-roll operations**; that is, it must be able to provide pump operations while moving. They generally have a water tank capacity of 100 to 500 gallons and pump capacities of between 100 and 500 GPM. In addition, almost all wildland fire apparatus have a Class A foam system. This extends the water supply and increases the water's ability to penetrate the fuel. Approximately 10% to 20% of new apparatus are equipped with a **compressed air foam system** (**CAFS**). Compressed air foam systems provide more extinguishing capability by lowering friction loss, increasing the reach of hose streams, and allowing the use of a lighter hose. In other respects, the designs of wildfire firefighting apparatus vary considerably because they are configured to meet the special needs of the areas they will serve. For example, fire departments in areas that have level terrain, with little variation in topography, do not need an apparatus with a side-slope stability system (a hydraulic system that allows the chassis of the truck to stay level on sloped terrain). However, in areas where the topography varies dramatically, the fire apparatus may be required to have wheels that are powered individually to allow for greater ground clearance (see Figure 5.6).[7]

pump-and-roll operation

The capability of an apparatus to pump water while it is in motion.

compressed air foam system (CAFS)

A high-energy foam-generation system consisting of an air compressor (or other air source), a water pump, and foam solution that injects air into the foam solution before it enters a hoseline.

Stay Safe

> NFPA 1906, *Standard for Wildland Fire Apparatus*
>
> *Notes:* NFPA standards 1901, 1902, and 1906 overlap when wildland fire apparatus are equipped with pump capacities of 250 to 500 GPM.
> How the apparatus is used will determine the standard that will be followed.

NFPA 1906 Standard for Wildland Fire Apparatus	Requirements for apparatus ground clearance, angle of approach and departure, side-slope stability, and pump-and-roll capabilities
NFPA 1901 Standard for Automotive Fire Apparatus NFPA 1902 Standard for Initial Attack Fire Apparatus	More stringent requirements in regard to hose, ladders, equipment, and water tank capacity

FIGURE 5.6 ◆ Wildland fire apparatus are designed for off-road operations. They are lighter and more maneuverable than structural fire apparatus.
Courtesy of James D. Richardson.

MOBILE WATER SUPPLY APPARATUS

Mobile water supply apparatus are designed to transport water to an incident scene for use by pumping apparatus. NFPA standards require that a mobile water supply apparatus have a minimum tank capacity of 1,000 gallons. Straight-chassis apparatus with tank capacities of 1,500 to 4,000 gallons must be of tandem rear-axle construction, whereas a tractor-trailer design is recommended for larger tank capacities. If the apparatus is equipped with full-size fire pumps as well as other standard tools and equipment of a fire pumper, it is classified as a tanker-pumper and can operate as either a mobile water supply apparatus or a pumper (see Figure 5.7).[8] Mobile water supply apparatus are also referred to as water tenders.

NFPA 1903, *Standard for Mobile Water Supply Fire Apparatus*

AERIAL APPARATUS

Aerial apparatus are vehicles designed to provide firefighters access to the upper levels of a structure that cannot be reached with ground ladders. These apparatus are primarily used for rescue and ventilation and for providing elevated master streams. The NFPA requires that aerial devices be power-operated and self-supporting with an elevation capability of at least 50 feet. (Working ladder height is determined by raising the ladder to its maximum elevation and extension and measuring from the ground to the highest rung on the ladder.) The International Fire Service Training Association classifies aerial apparatus into four categories:

- ◆ Aerial ladder apparatus
- ◆ Aerial ladder platform apparatus

FIGURE 5.7 ◆ Water tenders vary in carrying capacity from several hundred to several thousand gallons and are the primary water supply for many rural fire ground operations. Courtesy of James D. Richardson.

- Telescoping aerial platform apparatus
- Articulating aerial platform apparatus.[9]

Aerial ladders range in length from 50 to over 135 feet and must be able to support 250 pounds of weight at the tip, when fully extended horizontally and at a 45° angle. The ladder is **hydraulically operated** but is required to have electrically and manually operated backup systems that can be used to lower the ladder if the hydraulic system fails. The apparatus must carry a large complement of ground ladders, forcible entry tools, rescue tools, and ventilation equipment. The support chassis for the aerial ladder may range from a two- or three-axle single chassis to a tractor-trailer apparatus, known as a tiller truck. The tractor-trailer model is equipped with a rear-wheel steering compartment. The steering mechanism in the compartment is operated by a firefighter, who is referred to as "the tillerman." The rear wheel steering allows the trailer to be more maneuverable on narrow streets and in traffic than straight-chassis aerial apparatus (see Figure 5.8).[10]

A variation of the aerial ladder is the **aerial ladder platform**. This apparatus consists of a structurally reinforced aerial ladder, ranging in height from 85 to 110 feet, with a platform attached to the tip that is usually mounted over the rear axles of a single-frame three-axle chassis.

The platform, or basket, is required to support 750 pounds of weight in any position in which it is placed.[11] It is constructed of steel or aluminum alloy and is usually equipped with a heat shield and shower nozzle to protect personnel. Electric, air, and hydraulic outlets are supplied. Many platforms are equipped with 1½- and 2½-inch outlets, which allow handlines to be supplied from the platform to upper levels of a structure when standpipe or sprinkler systems are not available or are inoperable. Normally, these also have a permanently mounted turret nozzle and a water supply system capable of supplying flow rates of 1,000 GPM or more. A hydraulically operated backup system and two primary operating systems, one at ground level and one in the basket, are required (see Figure 5.9).

hydraulically operated

Functioning by utilizing pressure created by compressing oil in a cylinder.

aerial ladder platform

A power-operated aerial device that combines an aerial ladder with a personnel-carrying platform supported at the end of the ladder.

FIGURE 5.8 ◆ The single frame ladder truck may have a single or dual rear axle, and the aerial ladder may be midship or rear mounted.

The **aerial platform apparatus** consists of a platform with a permanently mounted turret nozzle and water supply system mounted on a **telescoping** or **articulating boom**. The boom may be of box-beam or tubular truss-beam construction. The only difference between the telescoping and articulating versions boom is

aerial platform apparatus

A fire apparatus that carries a hydraulically operated elevating platform that can be utilized for rescue, fire, or observation operations.

telescoping boom

Aerial device raised and extended through sections that slide within each other.

articulating boom

A boom constructed of several sections that are connected by a hinged joint that allows the boom to function much like the human elbow.

FIGURE 5.9 ◆ The aerial ladder platform allows the apparatus to be used as an observation post, for heavy stream operations, or for aerial ladder operations.
Courtesy of James D. Richardson.

how they operate. The telescoping boom is sectioned in a way that allows the sections to telescope into each other in a straight line. The articulating boom is constructed of several sections that are connected by a hinge that operates much like a human elbow. The functions of the aerial platform apparatus are basically the same as those of the aerial ladder platform (access to upper levels, rescue, ventilation, and master stream operations), but aerial platforms do not have aerial ladder operations capabilities. The boom of the telescoping platform does have a small ladder mounted on it, but its only function is to provide an emergency escape route for firefighters who are trapped on the platform.[12]

Each fire department determines the apparatus type and method of operation that will meet its needs. Articulating platforms are more maneuverable in confined areas, such as narrow streets or alleys. However, these types of aerial apparatus are more unstable than telescoping type apparatus if they are extended past their designed operating limits (which may occur during fire rescue operations). Telescoping platforms and ladder platforms are more stable and provide a means of escape for firefighters if they must abandon the platform. However, telescoping aerial platforms or ladder platforms are less maneuverable in confined areas.

NFPA 1904, *Standard for Aerial Ladder and Elevating Platform Fire Apparatus*

◆ SPECIALIZED APPARATUS

Prior to 1970, the main responsibility of a community's fire department was to put out fires, teach fire prevention, and administer some basic first aid. A few departments provided medical services, but in most communities, medical response and transport were provided by private agencies or funeral homes. Response times were long, and in some instances, transportation was actually denied because the victim could not pay for the service. In response to the public outcry, community leaders and emergency response professionals began looking for a practical way to provide faster, less expensive emergency medical service The solution they found was based on the medical response and transport techniques that were already in place in a few urban centers such as Los Angeles County and Jacksonville, Florida, as well as on lessons learned during the Vietnam War. Many returning veterans were now working in the emergency services, and their wartime experience in efficient methods for treating and transporting seriously injured patients proved invaluable. One example was the use of pressurized garments (MAST trousers) to maintain blood pressure in vital organs of patients who had low blood volume or were in severe shock. Other examples were the development of air ambulance helicopters and the use of advanced intravenous and drug techniques by paramedics. As a result of the input gathered from these groups and the vision and creativity of those who expanded on it, the United States today has one of the most sophisticated public emergency medical services in the world. Although some communities have placed the responsibility for this service on other emergency response agencies, such as the police department or a separate emergency medical services department, most communities have turned to their fire departments to implement this service in their 9-1-1 system.

Another consequence of the growing sophistication of the emergency medical services is that fire departments are now able to provide a **heavy rescue squad**.

heavy rescue squad

A team of firefighters who are highly skilled in various types of rescue operations that require the use of highly specialized equipment and training.

Many fire departments had light rescue squads capable of performing tasks involving extrication and medical treatment. However, these units were often also called for nonrescue tasks, such as providing extra help during fire suppression operations. In the 1970s, fire departments began to see the value of a specially trained team whose sole responsibility would be to provide rescue services in a variety of situations. This team would need extensive training in industrial and farm equipment extrication, trench and tunnel rescue, building collapse, high- and low-angle rope rescue, confined space and swift and rapid rising water rescue, and a designated vehicle to carry their equipment.

Between the years of 1977 and 1979 there were a series of devastating railroad tank car explosions in the communities of Jacksonville, Florida, Nashville, Tennessee, and Houston, Texas. In 1977, Jacksonville, Florida Fire Department trained the first fire department hazardous materials (HazMat) response team in the nation. By 1979, Houston, Texas and Nashville, Tennessee fire departments, followed Jacksonville's lead and established their HazMat response teams. These three fire departments set the standard for hazardous materials training, equipment use, and response. Although the first teams were ill-equipped and learned as they faced situations, today's hazardous materials teams possess some of the most sophisticated equipment and highly trained personnel in the world. After the events of September 11, 2001, hazardous materials teams are required to respond not only to chemical incidents but also to incidents that involve biological, radiological, and nuclear agents.[13]

AMBULANCE APPARATUS

An ambulance is a vehicle specially designed for delivering emergency medical personnel to an incident scene and transporting the sick and injured to a medical facility. The Type I ambulance is a modular unit that is mounted on a light- or medium-duty truck chassis. It has a window between the module and the cab portion of the apparatus to allow communications between the driver and caregivers. Occupants must exit the vehicle to move from the module to the cab. The Type II ambulance is a standard full-size van body with a raised roof to allow for more interior head clearance. The interior space is one complete unit, so occupants are able to move from the cab to the treatment area without exiting the vehicle. The Type III ambulance is a modular unit on a van chassis, but it has walk-through capability to allow occupants access to all areas of the vehicle. Ambulances must provide enough floor space to accommodate a stretcher for the patient, a work area for medical personnel, and adequate storage for all medical supplies (see Figure 5.10).

Fire department ambulances may be operated as **basic life support units** (**BLS**) or **advanced life support units** (**ALS**). Most BLS units provide first aid and nonemergency transportation to the hospital. Personnel are usually trained only to the Emergency Medical Technician Basic Level. ALS units provide advanced medical services, including cardiovascular monitoring, and under certain circumstances are allowed to administer drugs intravenously. ALS personnel are trained to the Emergency Medical Technician Paramedic level.

In addition to the standards set forth in NFPA 1901, ambulances must meet the federal requirements set forth by the General Services Administration Standard KKK-A-1822, *Federal Specifications for Ambulances*. The vehicle portion of the apparatus must also meet requirements set forth by the *Federal Motor Vehicle Safety Standards*.

basic life support units (BLS)

Medical units that provide medical and transportation needs for patients who do not have life-threatening injuries or illness.

advanced life support units (ALS)

Medical units that provide advanced medical treatment and transport to the severely injured or ill.

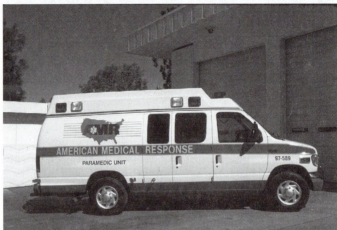

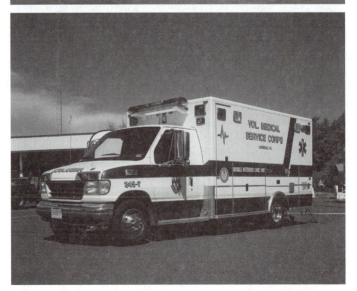

FIGURE 5.10 ◆ Emergency service organizations have a choice of the type of ambulance they wish to use.

RESCUE APPARATUS

Rescue apparatus are used to move specially trained rescue teams and equipment to the scene of an incident, where the team's job is to remove victims to a place of safety. These vehicles are not intended for patient transport. The number of assigned personnel and the type of equipment that is carried on the rescue vehicle varies, based on the needs of the community. Rescue apparatus are classified as light, medium, and heavy duty. Light- and medium-duty rescue apparatus are used to perform basic rescue operations that require a limited range of tools and equipment, such as hydraulic spreaders and cutters, jacks, saws, ropes and rigging, oxyacetylene cutting equipment, or air bag lifting equipment. These units may also carry firefighting tools and equipment and may be utilized as dual-purpose apparatus. Heavy rescue apparatus are single-purpose units, carrying equipment and personnel whose only purpose is to extricate victims from dangerous situations. These include a wide range of possible operations, including vehicle and industrial accident extrication, as well as high-angle, confined space, swift and rapid rising water, trench, and tunnel rescues. These apparatus consist of modular units that are mounted on a heavy-duty truck chassis with two or three axles because of the massive amounts of equipment they must transport (see Figure 5.11).[14] There is no NFPA standard specifically for rescue vehicles.

Stop and Think 5.1

If the type of fire apparatus purchased by a fire department is determined by the priority of threats and a needs analysis of the community that it serves, what kind of specialized apparatus would a rural community need?

FIGURE 5.11 ◆ Heavy rescue apparatus carries a variety of tools and equipment for specialized rescue operations.
Courtesy of James D. Richardson.

HAZARDOUS MATERIALS APPARATUS

A hazardous materials (HazMat) apparatus may be an enclosed trailer or a module on a three-axle heavy-duty truck chassis. These units carry specialized equipment for detecting hazardous materials and mitigating the damage they cause. This includes patching kits for various types of containers and agents; special protective clothing and breathing apparatus; air, water, and soil testing and monitoring devices; chemical, nuclear, radiological, and biological detection equipment; special communication devices; and weather monitoring equipment (see Figure 5.12).

OSHA

Occupational Safety and Health Administration.

OSHA Code of Federal Regulations 29 CFR 1910.120

The federal code that mandates the level of training for anyone working with hazardous materials.

The Occupational Safety and Health Administration (**OSHA**) establishes the levels of training for all public and private agencies that respond to hazardous materials incidents. The code governing these is **OSHA Code of Federal Regulations 29 CFR 1910.120** (also known as HAZWOPER, Hazardous Waste Operations and Emergency Response).

AIRCRAFT RESCUE AND FIREFIGHTING APPARATUS (ARFF)

These apparatus are specialized pieces of equipment that are used only for fighting aircraft fires. The units are very large with water tank capacities ranging from 1,500 to 6,000 gallons and foam storage tanks that carry up to 1,000 gallons of foam concentrate. Pump capacities can be as high as 2,000 GPM.[15] The National Fire Protection Association defines three categories of ARFF apparatus: major firefighting vehicles, rapid intervention vehicles (RIVs), and combined-agent vehicles. RIVs are designed to respond quickly to the emergency scene and initiate rescue and initial firefighting measures until the major firefighting vehicles arrive. Major firefighting vehicles are large apparatus that carry in excess of 3,000 gallons of firefighting foam.

FIGURE 5.12 ◆ Hazardous materials teams' equipment trailers carry a wide variety of specialized equipment and materials for mitigating chemical and other HazMat incidents.
Courtesy of James D. Richardson.

Combined-agent apparatus are similar to the major firefighting vehicles but are capable of simultaneously delivering firefighting foam and dry-powder extinguishing agents on a fire. All are required to have pump-and-roll and off-road capability. It is recommended that they have positive drive to each wheel in order to achieve maximum chassis clearance (refer to Figure 5.5). [16]

NFPA 414, *Standard for Aircraft Rescue and Fire-Fighting Vehicles*

The size, number, and type of ARFF apparatus needed are determined by Federal Aviation Administration (FAA) **regulations**.

regulations
Mandated by law.

FIREBOATS

Most port communities have fireboats to combat fires involving watercraft, wharves, and marinas. These apparatus range in size from high-speed shallow-draft vessels to large oceangoing tugboats. Fireboats can provide high volume pump capacities of over 20,000 GPM through master stream turret nozzles that flow up to 3,000 GPM each (see Figure 5.13). [17]

FIREFIGHTING AIRCRAFT

Aircraft, both fixed wing and rotary blade (helicopters), are used extensively in large wildland firefighting operations. Their primary duty is to support ground operations by dropping extinguishing agents or fire retardants on burning vegetation. The

FIGURE 5.13 ◆ Fireboats are capable of delivering thousands of gallons per minute.

fixed-wing aircraft are usually propeller-driven military or commercial aircraft that have been converted to serve firefighting functions. Their carrying capacity varies from 700 to 3,000 gallons of agent. Most firefighting helicopters do not have the carrying capacity of fixed-wing aircraft, but they do have the ability to hover over a fire and more accurately place the retardant or extinguishing agent on the target. The majority of helicopters transport their payload in a bucket that is suspended under the aircraft. This bucket is referred to as a "**Bambi Bucket**" and has carrying capacities ranging from 300 to 1,000 gallons. Heavy-lift helicopters, using a special "Twin Bambi Bucket" configuration, have demonstrated the ability to deliver up to 5,200 gallons of retardant or water.[18]

Bambi Bucket

A brand of collapsible bucket.

◆ PERSONAL PROTECTIVE EQUIPMENT (PPE)

Firefighters are asked to respond to many different types of emergencies, each presenting its own range of challenges and dangers, and each calling for a different kind of **personal protective equipment** (**PPE**). We will look next at specialized outfits, called ensembles, that are designed to respond to these diverse needs. These include the station/work uniform, as well as ensembles suited for wildland firefighting, structural firefighting, and proximity firefighting. Also discussed are ensembles designed to defend against very specific hazards: dangerous vapors, liquid splashes, and chemical or biological agents. We will also describe **self-contained breathing apparatus** (**SCBA**) and **personal alert safety systems** (**PASS**).

self-contained breathing apparatus (SCBA)

Wearable respirator supplying a breathable atmosphere that is carried by or generated by the apparatus and is independent of the atmosphere.

personal alert safety system (PASS)

Electronic lack-of-motion sensor that sounds a loud tone when a firefighter becomes motionless.

NFPA 1500, *Standard on Fire Department Occupational Safety and Health Program*, addresses the use, care, and maintenance of most types of fire PPE.

CLOTHING
Station Work Uniform
Station work uniforms are not designed to protect personnel from hazardous atmospheres, but they do offer some degree of protection when worn under other fire protection ensembles (see Figure 5.14).[19]

NFPA 1975, *Standard on Station/Work Uniforms for Fire and Emergency Services*

Structural Firefighting Ensemble
Protective clothing used during structural firefighting includes coats, trousers, helmets, gloves, footwear, and hoods. The coat and trousers construction consists of three layers: a thermal barrier, a vapor barrier, and an outer shell. Federal regulations forbid the removal of the protective barriers from the outer shell when being worn during a response. The minimum thermal protective standard of the structural firefighting ensemble coat and trousers requires that the wearer be protected from second-degree burns during the first 35 seconds of heat exposure (see Figure 5.15).[20]

NFPA 1971, *Standard on Protective Ensembles for Structural Fire Fighting and Proximity Fire Fighting*

FIGURE 5.14 ◆ A firefighter's work uniform affords little personal protection.
Courtesy of James D. Richardson.

Wildland Firefighting Ensemble

The wildland firefighting ensemble consists of a lightweight jacket and trousers, high-top boots, a helmet, gloves, web gear (a utility belt for carrying survival equipment and tools), and a fire shelter. The fire shelter is an aluminized sheet of fire-resistant material used during an emergency to protect an entrapped individual if his or her

FIGURE 5.15 ◆ The structural firefighting ensemble does not protect the wearer from toxic products that can be absorbed through the skin.
Courtesy of Dianne Richardson.

position is overrun by fire. The individual lies in a prone position on the ground and covers himself or herself with the sheet. The aluminized outer coating will reflect a large amount of the heat and help protect the individual until the fire has passed. The jacket and trousers must be constructed of fire-resistant cotton or fire-resistive material and are designed to be worn over undergarments of 100% cotton or fire-resistive materials.[21]

NFPA 1977, *Standard on Protective Clothing and Equipment for Wildland Fire Fighting*

Proximity Ensemble

The proximity ensemble is designed to be worn in situations where the firefighter may be working in close proximity to high radiated heat, such as that created by petroleum-based fires or during the burning of other liquid fuels. The coat, trousers, gloves, helmet, and hood have an aluminized coating, which increases heat reflection. The helmet is equipped with a face shield covered with an anodized gold material to protect the wearer from facial burns (see Figure 5.16).[22]

Hazardous Materials Ensembles

Hazardous materials operations require a variety of personal protective ensembles, depending on the degree and nature of the hazard. These ensembles are classified at four different levels, each providing a different degree of protection and each governed by a different NFPA standard.

FIGURE 5.16 ◆ The proximity suit allows the firefighter to advance closer to, but not enter, a high-intensity fire.
Courtesy of TEEX-Texas A&M System.

FIGURE 5.17 ◆ The Level A ensemble protects the wearer against toxic vapors and liquids but offers no fire protection.
Courtesy of Dianne Richardson.

The **Level A ensemble** offers the highest protection to personnel. It fully encapsulates and separates the wearer from the environment. This ensemble provides vapor protection, as well as splash protection. A self-contained breathing apparatus must be worn with the Level A ensemble (see Figure 5.17).

NFPA 1991, *Standard on Vapor-Protected Ensembles for Hazardous Materials Emergencies*

Level A ensemble
Highest level of skin, respiratory, and eye protection that can be afforded by personal protective equipment.

The **Level B ensemble** is very similar to the Level A suit except that it does not offer vapor protection. It is designed for splash protection only. It comes in two basic configurations: an overall type and an encapsulated type. A self-contained breathing apparatus must be worn with a Level B ensemble (see Figure 5.18).

NFPA 1992, *Standard on Liquid Splash-Protective Ensembles and Clothing for Hazardous Material Emergencies*

Level B ensemble
Personal protective equipment that provides the highest level of respiratory protection but a lesser level of skin protection.

The **Level C ensemble**, which comes in a coverall or two-piece style, is worn in areas where there is a danger of splashes, but respiratory hazards are less likely. Respiratory protection can be provided by an air-purifying respirator instead of a self-contained breathing apparatus (see Figure 5.19).

Note: If Level A, B, or C ensembles are used during a possible terrorist incident response, they must also conform to the standard set forth in NFPA 1994, *Standard on Protective Ensembles for Chemical/Biological Terrorism Incidents.*

Level C ensemble
Level B personal protective ensemble with filtered respirators for respiratory protection.

FIGURE 5.18 ◆ The Level B ensemble offers splash protection against toxic liquids. A self-contained breathing apparatus (SCBA) must be worn with this ensemble.
Courtesy of James D. Richardson.

Level D ensemble

The standard work uniform worn at the fire station.

The **Level D ensemble** is the regular station work uniform and provides no chemical protection. This ensemble is shown in Figure 5.14.

NFPA 1975, *Standard on Station/Work Uniforms for Fire and Emergency Services*

FIGURE 5.19 ◆ The Level C ensemble is similar to that for Level B. It offers respiratory protection with a filter mask and cannot be worn in an oxygen-deficient environment.
Courtesy of James D. Richardson.

SELF-CONTAINED BREATHING APPARATUS

The self-contained breathing apparatus (SCBA) is perhaps the most important piece of safety equipment available to the firefighter. SCBA must be worn when the work environment is determined to be immediately dangerous to life and health (IDLH). Common IDLH atmospheric conditions encountered by firefighters include the following:[23]

- ◆ Areas that are oxygen deficient. Any atmosphere in which the oxygen concentration is below 19.5%.
- ◆ Temperatures above 130° F. If the air temperature is above 130°, inhaled air will cause pulmonary edema and death by asphyxiation.
- ◆ Smoke. Consists of unburned particles and products of combustion, such as tar, carbon, and toxic gases that will cause irritation to the eyes and lungs, or death.
- ◆ Toxic atmospheres. Carbon monoxide, hydrogen sulfide, hydrogen cyanide, and carbon dioxide are toxic products of the combustion process. Some combustion byproducts are carcinogenic.

The maintenance, testing, and use of SCBA are established by federal regulations and the NFPA. Only 22 states, Puerto Rico, and the Virgin Islands are required to have respiratory protection plans that meet OSHA's 29 CFR 1910.134 requirements. It is recommended that those states that are not required to follow the OSHA regulations should adopt the nationally recognized standards set forth by NFPA 1500, NFPA 1404, and NFPA 1981 (see Figure 5.20).[24]

> NFPA 1500, *Standard on Fire Department Occupational Safety and Health Program.* NFPA 1404, *Standard for a Fire Department Self-Contained Breathing Apparatus Program.* NFPA 1981, *Standard on Open-Circuit Self-Contained Breathing Apparatus for Fire Service.*

PERSONAL ALERT SAFETY SYSTEM (PASS)

The personal alert safety system (PASS) is another critical piece of personal protective equipment. A PASS may be a separate device that is attached to the wearer's

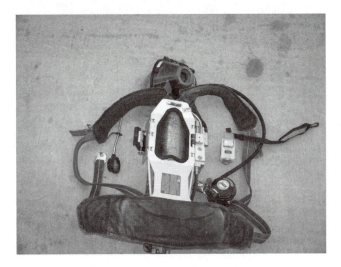

FIGURE 5.20 ◆ Self-contained breathing apparatus (SCBA) and personal alert safety systems (PASS) life safety tools should be used at any emergency scene.
Courtesy of James D. Richardson.

protective ensemble, or it may be incorporated into the SCBA. The PASS system contains a motion detection device. It is designed to emit a loud audible alarm automatically if the firefighter remains motionless for 30 seconds. The device can also be activated manually by the firefighter if he or she becomes trapped but remains conscious. Some PASS devices also incorporate a bright strobe light that serves as a location beacon.

Stop and Think 5.2

What is the major reason why a firefighter should never enter a hazardous area without proper protective clothing and breathing apparatus?

◆ TOOLS AND OTHER EQUIPMENT

Firefighters have access to a large variety of tools and equipment adapted for almost any imaginable use. Some are standard tools, such as the hand or hydraulically operated tools used in other work situations for such actions as striking, prying and spreading, cutting, and digging (see Table 5.1 and Figure 5.21). Others are technologically advanced devices that serve specialized firefighting functions.

THERMAL IMAGING CAMERAS

The safety and efficiency of fire ground and emergency response operations have improved immeasurably—and will continue to improve—as a result of technological advances. A notable example is the **infrared thermal imaging camera** (**TIC**). Originally designed for locating victims in search and rescue operations on the fire ground, TICs are now used in many other operations, such as locating fires in concealed

infrared thermal imaging camera (TIC)

Gives firefighters the ability to see heat generated by an object or person.

TABLE 5.1	Common tools and their uses for fire and rescue operations
Tool	*Use*
Pick-head ax	Chopping, cutting, puncturing
Flat-head ax	Chopping, cutting, hammering
Halligan tool	Prying
Hydraulic spreaders, cutters, and rams	Prying, pushing, and cutting
Pike pole	Pulling and puncturing
Circular saw	Cutting
Chain saw	Cutting
Adzehoe	Cutting and smoothing operations when clearing a fire break
Mattock	Used much as a pickax. It has a broad blade on one end and a pike on the other end.
Pulaski	Combines the axe with a mattock

FIGURE 5.21 ◆ Firefighters use a variety of tools that are designed for digging, cutting, pushing, pulling, and chopping.
Courtesy of James D. Richardson.

spaces, observing a fire's path of travel, and locating lost victims in rural areas or at accident scenes. Personnel responsible for operating the unit should be well trained in interpreting the data projected on the camera's imaging screen and should be aware of the limitations as well as the benefits of the device. A TIC should not be used as a navigation tool, for example, because the firefighter could become disoriented if the unit should malfunction (see Figure 5.22).

MOBILE DATA TERMINAL

The **mobile data terminal** (**MDT**) gives emergency dispatchers the ability to remain in direct contact with on-scene personnel. MDTs range in complexity from computerized

mobile data terminal (MDT)

Mobile computer that communicates with other computers on a radio system.

FIGURE 5.22 ◆ The thermal imaging camera allows firefighters to see through thick smoke to detect victims. It can also be used to find hidden fires in enclosed areas.
Courtesy of James D. Richardson.

devices mounted inside the emergency vehicle to handheld units. They consist of a screen that displays data sent to the terminal and a keyboard for entering information to be sent to the dispatcher. Some MDTs use the global positioning system (GPS), which provides the dispatcher with the location of each fire apparatus. Geographical information systems (GIS) may also be integrated into the mobile data terminal, allowing the dispatcher to transfer information such as street maps, addresses, telephone numbers, and building information to firefighters on the scene.

◆ SUMMARY

Regardless of its size, every fire department must have access to the basic resources it needs to protect its community. These resources include designated facilities for administrative support, training, and emergency response, as well as the vehicles, tools, and equipment needed to respond safely and efficiently to any predictable emergency. It is humanly impossible to plan for every contingency, but fire departments should be in a constant state of readiness to face incidents that are likely to occur in their jurisdictions.

On Scene

Urban flight caused the small rural community of Hill to experience a steady increase in population. With corresponding commercial growth and new residential housing, emergency calls became more than Hill's volunteer fire department could handle. The department had adequate equipment, but response times became longer because personnel were not always available to respond. This prompted the local government to begin the process of converting the volunteer department into a fully paid department. Volunteers grew apprehensive, fearing that their department, which had been serving the community for over 100 years, might be phased out as the paid department grew. A fire chief was hired to implement this transition. On his first day of duty, the new chief began to inventory the community's firefighting resources. He found that the fire station and all of its land, buildings, firefighting apparatus, and equipment were owned by the volunteer department. His first concern was to ensure that the volunteer department would give the paid department access to their resources until the paid department could establish its own resource pool.

Ensuring the availability of adequate resources is a primary function of a fire chief.

1. If you were the chief, how would you deal with this situation?
2. What course of action would you take to provide a smooth transition to a paid department?

Review Questions

1. What are the advantages to having the fire department's headquarters office located in the fire station?

2. What is the difference between a Type I pumper and an initial attack pumper?

3. By what methods do aerial platforms operate? How do they differ from aerial ladder platforms?

4. List four types of specialized apparatus.

5. Describe the difference between a Level A and a Level B PPE ensemble.

6. Explain how a PASS device operates.

■■■

Notes

1. *Fire service orientation and terminology.* (1995). (3rd ed.). Stillwater, OK: International Fire Service Training Association.

2. Forsman, Douglas. (2003). Training fire and emergency services. *Fire protection handbook.* (19th ed.) Quincy, MA: National Fire Protection Association.

3. Cricenti, Nicholas. (2003). Fire department facilities and fire training facilities. *Fire protection handbook.* (19th ed.). Quincy, MA: National Fire Protection Association.

4. Tutterow, Robert. (2003). Fire department apparatus and equipment. *Fire protection handbook.* (19th ed.). Quincy, MA: National Fire Protection Association.

5. *Fire service orientation and terminology.* (1995). (3rd ed.). Stillwater, OK: International Fire Service Training Association.

6. Tutterow, Robert. (2003). Fire department apparatus and equipment. *Fire protection handbook.* (19th ed.). Quincy, MA: National Fire Protection Association.

7. Cavette, Chris. (February 1, 2002).Wet and wild. *Fire Chief.* Retrieved March 10, 2006, from http://firechief.com/mag/firefighting_wet_wild/.

8. *Fire service orientation and terminology.* (1995). (3rd ed.). Stillwater, OK: International Fire Service Training Association.

9. Ibid.

10. Ibid.

11. Tutterow, Robert. (2003). Fire department apparatus and equipment. *Fire protection handbook.* (19th ed.). Quincy, MA: National Fire Protection Association.

12. Ibid.

13. Burke, Robert. (July 20, 2005). *HazMat team spotlight: Nashville, TN fire department.* Retrieved March 3, 2006, from Firehouse.com website: http://cms.firehouse.com/content/article/article.jsp?sectionId=18&id=41832.

14. *Fire service orientation and terminology.* (1995). (3rd ed.). Stillwater, OK: International Fire Service Training Association.

15. Ibid.

16. Tutterow, Robert. (2003). Fire department apparatus and equipment. *Fire protection handbook.* (19th ed.). Quincy, MA: National Fire Protection Association.

17. *Fire service orientation and terminology.* (1995). (3rd ed.). Stillwater, OK: International Fire Service Training Association.

18. Ditmore, Chris. (June 4, 1997). Twin Bambis deliver, at 40,000 lbs. *SEI Industries.* Retrieved March 19, 2006, from http://www.sonnet.com/usr/wildfire/mi26t.html.

19. Dodson, David. (2004). *Firefighter's handbook: Essentials of firefighting and emergency response.* (2nd ed.). Clifton Park, NY: Delmar Publishing.

20. Ibid.

21. Rutledge, Marty. (2004). *Firefighter's handbook: Essentials of firefighting and emergency response.* (2nd ed.). Clifton Park, NY: Delmar Publishing.

22. Ibid.

23. Teele, Bruce W., and Foley, Stephen N. (1997). Fire department apparatus and equipment. *Fire protection handbook.* (19th ed.). Quincy, MA: National Fire Protection Association.

24. Ibid.

■■

Suggested Reading

Burke, Robert. (July 20, 2005). *HazMat team spotlight: Nashville, TN fire department.* Retrieved March 3, 2006, from Firehouse.com website: http://cms.firehouse.com/content/article/article.jsp?sectionId=18&id=41832.

Cavette, Chris. (February 1, 2002). Wet and wild. *Fire Chief.* Retrieved March 10, 2006, from: http://firechief.com/mag/firefighting_wet_wild/.

Cricenti, Nicholas. (2003). Fire department facilities and fire training facilities. *Fire protection handbook.* (19th ed.). Quincy, MA: National Fire Protection Association.

Ditmore, Chris. (June 4, 1997). Twin Bambis deliver, at 40,000 lbs. *SEI Industries.* Retrieved March 19, 2006, from http://www.sonnet.com/usr/wildfire/mi26t.html.

Dodson, David. (2004). *Firefighter's handbook: Essentials of firefighting and emergency response.* (2nd ed.). Clifton Park, NY: Delmar Publishing.

Fire service orientation and terminology. (1995). (3rd ed.). Stillwater, OK: International Fire Service Training Association.

Forsman, Douglas. (2003). Training fire and emergency services. *Fire protection handbook.* (19th ed.). Quincy, MA: National Fire Protection Association.

Police and emergency vehicles. (October 2, 2006). *Neumann does communications.* Retrieved October 2, 2006, from Neumann Communications website: http://www.neumanncomm.com/police_and_emergency_vehicles.htm.

Fire Dynamics

6 CHAPTER

Key Terms

absolute temperature, p. 100
absolute zero, p. 99
atom, p. 94
backdraft event, p. 112
British thermal unit, p. 99
caloric (or calorific)
   value, p. 102
chain reaction of
   self-sustained
   combustion, p. 96
electron, p. 95
endothermic
   reactions, p. 103
energy, p. 99
exothermic, p. 96
fire flow, p. 102

fire load, p. 102
fire tetrahedron, p. 111
flash point, p. 97
flashover (rapid fire
   progress event), p. 112
heat of combustion, p. 102
hydrophoric materials, p. 104
latent heat, p. 100
latent heat of fusion, p. 100
latent heat of
   vaporization, p. 100
matter, p. 95
molecule, p. 95
neutron, p. 94
normal atmospheric
   pressure, p. 97

oxidation, p. 95
proton, p. 94
pyrolysis, p. 96
rate of heat release, p. 102
resistance, p. 106
specific gravity, p. 98
specific heat, p. 100
static electricity
   heating, p. 108
stoichiometric
   reaction, p. 102
Type I combustion, p. 110
Type II combustion, p. 110
vapor density, p. 97
vapor pressure, p. 97

Objectives

After reading this chapter, you should be able to:

- Describe the components of an atom.
- Explain how an atom becomes reactive.
- Describe the process of combustion at the molecular level.
- Describe the three forms of matter.
- Explain how the process of combustion affects the different forms of matter.
- Identify and describe the different types of heat-producing energy.
- Describe the four stages of fire evolution.
- Identify the four classifications of fires and the principles involved in their extinguishment.

◆ INTRODUCTION

Fire inspires both fascination and fear. When controlled, it can be the most productive force on earth, but when uncontrolled, it has enormous destructive potential. In this chapter, we will consider the scientific principles behind this awesome force of nature.

Fire is defined as a rapid, self-sustaining oxidation process accompanied by the evolution of heat and light of varying intensities.[1] This definition is accurate, brief, and readily accepted by the fire service, but there is much that it does not explain. For example, what is oxidation, and why is the oxidation process self-sustaining? What causes the evolution of heat and light? To answer these questions, the student of fire science must understand the chemistry, thermodynamics, and behavior of fire. An old military adage has it that good generals know and understand their enemies. The same holds true of the good firefighter. To outsmart the enemy, he or she must understand what fire is and how the environment in which it develops determines its behavior.

◆ ATOMS AND MOLECULES

atom

The smallest particle that constitutes the matter of an element.

proton

A particle in the nucleus of an atom that carries a positive electrical charge.

neutron

A particle in the nucleus of an atom that has no electrical charge.

The **atom** is the smallest unit of a chemical element. It consists of three components: protons, neutrons, and electrons. The core of the atom, called the nucleus, is usually composed of protons and neutrons. The **protons** have a positive electrical charge, and the **neutrons** have no electrical charge. Orbiting the nucleus at different levels are

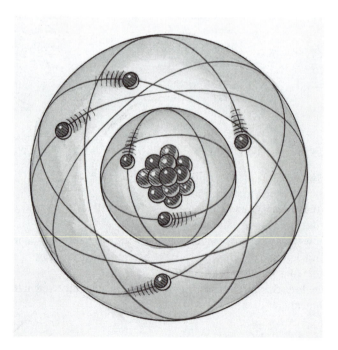

FIGURE 6.1 ◆ The atom is the smallest particle of matter. When two or more atoms are joined, a molecule is formed.

negatively charged **electrons**, usually equal in number to the positively charged protons. The number of electrons in the outermost orbit of the atom determines if the element will react chemically with other elements. The innermost orbit holds at most two electrons (except in the case of the element hydrogen, which has only one electron). The second orbit can hold a maximum of 8 electrons; the third, a maximum of 18. The oxygen atom has a total of eight electrons—two in the inner orbit and the remaining six in the outer orbit. Because its outer orbit does not contain the full eight electrons, oxygen bonds chemically with other elements in a process called **oxidation**, which will be explained more fully because it is crucial to fire chemistry.[2]

A **molecule** is formed when two or more atoms bind together in fixed proportions. The atoms that make up a molecule share pairs of electrons to form a chemical bond. A molecule may consist of atoms of the same chemical element—for example, the oxygen molecule (O_2) comprises two oxygen atoms. A molecule may also be composed of different elements—for example, the water molecule consists of two atoms of hydrogen and one atom of oxygen (H_2O). Molecules are in a constant state of motion. The amount of motion is determined by the physical state of the matter (see Figure 6.1).

electron

Negatively charged particles that orbit the nucleus of an atom.

oxidation

Chemical process that occurs when a substance combines with oxygen.

molecule

Two or more atoms joined together in a fixed proportions.

◆ MATTER, ENERGY, AND THE COMBUSTION PROCESS

Matter can exist in three physical states: solid, liquid, or gas. Because it occupies space and has weight, matter is quantifiable; that is, it can be measured in dimensions, volume, and mass (weight). It has properties that give it a physical description (see Figure 6.2). Matter can be changed chemically or physically. Physical changes are usually reversible. The physical state of matter is dependent on temperature and pressure. For example, at a normal sea-level atmospheric pressure of 14.7 pounds per square inch (psi) and at temperatures between 33°F and 211°F, water exists in a liquid state. At 32°F, water begins to solidify into ice; at 212°F, water will boil and convert to a gas. The chemical state of the water is not changed. Ice is solidified H_2O, and steam is vaporized H_2O. These processes can be reversed by melting the ice and condensing (cooling) the gas.

Chemical changes in matter are usually irreversible because the physical state of the matter changes as well as its molecular structure. New chemical substances are created. A prime example of this is the process of combustion (burning). We will next consider what happens in combustion at a molecular level.

matter

The substance or substances of which all physical things are composed. It occupies space, has weight, and can be measured. It exists as a solid, liquid, or gas.

FIGURE 6.2 ◆ Matter can exist in a solid, liquid, or gaseous state.
Courtesy of James D. Richardson.

SOLID MATTER

Solid matter is formed when molecules are so close together that they cannot change places. Free movement is not possible; the molecules can only vibrate in place. This bond gives the matter a definite form and shape. The closer the molecules are packed together, the denser the matter. Form and shape are maintained until a reactive force (such as heat) breaks the molecular bond.

When heat is applied to combustible solid matter, it is absorbed and dissipates throughout the mass. If heat is applied faster than the mass can dissipate it, the molecules begin to move faster and faster, collide with each other, and break apart. Some of the loose molecules bond with the oxygen in the atmosphere (oxidation), while other molecules remain in a free state. The oxidization process is an **exothermic** reaction, which means that heat is released when the bonding occurs. This heat in turn oxidizes more free molecules, which release more heat and cause more molecules in the solid to break away and oxidize, and so on. This process is referred to as the **chain reaction of self-sustained combustion**.[3] An example of this process can be seen by observing the combustion process in wood, the most common fuel that feeds the fires that firefighters face.

Wood is composed of cellulose. Being a living organism, cellulose contains moisture in its cellular structure. When wood is exposed to an ignition source, its molecular structure begins to change. As the surface temperature increases to 212°F, the moisture in the wood boils and converts to steam. As the temperature increases even more and penetrates deeper into the wood, more moisture is converted to steam. This process effectively dries the wood. When the surface temperature reaches 482°F, the wood begins to pyrolyze. **Pyrolysis** is the chemical decomposition of matter through the action of heat.[4] When pyrolysis is reached, wood releases combustible gases, and a layer of char forms over the heated surface. At this point, if a continuous source of ignition (a pilot flame) is present, the rapidly released combustible gases will ignite. As further heating is imposed on the wood surface, more combustible gases are released, and self-sustained combustion occurs. Many other factors that affect this process will be discussed later in this chapter.

LIQUID MATTER

When matter is in the liquid state, its molecules are in proximity to one another, but they are free to move in all directions, sliding past each other and changing places. This explains why liquids, unlike solids, have no characteristic shape. Instead, a liquid assumes the shape of the container in which it is confined and flows downward to occupy the lowest part of the container. The speed at which the molecules move is directly related to the temperature of the liquid. The higher the temperature, the faster the molecular movement.

The molecular movement in all liquids allows for the escape of molecules through the surface layer in the form of vapor. If the liquid is in an open container, all of the molecules eventually escape, and the liquid evaporates into a gaseous or vapor state. The speed of this process depends on the temperature of the liquid. Increasing the temperature increases the molecular movement, which increases the speed of evaporation.[5]

If the liquid is in a closed container and the temperature is kept constant, some of the molecules that escape from the surface collide with one another or the walls of the container and return to the liquid. An equilibrium is established when the rate of molecules leaving the surface of the liquid balances the rate of molecules returning to the liquid. When this equilibrium point is reached, the pressure exerted on the walls

exothermic

Chemical reaction between two or more materials that changes the materials and produces heat.

chain reaction of self-sustained combustion

When a fuel is heated to the point that it releases flammable vapors, which burn and generate more heat on the fuel, which releases more flammable vapors.

pyrolysis

The chemical decomposition of a substance by heat.

FIGURE 6.3 ◆ When the vapor pressure of a liquid is greater than the atmospheric pressure, the liquid will boil until all of the liquid is converted into vapor.
Courtesy of James D. Richardson.

of the container by the molecules remaining in the gaseous state is referred to as the **vapor pressure** of the liquid. Vapor pressure is dependent on the molecular characteristics of the liquid and its temperature. The stronger the bond of the molecules, the lower the vapor pressure of the liquid at any given temperature. As the temperature increases, the molecular movement increases, and vapor pressure increases proportionally. When the vapor pressure of the liquid equals atmospheric pressure, the liquid boils (see Figure 6.3).

It is important for firefighters to understand the concept of temperature and its effect on the vapor pressure and evaporation rates of different liquids. Remember, a liquid does not burn. The gases that the liquid generates when heated to a given temperature create the flammable atmosphere. In this context, "heat" is a relative term, because some liquids generate vapors at very low temperatures. For example, gasoline has a vapor pressure range of 7 to 14.5 psi. It can generate flammable vapors at −40°F, which is its **flash point**.[6]

Remember that the atmospheric pressure at sea level is 14.7 psi. If the vapor pressure of a liquid is higher than 14.7 psi, it can exists as a liquid only when it is confined in a vessel or when the temperature of the liquid is sufficiently lowered. For example, propane, a flammable gas that is used extensively for heating and cooking, has a very high vapor pressure (24.5 psi at 0°F). It is sold to consumers in a liquid state in a pressurized container. When the valve on the container is opened, the pressure inside the vessel is lowered; the liquid begins to boil and vaporize; and it is converted into a gas. For propane to exist as a liquid at **normal atmospheric pressure**, the temperature would have to be lowered to approximately −20°F. At that temperature, the vapor pressure of propane is 11.5 psi.[7]

GASEOUS MATTER

When a substance is in a gaseous state, the molecules move at high speeds. They collide with one another and the walls of the container (if the gas is confined). When a substance is in a gaseous or vapor state, it occupies much more space than it would in a solid or liquid state. Its **vapor density** is determined by its molecular weight in comparison with the molecular weight of normal air, which is 29. If the vapor has a higher

vapor pressure
The pressure exerted on the walls of a closed container by a liquid when the number of molecules escaping from the liquid reaches an equilibrium with the number of molecules returning to the liquid.

flash point
Minimum temperature at which a liquid gives off enough vapors to form an ignitable mixture with air near the liquid's surface.

normal atmospheric pressure
The pressure of the atmosphere at sea level is 14.7 pounds per square inch.

vapor density
Weight of a given volume of pure vapor gas compared with the weight of an equal volume of dry air at the same temperature and pressure.

FIGURE 6.4 ◆ If the vapor density is greater than 1.00, the gas will seek lower levels.
Courtesy of James D. Richardson.

molecular weight than that of air its vapor density is greater than 1.00 (the vapor density of air), and thus it tends to seek lower levels. If the vapor has a lower molecular weight than that of air, its vapor density is less than 1.00, and it thus it tends to rise in the atmosphere (see Figure 6.4).

Vapor density can be determined through a very simple formula:

$$\text{vapor density} = \frac{\text{molecular weight of the gas or vapor}}{29 \ (\text{molecular weight of air})}$$

The vapor density of propane can be determined as follows:

$$\text{vapor density of propane} = \frac{44.096}{29} = 2.004$$

The vapor density of propane is twice that of normal air. Therefore, propane vapors are heavier than air and will seek lower levels when released from confinement.

However, when we look at the **specific gravity** (its density compared with that of water) of the same substance in a liquid state, we may find that its behavior in the liquid state cannot be predicted by analogy to its behavior as a gas. For example, gasoline in a liquid state has a specific gravity of 0.72 to 0.76 and will "float" on top of water, which has a specific gravity of 1.0. Its vapor density, however, is 3.5. This means that when converted from a liquid to a gas, gasoline vapors will seek the lowest level possible. Its vapors are 3.5 times the weight of air. The liquid is lighter than water, but the vapor is heavier than air (see Figure 6.5).[8]

specific gravity

Weight of a substance compared with the weight of an equal volume of water at a given temperature.

Stop and Think 6.1

Why should firefighters be concerned with the form of matter with which they are faced on the emergency scene?

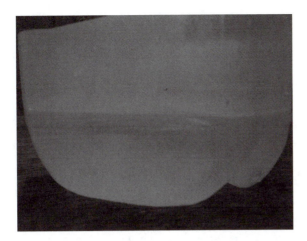

FIGURE 6.5 ◆ If the specific gravity of a liquid is less than 1.0, it will float on top of the water's surface.
Courtesy of James D. Richardson.

Energy is defined as the capacity or ability to do work. Energy itself cannot be observed. We can only perceive and measure the work that it performs. The First Law of Thermodynamics (also known as the Law of the Conservation of Mass and Energy) states, "Energy in all ordinary transformations is neither created nor destroyed, but is merely changed in form."[9] Energy exists in two forms, potential energy (stored and waiting to be released) or kinetic energy (active or in motion). It can take a variety of forms, including heat, light, electrical, mechanical, nuclear, and chemical energy. Chemical, electrical, and mechanical energies are the three most common sources of the fires encountered by firefighters. These energies generate thermal or heat energy, which is why it is crucial for firefighters to understand the measurement and physics of heat. We will consider these four types of energies and their application to fire science.

energy

The capacity to do work.

◆ HEAT ENERGY

Heat energy is a product of the other types of energy. It transfers from the hottest to the coldest substance and continues to do so until temperature equilibrium occurs. Heat is quantifiable, meaning it can be measured. For this purpose, in the United States, the fire service uses the English Customary system of measurement instead of the internationally recognized SI system (Système International), which is based on the metric system. Even though the use of the Customary system is discouraged by the scientific community, it is used in this text because of its wide use and acceptance by the fire service in the United States.

British thermal unit

The amount of heat required to raise the temperature of one pound of water at 60°F by 1°F at constant pressure.

FAHRENHEIT, RANKINE SCALES, AND BRITISH THERMAL UNITS (BTUs)

In the Customary System, temperature is measured using the Fahrenheit and Rankine scales, and heat units are measured in **British thermal units** (BTUs). The Fahrenheit (F) scale establishes a degree as a 1/180th differential between the melting point of ice and the boiling point of water. On this scale, water freezes at 32°F and boils at 212°F. A degree on the Rankine scale is equal to a degree on the Fahrenheit scale, but temperature is measured from absolute zero. **Absolute zero** is defined as the temperature at which all

absolute zero

The temperature at which all molecular movement ceases. This is established at −459.67° Fahrenheit, 0° Rankine (based on the Fahrenheit scale), −273° Celsius, and 0 Kelvin (based on the Celsius scale).

absolute temperature

Temperature that is measured from absolute zero.

specific heat

The amount of heat required to raise the temperature of a specified quantity of material by 1°C.

latent heat

The amount of heat energy absorbed or released by a substance when it changes physical states without a change in temperature.

latent heat of fusion

The amount of heat absorbed by a substance when converting from a solid to a liquid at the same temperature.

latent heat of vaporization

The amount of heat absorbed by a substance when converting from a liquid to a gas or vapor at the same temperature.

molecular movement ceases. **Absolute temperature** is measured from absolute zero. This has been established as −459.67°F or 0°R. A BTU is defined as the amount of heat needed to raise the temperature of one pound of water at 60°F by 1°F at constant pressure.[10]

SPECIFIC HEAT AND LATENT HEAT

Heat energy can be identified as specific heat or latent heat. **Specific heat** is the amount of heat absorbed by a substance as its temperature increases. **Latent heat** is the amount of heat absorbed or given up by a substance when it changes physical states without a change in temperature. For example, the specific heat of water is 1.0. It takes one BTU of heat to raise the temperature of one pound of water at 60°F by 1°F; whereas it takes 143.4 BTUs of heat to convert one pound of 32°F ice to 32°F water. The energy absorbed in this process is known as the **latent heat of fusion**. It takes 970.3 BTUs of heat to convert one pound of 212°F water to 212°F steam. The energy absorbed in this process is referred to as the **latent heat of vaporization**. This ability to absorb large quantities of heat during vaporization makes water a very effective extinguishing agent.

CONDUCTION, CONVECTION, AND RADIATION

Heat can be transferred in three different ways: by conduction, by convection, and by radiation.

Conduction

Conduction is the ability of heat to move through matter. The matter can be solid, liquid, or gas. Although heat conductivity can occur in matter that is in the liquid or gaseous state, it is most recognized as the means of heat transfer through solid matter. The heat conductivity of matter is directly related to its density. When a solid is heated, molecules become agitated and begin to transfer that movement to other molecules throughout the matter in an effort to dissipate the heat. The ability of the matter to dissipate the heat is directly related to its density. The closer the molecules are to one another, the denser the matter and the higher its conductivity. Metals have high heat conductivity because they are the densest substances on earth (see Figure 6.6). Concrete and masonry are also good conductors of heat because they have the ability to accumulate heat faster than it can be dissipated.[11]

FIGURE 6.6 ◆ Conduction is the transfer of heat through a solid object.
Courtesy of James D. Richardson.

FIGURE 6.7 ◆ Convection is the transfer of heat through circulating air or liquid.

Convection

Heat transfer by convection usually occurs in gases or liquids. When a gas is heated, its molecules become very active and begin to move freely away from one another so that there are fewer molecules in a given volume of the heated gas than in the same volume of cool gas. Because there are fewer molecules, the gas becomes lighter and begins to rise to a higher elevation and to transfer its heat to the surrounding areas in the upper levels. This process is known as convective heat transfer (see Figure 6.7).[12]

Radiation

Heat transfer by radiation does not require direct contact with the heat source or with convective heat currents. Radiant energy is transmitted as an electromagnetic wave that does not need to travel through any medium. The radiated heat is in the form of infrared rays, which are invisible to the human eye but behave in the same manner as visible light rays. Infrared heat rays travel in a straight line until they are absorbed or reflected by an object. They can pass through any object that is transparent. The color, opacity, and exposed surface area of a substance determine the degree of absorption of radiated heat. Infrared rays are readily absorbed by black or dark colors and by solid matter that has a large exposed surface area. White, light-colored, or mirrored surfaces reflect heat or infrared rays. Radiated heat is a major concern to firefighters when defending exposed surfaces. It has been estimated that 30% to 50% of the total amount of heat energy released during large fires of ordinary fuels is in the form of radiant energy.[13]

◆ CHEMICAL ENERGY

We have noted that chemical energy is a result of oxidation reactions. When a sufficient amount of heat is applied to a substance, it causes the breakdown of the molecular structure of a substance (fuel). The molecules that break free combine with oxygen in a chemical reaction called oxidation. Oxidation is exothermic, meaning that it produces heat, which in turn maintains the fuel at a temperature necessary for further oxidation to continue. This chemical process produces combustion and is the most common source of the fires dealt with by the fire service.

HEAT OF COMBUSTION OR CALORIC VALUE

heat of combustion

Total amount of thermal energy that could be generated by the combustion reaction if a fuel were completely burned.

caloric (or calorific) value

The amount of heat necessary to raise the temperature of one gram of water by 1° Celsius by one degree.

stoichiometric reaction

A reaction in which all reactants are present in fixed, definite proportions for the reaction to go to completion. The most violent of all combustion processes occurs when the fuel and oxygen proportions provide complete combustion, leaving no fuel or oxygen residue.

fire load

The amount of fuel in a compartment expressed in pounds per square foot obtained by dividing the amount of fuel present by the floor area.

fire flow

Quantity of water available for firefighting in a given area.

rate of heat release

The speed at which a fuel releases its caloric value when burning.

The amount of heat energy that a substance releases during the combustion process is known as its **heat of combustion** or **caloric (or calorific) value**. This varies with the type of substance and its mass. Calculation of the heat of combustion is based on a **stoichiometric** (complete combustion) **reaction**. This can occur only when the amount of fuel and the amount of oxygen are perfectly apportioned to provide complete combustion, leaving no fuel or oxygen at the end of the process. Although this type of combustion reaction can occur only in laboratory conditions, it is used in the fire service to calculate fire loads.[14]

Fire Load Computation

A **fire load** is defined as the potential heat energy of a fire and is calculated by multiplying the caloric value of a unit mass of the fuel feeding a fire by the amount of fuel available. In SI units caloric value is expressed in joules per gram; the English Customary system, in BTUs per pound, or per cubic foot if the substance is a gas or vapor.

In the United States, the fire service uses a rule-of-thumb estimate in computing caloric value. If the substance is organic (such as cotton, wood, paper, or plant matter), the estimated caloric value is 8,000 BTUs per pound. If the substance is a plastic or hydrocarbon, the estimated calorific value is 16,000 BTUs per pound.[15] These are theoretical figures because, as we have noted, complete combustion does not occur under real fire conditions. However, they are useful "ball-park" figures to work from when estimating fire loads. For example, 1,000 pounds of cotton and 1,000 pounds of foam plastics produce an estimated total fire load of 24,000,000 BTUs of heat (1,000 × 8,000 + 1,000 × 16,000 = 24,000,000).

Rate of Heat Release

Fire load, the total amount of heat that can be released by a fire, is only part of the equation that must be used when computing **fire flow**, the amount of water measured in gallons per minute (GPMs) that will be needed to extinguish a fire. The firefighter must also be prepared to estimate the rate of heat release. The **rate of heat release** is the speed at which heat energy is released from the substance. A one-pound block of wood and a one-pound pile of wood chips both have the potential to generate approximately 8,000 BTUs of heat. However, the one-pound block of wood will burn slower and release heat over a longer period of time than the burning pile of wood chips.

Other chemical energy reactions that are of concern to firefighters are spontaneous heating, the heat of decomposition, the heat of solution, and the heat of reaction.

SPONTANEOUS HEATING

Spontaneous heating takes place when a substance increases in temperature without the assistance of an outside ignition source. When oxygen combines with an organic substance, oxidation occurs, and heat is generated. If this heat cannot escape, the temperature will increase. The increase in temperature increases the rate of the chemical chain reaction process, doubling the rate with every 50°F increase in temperature. When the heat energy is generated faster than it can be dissipated, spontaneous ignition occurs. This process causes the fires that erupt when rags have been soaked in vegetable oils and

FIGURE 6.8 ◆ If heat is not allowed to dissipate, spontaneous heating can occur.
Courtesy of James D. Richardson.

improperly stored. Similarly, clothing has been known to ignite if it is stored in a closed bin immediately after being heated in a clothes dryer (see Figure 6.8). Bacterial activity in agricultural products can also generate enough heat to cause spontaneous ignition. In order for spontaneous ignition to occur, the following series of events must take place:

- The generation of heat through oxidation must be high enough to raise the temperature to the substance's ignition temperature.
- There must be proper ventilation to provide an adequate air supply to support combustion.
- The substance around the fuel must provide insulating properties that prevent heat dissipation.[16]

HEAT OF DECOMPOSITION

The heat of decomposition is generated during the breakdown of unstable compounds that were formed by **endothermic reactions** (reactions that absorb heat instead of giving off heat). These compounds have molecular structures that hold the fuel and the oxidizer in the same molecules. Once decomposition begins, it will generate large amounts of chemical heat energy.[17] In the early years of the movie industry, film was made of cellulose nitrate, an unstable compound. If this film was not stored in a cool, dark, dry environment, it would readily decompose and could spontaneously burst into flames. This fact was realized only after film-locker fires destroyed many classic and historic films. Even films that were not actually burned have been badly damaged by chemical decomposition (see Figure 6.9).

endothermic reactions

Chemical reactions in which substances absorb heat energy.

HEAT OF SOLUTION

The heat of solution is the heat released when a substance is dissolved in a liquid. The reaction can be endothermic or exothermic. If the reaction is endothermic, the substance will absorb heat and the solution will become cool. If the reaction is exothermic,

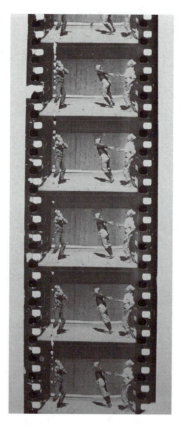

FIGURE 6.9 ◆ Cellulose nitrate film will decompose if not stored properly.

the temperature of the solution will increase and release heat. This type of reaction is usually insufficient to cause ignition.

HEAT OF REACTION

The heat of reaction is the heat released when a substance comes in contact with a liquid or with air. The intensity of the heat released during this reaction is capable of causing ignition. For example, magnesium is relatively stable when placed in cold water. However, when it is exposed to boiling water or steam, the heated metal reacts with the steam or boiling water, producing hydrogen, which ignites. Other metals, such as lithium or lithium hydride, react with water to form flammable gases. Water-reactive (a) substances are classified as **hydrophoric materials**. Pyrophoric materials are substances that can spontaneously ignite when exposed to air. Many pyrophoric materials also react violently with water (see Figure 6.10).

hydrophoric materials

Materials that react with water

Stop and Think 6.2

What is the only way firefighters can protect themselves against an unexpected event on the emergency scene?

FIGURE 6.10 ◆ The intense heat produced when combustible metals burn can reduce water to its constituent elements of oxygen and hydrogen, which become additional fuel that increases combustion.

◆ ELECTRICAL ENERGY

Electrical energy is the passing of electrons from one atom to another. Some atoms have electrons that are loosely attached. These electrons are in the outermost orbit or shell of the atom, farthest from the nucleus, and therefore have less attraction to the protons than electrons in closer orbits or shells. When an atom loses an electron, it has more protons than electrons and becomes positively charged. All atoms want to be "balanced." The positively charged atom has a strong attraction for free negatively charged electrons. Negatively charged electrons also seek positively charged atoms. The attraction increases in intensity as the number of positively charged atoms and negatively charged electrons increase. This attraction is called the *charge*. When electrons move through matter from one atom to another, one electron is gained, and another is lost. This movement is called a *current*. The matter that the electrons move through is called the *conductor*. Some matter is better suited for electrical conduction than others.

FORMS OF ELECTRICAL HEAT ENERGY

The six forms of electrical heat energy that are important to a firefighter are resistance, arcing, sparking, static, lightning, and induction.

Resistance Heating

resistance

The opposition to the flow of an electric current in a conductor component.

The opposition that a substance offers to the passage of an electric current is called its **resistance**. If the electrons in a substance are tightly held, a current cannot move through it very well. This type of substance is called an *insulator*. Materials such as rubber, glass, and dry air are excellent insulators and have very high resistance qualities. Copper, silver, aluminum, and steel—in fact, most metals—are excellent conductors because they hold their electrons loosely, enabling a current to pass readily through the material.[18]

The resistance in a conductor depends on its thickness and length in relation to the force of current flowing through it. As the speed of the electron movement from atom to atom increases (increased current flow), some electrons collide with the conductor's molecules, and the collision causes the molecules to break down. The energy that held the molecules together is released in the form of heat. This process is referred to as *resistance heating*. Resistance heating, when controlled, is very useful. It is used in electric appliances to provide heat for cooking (see Figure 6.11) and for heating structures. The incandescent lightbulb is a perfect example of resistance heating. However, when it is not controlled, resistance heating can produce extremely high temperatures, causing failure of the conductor or ignition of materials around it.

Arcing

Arcing heating is caused when an electric current is interrupted. This can occur intentionally when contacts are separated in a switch (see Figure 6.12) or accidentally when electrical contacts are loosely connected. In either case, electrical energy is discharged across an air gap, and the heat of that energy can increase until it is sufficient to ignite insulation, nearby combustible materials, or even a flammable atmosphere.

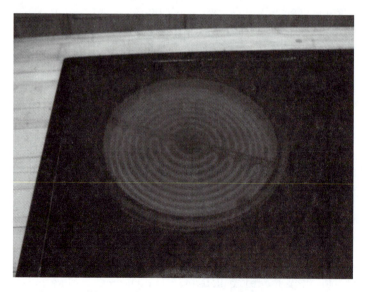

FIGURE 6.11 ◆ When controlled, heat of resistance can be used in many beneficial ways. Courtesy of James D. Richardson.

FIGURE 6.12 ◆ When an electrical contact is broken, an arcing spark will be created.
Courtesy of James D. Richardson.

Sparking

Sparking heating is different from arcing heating in that it is a short-term, one-time event in which there is a high-voltage discharge but the total energy output is low. Under normal environmental conditions, this type of heating is usually not capable of causing ignition because the heat energy is not sustained long enough. However, sparking heating is very dangerous in flammable atmospheres where the vapor/air mixtures are within their flammable range (see Figure 6.13).

FIGURE 6.13 ◆ Sparks created from metal work operations may ignite explosive atmospheres.
Courtesy of Dianne Richardson.

FIGURE 6.14 ◆ Static sparks have been known to ignite volatile atmospheres and explosive devices.

Static

static electricity heating

The heat generated by static electricity arcing.

Static electricity heating is related to sparking heating. It can occur when two materials are brought together and then separated, when fluids move through a conduit, or when finely divided solid (dust) particles touch as they pass each other. During these occurrences, it is possible for one surface to pick up electrons from the other. The surface that gains electrons becomes negatively charged, while the surface that loses electrons becomes positively charged. If proper bonding or grounding is not provided, sufficient electrical charge may accumulate to produce a spark. This spark is short term and is usually not hot enough to ignite ordinary combustible material around it. However, static sparks have been known to ignite flammable vapors, gases, certain explosives, and dust clouds (see Figure 6.14).[19]

Lightning is a form of static electricity that occurs when the action of water and ice crystals within clouds causes an imbalance in the electrical charge of the clouds. As negatively charged clouds pass over the ground, the ground becomes positively charged. Electrons will jump from cloud to cloud or to the ground and back in the form of a large static spark, which is what we know as lightning. The heat that is generated by this phenomenon will readily ignite any combustible material that it encounters. It can even cause masonry to explode. The only way to protect against this powerful type of electrical energy heating is by dissipating the charge to the ground and diverting it away from combustible materials through large metal cables and conductors (see Figure 6.15).

Induction

If a conductor is subjected to an alternating magnetic field, a flow of current is produced. As the frequency of the alternating magnetic field increases, the flow of the

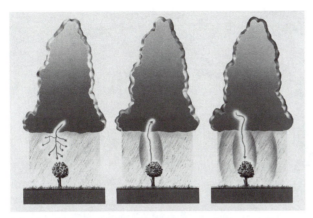

FIGURE 6.15 ◆ Lightning is a very high voltage static spark that can cause severe injury or death.

current increases and produces resistance heating. This is called *induction heating*, and it is the principle behind microwave ovens.

Stop and Think 6.3

What are the various types of electrical energy that can pose a threat to firefighters on the emergency scene?

◆ MECHANICAL ENERGY

Mechanical heat energy, which can be generated by either friction or compression, is responsible for a large number of fires each year. The vast majority of these fires are a result of friction heat.

FRICTION

The National Fire Protection Associations's *Fire Protection Handbook* defines friction heat as the mechanical energy used in overcoming the resistance to motion when two solids are rubbed together.[20] Friction is a part of everyday life, but when it is generated faster than it can be dissipated, it becomes a hazard. Friction between two metal objects, for example, can generate sparks ranging in temperature from 2,500°F to 5,400°F, depending on the types of metal involved. Although these temperatures are high, the total heat content of the spark is usually low, and ignition of combustible materials is unlikely unless flammable liquids, gases, or vapors are present.

COMPRESSION

Heat of compression occurs when a gas is compressed suddenly. As the molecules in the gas are forced closer together, they begin to collide and break apart, releasing heat. If the ambient vapor/air mixture is in the flammable range of the gas, combustion will take place. The diesel engine operates on this premise.

FIGURE 6.16 ◆ The burning of charcoal is a Type I combustion process. Courtesy of James D. Richardson.

COMBUSTION PROCESS

Type I combustion

Direct oxidation. Pyrolysis or decomposition of the matter is not required to occur in order to sustain combustion.

Type II combustion

Sequential oxidation. The molecular structure of the substance must be broken down to produce combustible gases, which will react with oxygen in the air.

The process of combustion can be separated into two types. **Type I combustion** is a process of direct oxidation, which means that pyrolysis or chemical decomposition is not necessary for the combustible material to burn. It combines directly with oxygen and does not need to convert to a gaseous or vapor state. Sulfur and charcoal are examples of materials that undergo Type I combustion (see Figure 6.16).

Type II combustion is the process of sequential oxidation, as in wood burning. The combustible solid or liquid material must go through the chemical decomposition process of pyrolysis before combustion occurs; that is, the molecular structure of the substance must be broken down to produce combustible gases, which will react with oxygen in the air. Type II combustion occurs in almost all liquids and solids that contain carbon or hydrogen in their molecular structure (see Figure 6.17).[21]

Previously, students of fire science were taught that the combustion process required three elements: fuel, oxygen, and heat. This was called the fire triangle. However, research has proven that another element is needed for combustion to take place: a self-sustained chemical chain reaction. In fact, the NFPA defines the combustion process as a self-sustained chemical reaction producing energy that causes more reactions of the same kind. By adding this fourth component, the fire triangle became known as the *fire tetrahedron*. If one of these components is removed from the process, combustion cannot take place (see Figure 6.18).

BURNING PROCESS

The burning process involves a predicable sequence of four stages: ignition, growth, full development, and decay.

Ignition Stage

The fuel is heated to a point at which flammable vapors are released and ignite.

FIGURE 6.17 ◆ The burning of wood is a Type II combustion process.

Growth Stage

At this point, the flame is small but begins to heat other areas of the fuel and combustibles in close proximity. The chain reaction process increases. The speed of the growth stage and size of the fire depend on the amount of oxygen and fuel that is

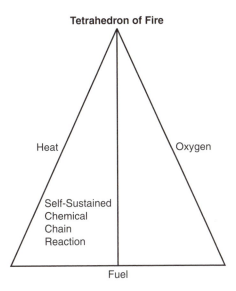

FIGURE 6.18 ◆ The combustion process can be stopped by removing any one part of the fire triangle or tetrahedron.
Courtesy of James D. Richardson.

available. Normal air contains approximately 21% oxygen. The combustion process can be maintained at oxygen concentrations as low as 14%. If the fire environment contains an oxygen concentration higher than 21%, it is considered an oxygen-enriched environment, and the growth of the fire will be more rapid.

Fully Developed Stage

At this point, all available fuels are burning. The intensity of a fire inside a structure depends not only on the amount of fuel available but also on the amount of oxygen available to support the combustion process. In an adequately oxygenated environment, the fully developed stage of an interior fire will culminate with a phenomenon known as a **flashover**, also referred to as a *rapid fire progress event*.

To understand rapid fire progress, consider a fully developed fire burning in a confined area, such as a room. Flammable gases and carbon by-products (including highly flammable carbon monoxide gas) formed by the combustion process collect at the ceiling and begin to travel laterally as the fire increases. If there is adequate oxygen in the room during the fully developed stage (15% to 21%), the gases are ignited and burn freely as they move along the ceiling. This event is referred to as a rollover and precedes flashover by a few seconds. The heat from these burning gases and by-products is radiated down and begins heating the nonburning contents of the room. The heated contents begin to pyrolyze and emit flammable vapors. Vapor emission increases with increased pyrolysis until the vapors reach their ignition temperature. This ignition or flashover is all-encompassing. It appears that the entire contents of the room ignite simultaneously. Temperatures can reach as high as 2,000°F. It is at this point that the fully developed stage peaks in its severity.

If the same enclosed room is underventilated, there is potential for a **backdraft event** to occur. In this case, heat and fuel are in plentiful supply, but there is not sufficient oxygen to support combustion. The burning fuels generate superheated gases that accumulate and fill the room. Three sides of the fire tetrahedron are present: heat, fuel, and a chemical chain reaction. Only oxygen is missing. However, since the gases are heated well above their ignition temperatures, if oxygen enters the room, the gases will ignite suddenly, and perhaps explosively.[22]

Decay Stage

This is the final stage of the combustion process. Decay occurs as the amount of fuel is diminished by the chemical decomposition of combustion. When the fuel is exhausted, the fire will extinguish itself.[23]

flashover (rapid fire progress event)
Stage of a fire at which all surfaces and objects within a space have been heated to their ignition temperature and flame breaks out almost at once over the surface of all objects in the space.

backdraft event
Instantaneous explosion or rapid burning of superheated gases that occurs when oxygen is introduced into an oxygen-depleted space.

◆ FIRE CLASSIFICATION

Fires are classified as Class A, B, C, D, and K. Each classification is based on the type of fuel that is expected to burn. The burning characteristics of each fuel class are different and require different methods of extinguishment.

CLASS A: ORDINARY COMBUSTIBLES

The fuels of Class A fires are considered ordinary combustibles and include wood, paper, plastic, and organic solids. The primary means of extinguishment for Class A fires is cooling with water, because water has a high heat absorption capability when it is converted into steam. One gallon of 60°F water can absorb over 9,300 BTUs of heat when it is completely converted into steam.

CLASS B: FLAMMABLE LIQUIDS

Class B fires involve flammable liquids, such as some oils, gasoline, and kerosene. The primary means of extinguishment for these types of fires is through smothering. The oxygen supply must be cut off from the fuel to interrupt the fire tetrahedron. Various types of water-based foam-extinguishing agents are used in this process. They function by creating a film over the top of the burning fuel to inhibit the production of flammable vapors. The water in the foam also acts to cool the liquid below its ignition temperature. Dry chemical agents are also used to provide extinguishment by disrupting the chemical chain reaction of combustion.

CLASS C: ENERGIZED ELECTRICAL FIRES

Class C fires are initiated by electricity. In the majority of cases, the electric current is only the original heat source that ignites surrounding material. When the electrical supply is removed, the fire continues to burn but is reclassified as a Class A or B fire.

CLASS D: FLAMMABLE METALS

Combustible metal fires are designated as Class D fires. Because metal fires are caused by chemical reactions, the usual method of extinguishment is disruption of the chemical chain reaction. There is no single extinguishing agent that can be used on all metals, and some agents are designed to be used on only one type. Water can be used only on certain types of metal fires, and only if extremely large volumes are available. Small quantities of water can cause a violent escalation in the fire. The intense heat generated by some burning metals has the capability to break down the molecular structure of water into its constituents of hydrogen and oxygen and burn it away in a high-intensity display of heat and light.[24]

CLASS K: HIGH-TEMPERATURE COOKING OILS

On July 1, 1998, NFPA 10, *Standard for Portable Fire Extinguishers*, mandated that all extinguishing systems designed for the protection of cooking appliances using vegetable oils or animal fats as a cooking medium be labeled as Class K fires. Before this standard went into effect, cooking oils or grease fires were classified as Class B fires. Dry chemical agents, such as sodium bicarbonate and potassium bicarbonate, were used for extinguishment. However, with the development of high-temperature cooking oils, these dry chemical agents became less efficient. Class K extinguishers are designed to use a wet chemical agent that is a mixture of potassium acetate and potassium citrate. It extinguishes the fire by chemically reacting with the fatty acids in the cooking oil or fat to form a layer of soap or foam over the top of the fuel, separating the oxygen from the fuel. This reaction is called *saponification*. The wet chemical agent also cools the cooking appliance and oil to reduce the possibility of reignition.

Uncontrolled fire is a living and breathing beast that seeks to consume everything in its path. It destroys matter by altering its chemical composition. Fire can reduce matter to its basic elements or create new compounds through the process of combustion. Knowledge of the chemistry and physics of fire is vital to the firefighter's success because it makes it possible to predict a fire's behavior and to initiate effective extinguishment strategies.

Heat is transferred by three methods: conduction, convection, and radiation. Conduction is the transfer of heat through matter. The heat conductivity of matter is directly related to its density; the more dense the matter, the higher the conductivity. Hence, conduction usually refers to the transfer of heat through solids. Heat transfer by convection occurs by the circulation of gases or liquids. The hotter the temperature of a gas, the fewer molecules there will be in a given volume, making the gas lighter than it would be in a cooler state. During combustion, the lighter, heated gas rises to higher levels and transfers its heat to surrounding areas in the upper levels. Radiated heat is an electromagnetic infrared wave and does not require a transfer medium. About 30% to 50% of the total amount of heat energy released during large fires of ordinary fuels is in the form of radiant energy.

Energy cannot be created or destroyed, only changed in form. Chemical, mechanical, and electrical energy are the most frequent sources of ignition.

Heat encountered by the fire service is commonly generated by chemical energy in the form of an oxidation reaction. In oxidation, oxygen in the air combines with the gases produced by a burning fuel to produce a new gas. For example, the carbon of burning wood is oxidized to produce carbon dioxide (CO_2). The reaction is exothermic, meaning that it produces heat, which in turn heats more fuel, setting off a chain reaction of sustained combustion. Every fuel has a caloric value. If the caloric value is known, an estimate of anticipated heat release can be calculated.

Electrical energy is produced by the passing of electrons from one atom to another. When an atom loses an electron, it becomes positively charges and seeks a negatively charged free electron. The movement of electrons from one atom to another is called current. There are six forms of electrical heat energy: resistance, arcing, sparking, static, lightning, and induction.

There are two types of mechanical energy: friction and compression. Friction is the major cause of most mechanical heat energy fires. Friction sparks can generate temperatures between 2,500°F and 5,400°F. These are extremely hazardous if flammable liquids, gases, or vapors are present. Gases that are compressed suddenly will release heat. If the heat reaches the flammable range of the gas, combustion takes place. The heat produced is known as the heat of compression.

There are two types of combustion. Type I combustion is direct oxidation. Pyrolysis or decomposition of the matter is not required to occur in order to sustain combustion. Type II combustion is sequential oxidation and does require pyrolysis. The molecular structure of the matter must be broken down to produce combustible gases that react with air. Four components are necessary for combustion to occur: fuel, oxygen, heat, and a chemical self-sustained chain reaction. If one component is absent, combustion cannot occur.

Fire develops in four stages: ignition, growth, fully developed, and decay. There are four fire classifications that are defined according to the type of fuel involved: Class A (ordinary combustibles: wood, paper, plastics, and so on), Class B (flammable

FIGURE 1.2 ◆ The fire rattle was used to alert the town of an emergency.

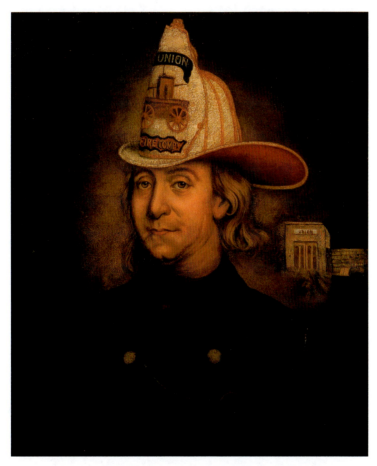

FIGURE 1.3 ◆ American Founding Father Benjamin Franklin wears a fire helmet in a portrait.

FIGURE 2.2 ◆ Cadets attending fire academy.
Courtesy of TEEX-Texas A&M System.

FIGURE 2.3 ◆ ARFF apparatus.
Courtesy of TEEX-Texas A&M System.

FIGURE 2.4 ◆ Water rescue training.
Courtesy of TEEX-Texas A&M System.

FIGURE 2.5 ◆ Collapsed structures training.
Courtesy of TEEX-Texas A&M System.

FIGURE 2.6 ◆ Simulated chemical complex fire training structure.
Courtesy of TEEX-Texas A&M System.

FIGURE 3.3 ◆ Potential candidate dressed in a suit for a first interview.
Courtesy of TEEX-Texas A&M System.

FIGURE 5.6 ◆ Wildland fire apparatus are designed for off-road operations. They are lighter and more maneuverable than structural fire apparatus.
Courtesy of James D. Richardson.

FIGURE 5.7 ◆ Water tenders vary in carrying capacity from several hundred to several thousand gallons and are the primary water supply for many rural fire ground operations.
Courtesy of James D. Richardson.

FIGURE 5.14 ◆ A firefighter's work uniform affords little personal protection.
Courtesy of James D. Richardson.

FIGURE 5.15 ◆ The structural firefighting ensemble does not protect the wearer from toxic products that can be absorbed through the skin.
Courtesy of Dianne Richardson.

FIGURE 5.16 ◆ The proximity suit allows the firefighter to advance closer to, but not enter, a high-intensity fire.
Courtesy of TEEX-Texas A&M System.

FIGURE 5.17 ◆ The Level A ensemble protects the wearer against toxic vapors and liquids but offers no fire protection.
Courtesy of Dianne Richardson.

FIGURE 5.18 ◆ The Level B ensemble offers splash protection against toxic liquids. A self-contained breathing apparatus (SCBA) must be worn with this ensemble.
Courtesy of James D. Richardson.

FIGURE 5.19 ◆ The Level C ensemble is similar to that for Level B. It offers respiratory protection with a filter mask and cannot be worn in an oxygen-deficient environment.
Courtesy of James D. Richardson.

FIGURE 5.22 ◆ The thermal imaging camera allows firefighters to see through thick smoke to detect victims. It can also be used to find hidden fires in enclosed areas.
Courtesy of James D. Richardson.

FIGURE 6.4 ◆ If the vapor density is greater than 1.00, the gas will seek lower levels.
Courtesy of James D. Richardson.

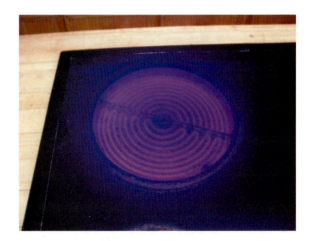

FIGURE 6.11 ◆ When controlled, heat of resistance can be used in many beneficial ways.
Courtesy of James D. Richardson.

FIGURE 6.13 ◆ Sparks created from metal work operations may ignite explosive atmospheres.
Courtesy of Dianne Richardson.

FIGURE 6.16 ◆ The burning of charcoal is a Type I combustion process.
Courtesy of James D. Richardson.

FIGURE 7.5 ◆ Old pier and beam foundations were made from tree trunks. This is very hazardous to firefighters when the piers are exposed to fire.
Courtesy of James D. Richardson.

FIGURE 7.7 ◆ There are several configurations of lateral bracing for walls.
Courtesy of James D. Richardson.

FIGURE 7.8 ◆ Gusset plates tie the truss components together. Gussets may fail under fire exposure, which then causes the truss to fail.
Courtesy of Lampasas Building Components of Texas.

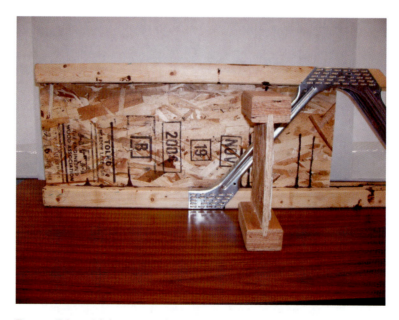

FIGURE 7.9 ◆ I-joists are engineered to perform equally as well as a solid wood beam of the same dimension. However, this lightweight support component will fail more often than a solid-wood component under fire conditions.
Courtesy of James D. Richardson.

FIGURE 7.13 ◆ Metal will fail quickly under fire conditions.
Courtesy of Universal City (TX) Fire Department.

FIGURE 7.15 ◆ Fire-resistive coatings can be sprayed on the steel structural components.
Courtesy of James D. Richardson.

FIGURE 7.20 ◆ Metal stars on the outside of a building should be a warning to firefighters that the masonry wall is unstable and will collapse if the rods connecting the metal stars fail.
Courtesy of James D. Richardson.

FIGURE 7.24 ◆ Used in post-and-beam construction, This type of connection is very strong but will readily fail under fire conditions if the tenon is exposed and burned.
Courtesy of James D. Richardson.

FIGURE 9.10 ◆ Water is supplied through the 4 inch steamer connection of the hydrant.
Courtesy of James D. Richardson.

FIGURE 9.15 ◆ If the stem of the OS&Y valve is extended out from the operating wheel, the valve is open.
Courtesy of James D. Richardson.

FIGURE 9.16 ◆ The post indicator valve is an operating device for a gate valve underground.
Courtesy of James D. Richardson.

FIGURE 9.17 ◆ The wall post indicator valve is an operating device for a gate valve inside the building.
Courtesy of James D. Richardson.

FIGURE 9.18 ◆ The post indicator valve assembly is a butterfly valve, the only indicator valve that is not a gate valve.
Courtesy of James D. Richardson.

FIGURE 9.19 ◆ The fire department is responsible for knowing which zones the fire department connections supply.
Courtesy of James D. Richardson.

FIGURE 9.20 ◆ An alarm valve may be designed for wet pipe, dry pipe, preaction, and deluge systems.
Courtesy of Tyco Fire and Building Products.

FIGURE 9.34 ◆ Dry chemical agents are nontoxic but can be highly irritating.
Courtesy of Ansul Incorporated.

liquids and gases), Class C (energized electrical equipment), Class D (combustible metals), and Class K (cooking oils and fats).

If a firefighter has a thorough understanding of the chemistry and physics of fire, he or she will be able to predict, with a reasonable amount of accuracy, how the fire will act and react under certain conditions. This knowledge is not only essential to success in fighting the fire but is also a basic ingredient in the recipe for the fire-fighter's survival. Firefighting is one of the most hazardous professions in the world; the more knowledge firefighters bring to the scene, the more likely they will return home safely.

On Scene

The Champion Forest Volunteer Fire Department, near Houston, Texas, responds to a call at a large area feed store. On their arrival, the firefighters find the building heavily involved in fire. The older part of the structure was built with heavy timber construction, which provides a very high fire load. A large metal-structure warehouse has been added to the original building. This warehouse is 100 feet wide, 200 feet long, and 24 feet high. So far, the fire is confined to the original part of the structure, which contains a variety of chemicals and insecticides used in ranching and farming. The exposed warehouse is full of bagged chemical fertilizers. Other exposed structures nearby include several bulk fuel storage tanks, area residences, and an elementary school that has scheduled class the next morning.

1. If you were the incident commander, what would be your first concern?
2. What hazards can be expected at this type of fire?
3. Would water be an effective extinguishing agent? If so, what would be the results of using water on this fire?
4. What protective measures would you provide for the firefighters?

Review Questions

1. What are the components of an atom? What identifying features do these components have?
2. What occurs at the molecular level during the process of combustion?
3. In what forms can matter be found? How do they differ molecularly?
4. Describe the different types of heat energy.
5. Identify the stages of a fire's evolution. Describe events that can occur in each stage.
6. Identify and describe the four main classifications of fire.

Notes

1. Tuve, Richard. (1976). *Principles of fire protection chemistry*. Boston: National Fire Protection Association.
2. Ibid.
3. Miale, Frank. (1999). *Fire fighter's handbook: Essentials of fire fighting and emergency response*. Clifton Park, NY: Delmar Publishing.

4. Drysdale, D. D. (2003). *Fire protection handbook*. (19th ed.). Quincy, MA: National Fire Protection Association.

5. Delpierre, G. R., & Sewell, B. T. (2002). Liquids and solids. *PHYSCHEM*. GRD Training Corporation. Retrieved October 7, 2005, from http://www.physchem.co.za/Kinetic/Liquids.htm.

6. Material safety data sheet number 401430ET - 2. *Product and company identification*. Equiva Services - MSDS. Retrieved October 16, 2005, from http://www.equivashellmsds.com/getsinglemsds.asp?ID=201791.

7. Flame Engineering, Inc. Propane technical information. *Propane & safety information*. Retrieved October 7, 2005, from http://www.flameengineering.com/Propane_Info.html.

8. Material safety data sheet number 401729E. *Product and company identification*. Equiva Services - MSDS. Retrieved October 16, 2005, from http://www.equivashellmsds.com/getsinglemsds.asp?ID=201791.

9. Skrotzki, Bernhardt. (1963). *Basic thermodynamics*. : New York, NY: McGraw-Hill.

10. Drysdale, D. D. (2003) *Fire protection handbook*. (19th ed.). Quincy, MA: National Fire Protection Association.

11. Miale, Frank. (1999). *Fire fighter's handbook: Essentials of fire fighting and emergency response*. Clifton Park, NY: Delmar Publishing.

12. Tuve, Richard. (1976). *Principles of fire protection chemistry*. Boston: National Fire Protection Association.

13. Drysdale, D .D. (2003). *Fire protection handbook*. (19th ed.). Quincy, MA: National Fire Protection Association.

14. Ibid.

15. Brannigan, Francis. (1999). *Building construction for the fire service*. (3rd ed.). Quincy, MA: National Fire Protection Association.

16. Hall, Richard, & Adams, Barbara (Eds.) (1998). *Essentials of fire fighting*. (4th ed.). Stillwater, OK: International Fire Service Training Association.

17. Drysdale, D. D. (2003). *Fire protection handbook*. (19th ed.). Quincy, MA: National Fire Protection Association.

18. California Energy Commission. (2003). Energy story. *What is electricity?* Retrieved October 30, 2005, from http://www.energyquest.ca.gov/story/chapter02.htm.

19. Drysdale, D. D. (2003). *Fire protection handbook*. (19th ed.). Quincy, MA: National Fire Protection Association.

20. Ibid.

21. Tuve, Richard. (1976). *Principles of fire protection chemistry*. Boston: National Fire Protection Association.

22. Grimwood, Paul. (February 2003). *Flashover—A firefighter's worst nightmare!* Retrieved November 3, 2005, from http://www.firetactics.com.

23. Miale, Frank. (1999). *Fire fighter's handbook: Essentials of fire fighting and emergency response*. Clifton Park, NY: Delmar Publishing.

24. Hall, Richard, & Adams, Barbara (Eds.). (1998). *Essentials of fire fighting*. (4th ed.). Stillwater, OK: International Fire Service Training Association.

Suggested Reading

Brannigan, Francis. (1999). *Building construction for the fire service*. (3rd ed.). Quincy, MA: National Fire Protection Association.

Bush, Loren, & McLaughlin, James. (1979). *Introduction to fire science*. (2nd ed.). Encino, CA: Glencoe Publishing Co., Inc.

California Energy Commission. (2003). Energy story. *Resistance and static electricity*. Retrieved October 30, 2005, from http://www.energyquest.ca.gov/story/chapter03.html.

California Energy Commission. (2003). Energy story. *What is electricity?*. Retrieved October 30, 2005, from http://www.energyquest.ca.gov/story/chapter02.htm>.

Drysdale, D. D. (2003). *Fire protection handbook.* (19th ed.) Quincy, MA: National Fire Protection Association.

Fire Equipment Manufacturers' Association. (September 8, 2006). Statement on class K extinguishers. *The life safety group*. Retrieved September 8, 2006, from http://www.femalifesafety.org.

Flame Engineering, Inc. Propane technical information. *Propane & safety information*. Retrieved October 7, 2005, from http://www.flameengineering.com/Propane_Info.html.

Grimwood, Paul. (February 2003). *Flashover—A firefighter's worst nightmare!* Retrieved November 8, 2005, from http://www.firetactics.com.

Hall, Richard, & Adams, Barbara (Eds.). (1998). *Essentials of fire fighting.* (4th ed.). Stillwater, OK: International Fire Service Training Association.

Miale, Frank. (1999). *Fire fighter's handbook: Essentials of fire fighting and emergency response*. Clifton Park, NY: Delmar Publishing.

Skrotzki, Bernhardt. (1963). *Basic thermodynamics*. New York, NY: McGraw-Hill.

Solids, liquids and gases—Their "molecular" structure. *Science for all*. Retrieved October 16, 2005, from http://www.apqj64.dsl.pipex.com/sfa/slg_structure_m.htm.

Tuve, Richard. (1976). *Principles of fire protection chemistry*. Boston: National Fire Protection Association.

Building Construction

7

Key Terms

axial loads, p. 120	gusset plates, p. 128	spalling, p. 133
compressive force, p. 122	load-bearing, p. 123	tensile force, p. 130
compressive stress, p. 120	non-load-bearing, p. 123	tensile stress, p. 120
corbels, p. 135	occupancy, p. 119	top plate, p. 142
dead-load, p. 126	shear forces, p. 132	torsional loads, p. 121
eccentric loads, p. 120	sole plate, p. 142	

Objectives

After reading this chapter, you should be able to:

- Describe the five categories of construction recognized by the National Fire Protection Association.
- Explain how each construction category reacts under fire conditions.
- Identify the various construction elements of each construction category.
- Describe the potential hazards that each construction poses to firefighters.
- Identify the various forces and loads that affect building integrity and explain how these forces and loads can cause building failure under fire conditions.
- Identify construction elements that may contribute to building collapse under fire conditions and describe the warning signs that may alert firefighters to a possible building collapse.

◆ INTRODUCTION

There is no such thing as an average fire. Each incident is unique. Different ignition sources, different fuels, and different construction types are a few of the variables that determine a fire's character. Although it is impossible to predict fully the course any fire will take, the more firefighters know about these variables, the more effective and safer they will be. In this chapter, we will consider building construction and how it affects fire behavior.

Building construction includes the design of a structure, the materials used to build it, and the configuration of structural elements in the design. Acquiring a comprehensive knowledge of building construction is one of the most challenging professional duties of today's firefighters. This requires knowing the properties of different building materials and how they react to fire. It also requires understanding basic principles of physics such as load and force.

Of course, in the heat of battle, firefighters cannot take the time to perform complicated calculations, but a basic knowledge of these principles can have immense benefits. It can help incident commanders decide whether to initiate offensive or defensive fire ground operations. It can teach firefighters to read the warning signs of life-threatening events such as rollovers, flashovers, and imminent building collapse. Firefighters may experience one or all of these phenomena in any type of structure.

Knowledge of building construction, coupled with all-important experience, can spell the difference between life and death.

◆ LOADS AND FORCES

A load is defined as "any of the forces that a structure is calculated to oppose, including any permanent force, any moving or temporary force, wind, or earthquake forces."[1] It is very important that firefighters understand how these forces affect a structure. Keep in mind that architects design and builders build with these forces in mind. They provide support elements adequate to withstand the forces that will be imposed on them. However, when these support structures are weakened by fire, the equations change, and the building can become unstable and subject to collapse. This scenario was played out in the early hours of February 14, 2000, when two members of the Houston (TX) Fire Department entered into a burning fast food restaurant. They were unaware that there was fire in the concealed space overhead that was formed by the wooden roof trusses. The fire weakened the trusses to the point that they could no longer support the weight of the commercial equipment on the restaurant's roof. The roof collapsed, fatally trapping the two firefighters.

When calculating building design, architects and builders also consider how the building will be used and what it will contain—in other words, its **occupancy**. What forces will the contents of a building and the people who live or work there exert on the structure? Loads and stresses must be calculated and structural support elements provided to counter these forces. If the occupancy changes and the building is not renovated or is improperly renovated to support the new occupancy, new loads and stresses placed on the building may exceed the original load design. This is a major concern for firefighters. Although the building may support the excess loads under normal conditions, there is a possibility of premature collapse if fire attacks the stressed structures. Firefighters who are unaware of new loads may expect a burning structure to hold up longer than it does. On June 5, 1998, two New York firefighters were killed and four were seriously injured while fighting a fire in a commercial/residential wood-frame structure when the rear of the second floor collapsed. After the investigation, one recommendation made by National Institute of Occupational Safety and Health investigators was to "ensure that modifications/renovations to buildings are in compliance with applicable

occupancy

General term for how a building structure or home will be used and what it will contain.

building codes, i.e., any renovations or remodeling does not decrease the structural integrity of the supporting members."[2]

LOAD IMPOSITION

As we have seen, structural support members are designed to withstand designated load impositions. When these support elements are attacked by fire, they may be weakened or distorted, which causes the load imposition to change. Load imposition can be classified as axial, eccentric, or torsional.

AXIAL LOADS

axial loads

Loads that are placed perpendicular to the central axis of the support structure and utilize the full support capabilities of that structural component.

Axial loads are proper loads, that is, the loads that structural elements were designed to support. They are evenly distributed and perpendicular to the central axis or plane of the supporting element. If load imposition is axial, the supporting elements are able to carry the maximum load capacity (see Figure 7.1).

ECCENTRIC LOADS

eccentric loads

Loads that are placed perpendicular to the support structure but do not pass through the central axis, thus causing the support structure to bend.

Eccentric loads are improper loads. They are not evenly distributed over the central axis or plane of the supporting structure. Instead, they are perpendicular to that central axis or plane. This off-center position creates a bending force on the structure. This places **tensile stress** on one side of the structure as it attempts to stretch and **compressive stress** on the opposite side as it contracts. The resulting imbalance weakens the support (see Figure 7.2).

tensile stress

Stress in a structure that tends to pull it apart.

compressive stress

The stress applied to materials that results in their decrease in volume.

FIGURE 7.1 ◆ Axial loads take advantage of the full load-carrying capacity of the column. Courtesy of James D. Richardson.

FIGURE 7.2 ◆ Columns will bend under eccentric loads.
Courtesy of James D. Richardson.

TORSIONAL LOADS

Torsional loads are improper loads that are offset from the central axis of the support structure but are in the same plane as the cross section. These loads create a twisting motion in the supporting member (see Figure 7.3).[3]

torsional loads

Loads that are placed only on the longitudinal axis of the structural component causing it to twist.

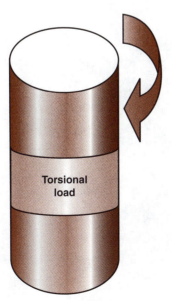

FIGURE 7.3 ◆ Torsional loads twist the column.
Courtesy of James D. Richardson.

Cataclysmic events, such as earthquakes or explosions, may subject a structure to both eccentric and torsional loads simultaneously. The resulting stresses are dynamic, three-dimensional, and highly intense. They radically increase the likelihood of rapid and serious damage.

Stop and Think 7.1

What are the most common loads and forces placed on a structure under fire conditions and why should firefighters be concerned?

◆ STRUCTURAL COMPONENTS

Most buildings have the same basic structural components: a foundation or floor, exterior walls or supports, a ceiling support structure, and a roof. It is essential to understand these structural components in order to make the best decisions in an emergency situation.

FOUNDATIONS

compressive force

A load that causes structural elements to shorten.

The foundation is the final recipient of all loads placed on a structure. It must be stable enough to accept the combined **compressive force** of the weight of the building and all loads delivered to the building. Today, most foundations are steel-reinforced concrete slabs. Concrete is a mixture of cement, sand or gravel, and water. It has a very high resistance to compression forces and continues to harden (cure) indefinitely.[4] The type of soil on which a building stands determines the structural design of the foundation, how much steel reinforcement is used, and the makeup of the concrete mixture (see Figure 7.4).

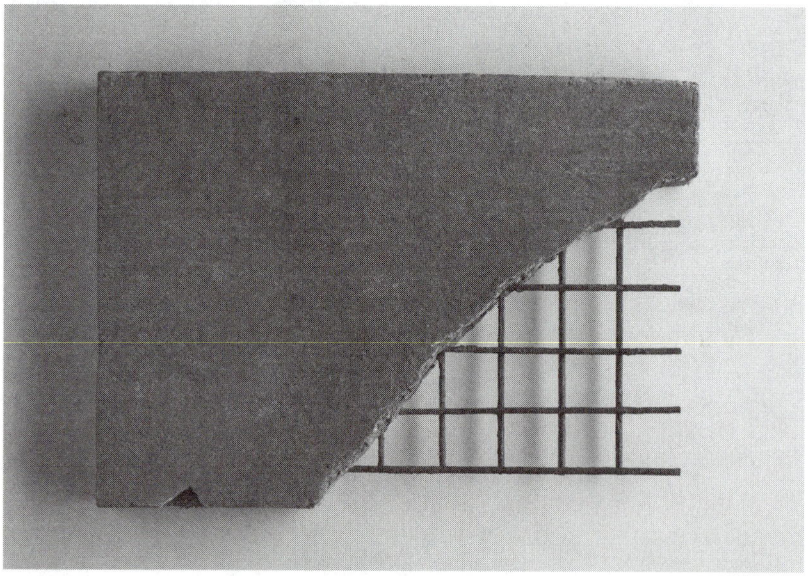

FIGURE 7.4 ◆ Concrete is designed for compression loads.

FIGURE 7.5 ◆ Old pier and beam foundations were made from tree trunks. This is very hazardous to firefighters when the piers are exposed to fire.
Courtesy of James D. Richardson.

Firefighters may encounter older structures with foundations built of wood joists and beams. This construction, sometimes referred to as *pier and beam,* consists of wooden floor joists supported by wooden girders, which are in turn supported by wood, concrete, or steel piers. The concealed space created by this type of construction provides a perfect environment for hidden fire to move around, over, and under firefighters, especially if the floor joists are of open-work truss construction. When fire weakens the support structures, they may well give way, resulting in floor or wall collapse (see Figure 7.5).

WALLS

Walls can be classified as bearing and nonbearing. **Load-bearing** walls support the roof and any upper floors. All loads placed on the structure are transmitted through the load-bearing walls to the foundation. **non-load-bearing** walls support only their own weight (see Figure 7.6). Structurally speaking, load-bearing walls function as columns. They are designed to accept vertical loads and to transmit those loads to the foundation. Bracing must be added to walls to give them resistance to lateral loads such as wind forces. There are a variety of brace configurations. The most common types are diagonal bracing, chevron bracing, and X bracing (see Figure 7.7 A, B, C).

WOOD TRUSSES AND I-JOISTS

Today, wood trusses are designed by computer engineering programs. All of the angles and lengths of the components are cut with precision and precisely placed in the web, the area between the top and bottom cord. However, the actual strength of

load-bearing

A component of a structure that is designed to bear the weight of the structure and loads and transmit this weight to the foundation.

non-load-bearing

A partition or wall that supports only itself or interior finish components.

FIGURE 7.6 ◆ (a) Load-bearing walls transfer loads from other support structures to the foundation. (b) Non-load-bearing walls transfer only their own weight to the foundation. Courtesy of James D. Richardson.

the wood truss does not rest completely with its components. It also depends on the connectors that tie the web components and cords together. A truss is an assemblage of building components that form a rigid framework. It consists of a top cord (top board), a bottom cord (bottom board), and interconnecting members that create a web. The strength of the truss is achieved by configuring the web components in triangles, which provide greater stability than right-angled configurations when stress is imposed. Trusses are replacing solid lumber joists and roof components in many wood-frame structures because they are lighter, easier, and faster to install. Also, truss construction for ceiling joists allows for spans as great as 80 feet without the need for support columns. Prior to 1952, truss components were connected with wood gussets, staples, nails, glue, or mortise-and-tenon joints. After that date, metal gussets came into use. Today's standard connectors are steel plates stamped from 16-, 18-, or 20-gauge structural steel and are coated with zinc to reduce the possibility of rusting. One side of the plate has teeth that can range in length from 5/16 to 9/16 inch. The plate contains eight teeth per square inch and is sized according to the size of the wood components and the amount of stress that will be transferred. Today, the connector plates are pressed into place by computer-operated machines. This creates a much stronger connection than hand-mounted connections. For example, a 3 × 4-inch connector plate, applied to both sides of a web joint, provides a working strength of 2,400 pounds. with a safety design load of 7,680 pounds.[5] This is very impressive when viewed from a construction standpoint. However, from the firefighter's viewpoint, there is a caveat: Under fire conditions, metal connectors will start to lose structural integrity when they are heated to 1,000°F. If the connectors fail, the entire truss is likely to collapse. Failure of the metal connectors due to heat

FIGURE 7.7 ◆ There are several configurations of lateral bracing for walls.
Courtesy of James D. Richardson.

FIGURE 7.8 ◆ Gusset plates tie the truss components together. Gussets may fail under fire exposure, which then causes the truss to fail.
Courtesy of Lampasas Building Components of Texas.

dead-load

The structure and any equipment or building components that are permanently attached to the building.

exposure is not the only reason wood trusses collapse under fire conditions. The trusses may also be weakened by:

- ◆ Damage caused by pests (termites, carpenter ants, and so on)
- ◆ Corrosion of the metal connectors
- ◆ Excessive **dead-load** weight beyond the design load
- ◆ Illegal modifications to the truss (see Figure 7.8)

The wood I-beam construction component came into use in the 1990s and continues to increase in popularity in today's housing industry. It makes possible open spans of up to 60 feet without support columns. This is less than the 80-foot spans that truss construction allows, according to the previous discussion. The I-beam was designed to take the place of solid-wood and wood-truss joists. In fact, construction professionals refer to it as an "I-joist."

Like the truss, the I-joist has three components: top and bottom cords (often referred to as flanges) and web. It offers a much higher strength-to-span ratio than dimensional lumber. This means that it can support more weight than a solid beam of equal length, width, and depth. Most I-joist components are held together by glue. However, wide I-joists (10 inches wide or more) may use a metal angle brace gusset for additional support. The flanges are constructed of laminated veneer or solid lumber and range in dimensions from $1\frac{1}{2} \times 1\frac{1}{2}$ inches to $1\frac{1}{2} \times 3\frac{1}{2}$ inches. The I-joist web is composed of plywood, laminated veneer lumber (LVL), or oriented strandboard (OSB). Instead of using the triangle principle for strength, the I-joist web takes advantage of the 90° cross grain of the veneered layers in the plywood, LVL, and the cross strands of the OSB.[6] Engineered structural components such as LVL and OSB are becoming increasingly more common in the housing industry (see Figure 7.9).

FIGURE 7.9 ◆ I-joists are engineered to perform equally as well as a solid wood beam of the same dimension. However, this lightweight support component will fail more often than a solid-wood component under fire conditions.
Courtesy of James D. Richardson.

CEILING JOISTS

Ceiling joists are beams that support ceiling loads and are themselves supported by larger beams, girders, or load-bearing walls. Ceiling joists tie the exterior and interior walls together and double as floor joists for the attic or for a second story. When designed loads are imposed onto ceiling joists, the load is distributed horizontally to the ends of the joists resting on the load-bearing walls. The load is then transmitted vertically through the walls to the foundation. Any load-bearing wall or column under the joists will also assume part of the load. Ceiling joists may be constructed of lightweight steel trusses, wooden trusses, I-beam steel, or solid-dimensional lumber. The length of the span and the load that must be carried determine the size of the joists. This is often specified in a jurisdiction's building code (see Figure 7.10).[7]

ROOF CONFIGURATIONS

There are two major kinds of roofs: flat or pitched.

Flat Roofs

Flat-roof construction utilizes the ceiling joists as support for the roof assembly. The ceiling joists in effect become the rafters. Truss construction is often found in flat-roofed buildings built since the 1950s and designed with large open areas. This type of construction provides large concealed spaces in the attic area in which fire can travel unnoticed by firefighters. Remember that the ceiling joists tie the exterior and interior walls together. If the trusses fail under fire conditions, not only will the roof collapse, but the walls that are tied together by these trusses may also fail.

FIGURE 7.10 ◆ Ceiling joists are beams that connect parallel walls and support the interior ceiling covering.
Courtesy of Pickett's Video & Training.

Pitched Roofs

Pitched roofs can be of conventional or prefab truss construction. Conventional pitched roofs are shaped and assembled on the construction site and are built one rafter at a time. This type of construction is referred to as *stick frame*. The stick-framed roof may be constructed with a truss design, but the majority of truss roof are prefabricated construction.

gusset plates

Metal plates used to unite multiple structural members of a truss.

Prefab trusses are assembled at fabrication shops by cutting and assembling the components, using **gusset plates** to hold them together. The trusses are carried to the work site and lifted into place onto the support structure. Because the bottom cord of the truss is also the ceiling joist, failure of the truss during a fire will result in the structural instability of the supporting walls (see Figure 7.11).

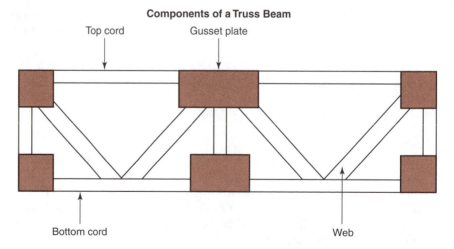

Components of a Truss Beam

Top cord   Gusset plate

Bottom cord   Web

FIGURE 7.11 ◆ If one component of a truss fails, the entire truss will fail.
Courtesy of James D. Richardson.

The arched roof is a form of pitched roof. It incorporates the truss design and is used to span large areas, such as gymnasiums, It is referred to as a *bowstring truss* because its design resembles an archer's bow.

Building design and construction are the major factors determining how a structure reacts under fire conditions. Firefighters knowledgeable about different building construction classifications have a definite edge when it comes to predicting a fire's behavior—knowledge that could save a life.

Although there are several model building codes that classify building construction, this chapter will focus on the building construction types recognized by NFPA 220, *Standard on Types of Building Construction*, and the new NFPA 5000, *Building Construction and Safety Code.*

There are five basic building classifications. NFPA 220 lists these classifications as follows:

- Type I, Fire Resistive
- Type II, Noncombustible
- Type III, Ordinary
- Type IV, Heavy Timber
- Type V, Wood Frame

These five classifications are used in the new NFPA 5000 building code. However, following the recommendations of the Board for the Coordination of the Model Codes (BCMC), the new building code removed the designations fire resistive, ordinary, and heavy timber, and so on, and simply designated Types I and II as noncombustible and Types III, IV, and V as combustible.[8]

In discussing these types of constructions, we will refer to fire-resistive ratings. A fire-resistive rating is based on the length of time (in hours) that a structural component can be expected to maintain structural integrity while being subjected to a standard fire exposure test, such as outlined in NFPA 251, *Standard Methods of Tests of Fire Endurance of Building Construction and Materials.* A fire rating of one hour means that the assembly will resist breakthrough for one hour under controlled test conditions.

TYPE I CONSTRUCTION

In Type I construction, the structural components are noncombustible and protected, usually by encasement in a heat-resistant material that provides temporary structural stability when the components are exposed to fire. Because the structural components are protected, they are given a fire-resistive rating. Although the structural components do not contribute fuel to a fire, some combustible materials are permitted for use in interior finishes, roof structures and coverings, doors and windows, and exterior trim.[9]

The new NFPA Building Code divides Type I construction into two subtypes: Type I-443 and Type I-332. Each numerical digit represents the number of hours of fire resistance required for certain components of the structure.

- The first digit identifies the fire-resistive rating for the load-bearing exterior walls as three or four hours, respectively.

- The second digit identifies the fire-resistive rating of interior load-bearing walls, columns, beams, girders, trusses, and arches as three or four hours, respectively.
- The third digit identifies the fire-resistive rating for floor and roof assemblies as two or three hours, respectively.

Although concrete block or masonry can be used in Type I construction, the most common structural materials used are steel-reinforced concrete or protected steel.[10]

Steel

Steel is a very strong structural material, unique in its ability to withstand compressive and **tensile forces** equally. Alloying steel with other metals gives it versatility and allows it to be used for a variety of construction purposes. However, strong as it is, steel begins to lose its structural integrity when subjected to fire temperatures of 1,000°F, which is not unusual in structure fires. In fact, tests have shown that when structural steel is heated above 1,200°F, "steel properties decrease so dramatically that they are of no structural interest." [11] The length of time between heat exposure and the structural failure of steel depends on the following factors:

- Mass of the Steel Component. The more density (mass) that a steel component has, the longer it will take to heat to 1,000°F. Steel components with less density absorb heat faster.
- Heat Release Rate of the Fire. The higher the heat release, the faster the steel is heated.
- Forces or Loads Exerted on the Steel Component. A steel component under stress will fail at a lower temperature than a component that is not load bearing.
- Type of Connections Joining the Steel Components. When structural steel members are welded or bolted together to form an assembly, they will maintain their structural integrity longer under fire conditions than if the components are simply supported (see Figure 7.12).[11]

tensile force

A force that causes the structural element to lengthen.

FIGURE 7.12 ◆ Structural collapse can occur if a connection fails.
Courtesy of James D. Richardson.

Structural steel becomes very animated under fire conditions, especially if it is under a load or restrained at its ends. If the beam is restrained, it will distort by bending or twisting when fire conditions reach or exceed 1,000°F. This can create excessive load imposition on support structures and cause structural failure. Steel has a memory when it is placed under stress. The memory allows the steel to return to its original shape after the stress has been removed. This is referred to as *flexibility*. The point at which the steel component will not return to its original shape after the stress is removed is called the *point of strain*. When water streams are directed onto a distorted beam, it will return to its original shape if it has not passed its point of strain. If it has passed its point of strain, the beam will remain in its distorted shape when cooled. When buckled beams are cooled, they can retract and fracture the connections at the ends. This occurs because the safety rating for bolts and welds is higher than for structural steel. When the retracting steel is torn away from the bolts and welds, structural components are weakened and collapse is possible.

Unrestrained beams also pose dangers for firefighters. An unrestrained steel beam 100 feet in length will elongate 9.5 inches when heated to 1,000°F. However, when that same beam is cooled to a normal ambient temperature, it will return to its original length. Elongation can cause a wall or support to be displaced. If this has occurred, when the beam is then cooled and retracts to its original length, it may dislodge from the support and initiate a structural collapse (see Figure 7.13).[12]

There are several methods of protecting structural steel components from the adverse effects of fire exposure. These methods are not designed to make the steel components fireproof but to give them a fire-resistive rating. One method of protecting structural steel is encasement. The fire-resistive ratings for encasements depend on the moisture content and the thickness of the materials used. Fire-resistive materials can be made of tile, gypsum, plaster and lath, or concrete. Each material has advantages and disadvantages. Tile is a good insulator, as is gypsum, which has the added advantages of being lightweight and high in moisture content. Both

FIGURE 7.13 ◆ Metal will fail quickly under fire conditions.
Courtesy of Universal City (TX) Fire Department.

FIGURE 7.14 ◆ Encasement is a method of giving structural steel fire resistance. Courtesy of James D. Richardson.

materials are, however, easily damaged. If they should break away, the structural steel that they are meant to protect would be exposed to potential fire damage. Lath and plaster is used in conjunction with gypsum board to encase structural steel. It is a good substitute for concrete because it is much lighter, but concrete remains the best, and most permanent, fire-resistive material. Its main disadvantages are its weight and expense (see Figure 7.14).[13]

A second method of protecting structural steel is by spraying the steel components with a fire-resistive coating. It is, however, difficult to gauge the thickness of a sprayed coating or to maintain an even application on the steel component. Consequently, the fire resistance of components treated in this way is not easy to determine. Moreover, although a cement-based spray coating is an excellent insulator, it is easily damaged and can be accidentally removed or damaged by maintenance personnel (see Figure 7.15).[14]

Concrete

Concrete is a mixture of cement, sand, and aggregate. Water is added to the mixture to transform it into a fluid state so that it can be molded into a desired configuration. Concrete building components are formed by pouring the liquefied mixture into a form or "falsework" and allowing it to harden or cure to a predetermined working strength. This material has a natural strength against compressive loads that makes it an ideal material for columns, beams, and floor construction. Concrete structural components can be made on the building site (cast in place) or at a remote location and delivered to the work site (precast).

There are several configurations in which concrete can be used to take advantage of its structural qualities:

shear forces

Loads that cause the structural element to deform or fracture in a direction parallel to the force, by sliding one section of the element along others.

- Plain Concrete. Composed of cement, sand, aggregate, and water, it is very strong against compressive forces. However, plain concrete is vulnerable to the tensile and **shear forces** (loads that cause the structural element to deform or fracture in a parallel direction to the force, by sliding one section of the element along others) that are placed on walls, beams, girders, columns, and floors.

FIGURE 7.15 ◆ Fire-resistive coatings can be sprayed on the steel structural components.
Courtesy of James D. Richardson.

- Steel-Reinforced Concrete. Steel bars or steel mesh are added to the concrete mixture when it is poured. Curing the liquefied concrete bonds to the steel. The strength characteristics of the two materials complement one another, creating a structural component that has compressive, shear, and tensile strength.
- Prestressed (Pretensioned) Concrete. Prestressed structural components are cast at a concrete plant and subjected to a tensioning process. Prestressing cables, referred to as *tendons*, are placed in the falsework, or casting bed, and tightened to about 30,000 pounds. of tension. The concrete is poured around the cables and allowed to bond with the steel as it cures. When the tension is released from the cable ends, the steel tries to retract, creating a compression force on the concrete component, which gives it resistance against tensile forces. Prestressed components can be identified by the cut cables on the ends of the casted units.
- Post-Tensioned Concrete. Post-tensioned structural components are cast in place. High-strength steel cables are placed in the forms prior to pouring the concrete. The cables are coated or placed inside tubes to prevent them from bonding with the concrete. After the concrete is cured to the desired strength, the cables are tensioned with hydraulic jacks placing compression forces on the component. Post-tensioning is used extensively for floors in high-rise buildings because it allows thinner structural members to carry heavier loads.

The moisture content of concrete makes it a very effective fire-resistive material, but moisture can contribute to its degradation under certain fire conditions. When it is exposed to intense or prolonged heat, the moisture in the pores of the concrete converts to steam and expands, causing the concrete to break apart. This phenomenon is called **spalling.** When spalling occurs, the concrete is broken away from reinforcement steel, exposing the steel to the fire environment. The heated steel, having bonded with the concrete, will try to expand and bend, causing more portions of the concrete to fail. Recently cured concrete will spall at a faster rate than older "aged" concrete because it has a higher moisture content (see Figure 7.16).[15]

spalling

The process in which parts of a concrete structure break or chip off when the structure is exposed to extreme heat.

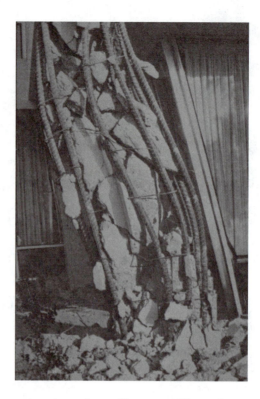

FIGURE 7.16 ◆ Exposure to extreme heat will cause spalling and expose the reinforcing steel. This causes the steel to expand and the concrete to crack and break away.

TYPE II CONSTRUCTION

Type II construction is described in the NFPA 5000 building code as noncombustible. The structural components are of noncombustible or limited-combustible materials. Type II construction is quite similar to Type I, but its fire-resistive rating is not as high, and protection is required in only two subtypes of this category: Type II-222 and Type II-111. These components must be protected and must have a two- or one-hour fire-resistive rating, respectively. They are considered adequate for residential, educational, institutional, business, and assembly occupancy structures. Structural components of Type II–000 are of unprotected noncombustible construction. These have no fire-resistive rating and do not provide structural stability under fire conditions. This type of construction is common and may be found on commercial and private properties in the form of business office buildings, garages, and barns (see Figure 7.17).

Stop and Think 7.2

What factors should firefighters be especially concerned about when fighting fires in structures of Type I or Type II construction?

TYPE III CONSTRUCTION

Type III construction is quite different from Types I and II. In Type III construction, the exterior walls are required to be constructed of noncombustible or limited-

FIGURE 7.17 ◆ Type II construction may be protected or unprotected.
Courtesy of James D. Richardson.

combustible building materials. They can be of stone or brick masonry, metal, concrete, or any combination of these. In old-style Type III buildings, the exterior bearing walls are usually of masonry construction. They are very thick at the lower levels and progressively decrease in thickness at each upper level. Ledges known as **corbels** are formed between stories and support the floor joists for the next floor (see Figure 7.18).

corbels
Pieces of stone jutting out of a wall to carry weight.

FIGURE 7.18 ◆ Corbels are ledges constructed on the inside of a masonry wall to support the floor joist of the upper level.
Courtesy of James D. Richardson.

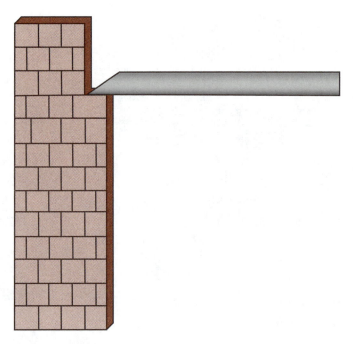

FIGURE 7.19 ◆ The fire cut allows the ends of the joist to burn faster and fall inward.
Courtesy of James D. Richardson.

Most of these buildings were built in the 19th century.[16] Their design poses a serious inherent threat to firefighters. Even though the exterior walls are non-combustible, the interior structural support components may be constructed partly, or entirely, of combustible material. Many of these buildings have ceiling and floor joists, floor and ceiling coverings, interior walls, girders, beams, and columns of wood. These components are not only combustible, but they are also assembled in a way that creates concealed spaces—for example, between floor and ceiling joists and between wall studs. Fire can move through these spaces un-noticed by firefighters, from room to room or from floor to floor. Moreover, in Type III construction, the floors are structurally engineered to collapse under fire conditions to prevent collapse of the exterior walls. The floor joists, supported by the corbelled exterior walls, are "fire cut" (cut on an angle) on each end. This al-lows the ends of the joists to burn through faster and ultimately drop away from the wall (see Figure 7.19).

The fire-cut design was actually intended to save firefighters' lives. During the 19th century, fires were fought from outside buildings. An interior collapse was much preferred over an exterior wall collapse. Exterior walls might still collapse, but it was fairly certain that the interior structures would go first. Modern firefighting strategy and tactics usually call for an aggressive interior attack. In older Type III buildings, such an attack could have fatal consequences. Firefighters must familiarize them-selves with buildings of this construction category in their response area.

There are several indicators of danger that firefighters should look for in Type III buildings. It is critical that these be identified during prefire planning to avoid unpleasant surprises during an actual firefighting operation:

- ◆ Metal Stars or Plates on the Exterior Masonry Walls. These plates indicate that the walls of the building are supported by being tied together with a metal rod or cable connected to

FIGURE 7.20 ◆ Metal stars on the outside of a building should be a warning to firefighters that the masonry wall is unstable and will collapse if the rods connecting the metal stars fail. Courtesy of James D. Richardson.

an opposite wall or a floor beam. Cold-drawn steel fails at 800°F. The rod or cable can fail under fire conditions and cause the wall to collapse (see Figure 7.20).

- ◆ Cracks in the Masonry and Mortar. In Type III construction, the load-bearing masonry wall is the support structure. Cracks in the masonry or mortar can be an indicator of wall movement, imposition of excessive load on the supporting elements, or foundation failure (see Figure 7.21).
- ◆ Sand Lime Mortar. Sand lime mortar is vulnerable to water erosion. Fire streams can cut the mortar from between the masonry and weaken a wall. Prefire inspection should identify

FIGURE 7.21 ◆ Cracks in a masonry wall may indicate foundation weakness. Courtesy of Bob Hopkins, Weatherford (TX) Fire Department.

FIGURE 7.22 ◆ Sand lime mortar is used in the restoration of historical buildings.
Courtesy of James D. Richardson.

deteriorating sand lime mortars.[17] Keep in mind that Type III buildings listed as historical structures are highly likely to have sand lime mortar, as historical preservation rules require that it be used in any reconstruction work (see Figure 7.22).

Like Type II construction, Type III construction has two subdivisions: protected and unprotected. In Type III-211 protected construction, the exterior load-bearing walls are required to have a fire-resistive rating of two hours. The combustible or limited-combustible interior structural support components must be protected and have a fire-resistive rating of one hour. These components include all structural frames, columns, girders, and beams supporting loads of more than one floor. Floor construction must have a one-hour fire-resistive rating.

In Type III-200 unprotected construction, the exterior load-bearing walls are required to have a two-hour fire-resistive rating. However, the interior support structures—columns, beams, and girders supporting loads from more than one floor—and the floor assemblies are constructed of unprotected combustible materials. NFPA 220 categorizes Type III construction as ordinary construction. This description has been dropped by the new NFPA 5000 building code, which designates these buildings as combustible (see Figure 7.23).[18]

TYPE IV CONSTRUCTION

In the mid-19th century, insurance companies demanded that the textile industry build structures that would provide greater fire resistance than was offered by ordinary construction. In response, builders developed the heavy-timber construction method, now designated as Type IV. The exterior load-bearing walls were of solid masonry (just as in Type III construction), but interior support timbers had to be of larger dimensions than was standard. Concealed spaces between the floors and roofs were eliminated.

Unfortunately, today there are different interpretations of what constitutes heavy-timber construction, which is also sometimes called *mill construction*. NFPA 220 uses the term heavy timber to identify Type IV construction, but due to the confusion, the new NFPA 5000 categorizes Type IV construction simply as combustible.

FIGURE 7.23 ◆ Unprotected Type III construction poses a threat to firefighters because it contains a large amount of concealed spaces in which fire can travel unseen.
Courtesy of James D. Richardson.

NFPA 220 defines Type IV construction as a structure with noncombustible or limited combustible exterior and interior walls. The interior structural components, such as columns, beams, arches, floors, and roofs, must be constructed with large dimensional solid or laminated wood and have no concealed spaces. The standard also requires that interior load-bearing walls have a fire-resistive rating of no less than one hour and exterior load-bearing walls have a fire-resistive rating of no less than two hours.[19] All wood columns used to support floor assemblies are required to be a minimum of 8 inches square (sawn posts) or 8 inches in diameter (round posts). The wood columns must be a minimum of 6 × 8 inches if they support a roof assembly. Floor decking must consist of 3-inch tongue-and-groove planks, and roof decking must be 2-inch tongue-and-groove planks. All wood beams and girders must be a minimum of 6 inches wide and 10 inches in depth if they support a floor assembly and 4 × 6 inches if they support a roof assembly. Even though the interior components are combustible, the large dimensions of the lumber give them a certain amount of fire resistance. In fact, wood over 6 inches thick can be more fire resistant than unprotected steel.[20] As the surface of the wood burns, it forms a layer (char) that insulates the unburned wood beneath. This process, called pyrolysis, reduces the burn rate of the wood and allows the wood to maintain structural integrity longer. In the United States, burn-rate testing of laminated and sawn timbers has established a char rate of 0.024 inch per minute.[21] For example, if a 2 × 4 inch solid wood board is exposed to heat and begins to pyrolyze, or char, at the point of heat contact, the 2 × 4 will be reduced in dimension by 0.024 inch per minute at the char site.

The load-carrying capability of structural components depends not only on the strength of the components but also on the connections that tie them together. The most stable way to connect two beams is to load the ends of the beams axially on top of a column or girder. When this is not possible due to height restrictions, beams may be tied to girders with mortise-and-tenon joints, or columns may be corbelled to

FIGURE 7.24 ◆ Used in post-and-beam construction, this type of connection is very strong but will readily fail under fire conditions if the tenon is exposed and burned.
Courtesy of James D. Richardson.

accept the ends of the beams. It is recommended that the connections have higher fire resistance than the structural components. However, this may not always be the case. Mortise-and-tenon joints and corbels are not as solid as heavy-timber beams, girders, or columns. These connecting elements can fail under fire conditions and cause premature collapse, even though the structural elements they connect have maintained their strength. Firefighters should be aware of this during prefire planning (see Figure 7.24).

TYPE V CONSTRUCTION

In Type V construction, structural components are entirely of wood. NFPA 220 refers to Type V structures as wood-frame construction. This classification has been dropped under the new NFPA 5000 building code and replaced with a combustible classification. Type V structures come in many different forms and all are very susceptible to fire. The NFPA places Type V construction under two subclassifications: Type V–111, which has a one-hour fire-resistive rating throughout the building, and Type V–000, which carries no resistive rating.[22]

Next, we will look at some aspects of Type V wood-frame construction and consider the challenges they present to firefighters.

Balloon-Frame Construction

Balloon-frame construction originated in the 1830s. It was a building method that required smaller-dimensional lumber than previous forms of wood construction. Mortise-and-tenon joints were replaced with machine-made nails. The defining feature in balloon-frame construction is that the wall studs run continuously, two or more stories, from the sill to the top plate of the building. The studs and the first-floor joists rest on the sill. The floor joists for the upper stories are tied to the supporting exterior walls by a ribbon strip, which is a board that is nailed to the interior edge of the studs. The joists rest on the ribbon strip, much the same as beams rest on a girder, and are nailed to the inside face of the studs.

The natural inclination of wood is to expand and contract across the grain, not with the grain. In balloon construction, most of the structural components are assembled so that there is less cross-sectional wood grain framing than in other types of wood-frame construction. Because the exterior wall studs are continuous from floor to attic, the wood grain is in one direction and unbroken. Therefore, there is less expansion and contraction of structural components in a balloon-framed structure than in conventional wood-frame construction. This quality makes balloon framing ideal for two-story structures with brick veneer, stone veneer, or stucco exterior walls. It is also the construction method of choice for interior load-bearing components in Type III masonry structures. Because there is less cross-sectional grain construction, the dimensional changes in the walls and the settling that occurs between the exterior walls and interior support structures are minimized.

The biggest drawback to balloon-frame construction is the large concealed spaces and voids that are created by its design configurations. If the exterior walls of a balloon-framed structure are not provided with proper firestops, fire can easily travel from the floor to the attic between the studs and to opposite load-bearing walls between the ceiling or floor joists. This fire movement may be undetected by firefighters until it breaks out above or behind them. Firestops are blocks of wood placed between studs at each floor level. They may also be placed between floor or ceiling joists. Their sole purpose is to inhibit fire spread throughout the building. Unfortunately, many older structures with this type of construction are not properly firestopped. Firefighters should try to identify these structures during prefire planning. One way to identify balloon construction in older buildings is to observe the location of upper- and lower-level windows and doors. In these older buildings, the upper story windows are usually aligned vertically with the lower floor windows. Exterior wall studs must be cut to accommodate window and door openings, and vertical alignment of the openings reduces the number of studs that must be cut. The widespread use of balloon-frame construction declined in the early to mid-20th century because the long structural framing members required became more difficult to obtain. However, some custom builders still occasionally use this construction method (see Figure 7.25).[23]

FIGURE 7.25 ◆ Balloon-frame construction provides many concealed spaces and voids for fire to travel through. If all the upper-level windows are aligned with the first-floor windows, it may indicate balloon-frame construction.
Courtesy of James D. Richardson.

Platform-Frame Construction

Platform framing gained popularity at the end of World War II. Platform framing allowed the use of shorter lengths of lumber, which reduced construction time and was easier to acquire at the time. This type of construction also reduced the amount of concealed spaces that were prevalent in balloon-frame construction. In platform-frame construction, wall studs extend only the height of each floor. The bottom of the first floor studs are nailed to a bottom board called a **sole plate**. The sole plate lies horizontally on a concrete foundation or on a wooden platform floor. The platform is constructed with floor joists covered by subflooring. The top of each stud on the first floor wall is nailed to a **top plate**. The top plate is constructed of two boards nailed together to form a beam. The top plate serves a dual purpose. It acts as a support structure for the ceiling or floor joists to rest on and also as a firestop to prevent fire spread if a fire develops inside the wall. When a second story is added, another floor platform is constructed of wood joists and subflooring, and a separate wall is erected on this platform. Basically, platform framing consists of a structure on top of a structure. There is no structural continuity between the floors. This makes the structure very susceptible to wind forces (see Figure 7.26).

Platform construction allows for many variations of design and structural components. From the 1940s until the late 1950s, builders used solid-dimensional lumber for ceiling joists, usually measuring 2×6 or 2×8 inches depending on the span, and 2×12-inch lumber for floor joists. To reduce cost and construction time, many builders have replaced solid-dimensional lumber with engineered-wood structural components. These lightweight components may be superior in strength under normal conditions, but they can readily fail under fire conditions. These structural components have less density, and because of their smaller dimensions, they have more surface area, which makes ignition easier. (See Chapter 6, "Fire Dynamics.")

sole plate

The bottommost component of a wall assembly. The bottom of the wall studs are connected to this component, which in turn is connected to the foundation of the structure.

top plate

Horizontal part between where the roof and studs finish.

FIGURE 7.26 ◆ Platform framing is the most common form of wood construction today.

Buildings are designed to carry specified loads and to endure specified forces. Renovations or changes in how a building is used (its occupancy) may alter the forces and loads in ways that exceed the building's capabilities. Firefighters must be able to identify the different types of loads and forces and to determine how a structure will react when subjected to a particular load or force.

A building consists of a foundation, floor, walls, and roof. The structural components are designed to transfer all loads to the foundation. The walls, columns, and foundation are designed to resist compression forces. Walls must be braced to strengthen them against lateral forces caused by wind. Beams and girders are designed to resist tensile and shear forces. Their function is to transfer loads horizontally to the support columns or walls.

The NFPA 220 and the NFPA 5000 building codes list five types of building construction. NFPA 220 designates Type I as fire resistive, Type II as noncombustible, Type III as ordinary, Type IV as heavy timber, and Type V as wood frame. NFPA 5000 eliminated these five designations and simply categorizes Types I and II construction as noncombustible and Types III, IV, and V as combustible.

Firefighters should study and be familiar with the characteristics of each construction category, focusing especially on those aspects of design and materials that would influence how a fire behaves. They should have a solid understanding of the structures in their jurisdiction and the special challenges these buildings may pose in the event of a fire. Keeping up to date on new, lightweight building components is also critical.

To put it simply: Firefighters should be ready for every possible situation. And then, after all their homework, painstaking preparation, and plans are in place, expect the unexpected!

On Scene

At 4:35 AM on February 6, 1997, firefighters of the Stockton (CA) Fire Department responded to a reported garage fire. On arrival at the scene, they found a single-family residence heavily involved with fire. It was reported that a victim was trapped in the residence. Search and firefighting teams entered the building, and a second alarm was ordered to increase the number of fire personnel and protect exposed residences in close proximity to the fire.

At 4:45 AM, the rear portion of the involved residence suddenly collapsed, killing two firefighters and severely injuring another. Firefighters later reported that there had been no warning signs of a potential collapse. The investigation that followed revealed that a two-story addition had been built onto the rear of the residence in the 1950s. This was a large open-area structure with no interior supporting walls. The exterior walls were the load-bearing supports for the second floor. When the structural integrity of the exterior walls was compromised by fire, the walls failed, and the entire second story collapsed to the first floor.

Further investigation revealed that the firefighters had not realized that the structure had a second level because the rear of the structure was obscured by smoke, fire, and darkness. They assumed that the structure was a one-story residence because there were no other two-story residences in the neighborhood.

1. What suggestions would you make to reduce the possibility that this type of incident would occur again?
2. How could this incident have been prevented?

Review Questions

1. List and categorize the types of construction designated by NFPA 220 and NFPA 5000.
2. What do the digits in the fire-resistive ratings represent?
3. What hazard does Type II construction pose to firefighters?
4. What types of forces are columns and foundations designed to withstand?
5. What hazard does old-style Type III construction pose to modern firefighting tactics and strategies?
6. What is the difference between pre-stressed and post-tensioned concrete?

Notes

1. Guralnik, David. (1982). *Webster's new world dictionary*. (2nd College ed.). New York: Simon and Schuster.
2. *Fire fighter fatality investigation report F98-17*. (June 1998). Retrieved April 7, 2008, from the NIOSH website: http://www.cdc.gov.niosh/fire/reports/face9817.html.
3. Prendergast, Edward. (1999). *Building construction related to the fire service*. (2nd ed.). Stillwater, OK: International Fire Service Training Association.
4. Brannigan, Francis. (1999). *Building construction for the fire service*. (3rd ed.). Quincy, MA: National Fire Protection Association.
5. Davis, Richard. (2003). *Fire protection handbook*. (19th ed.). Quincy, MA: National Fire Protection Association.
6. Feirer, John, & Hutchings, Gilbert. (1986). *Carpentry and building construction*. (3rd ed.). New York: Popular Science Books. Retrieved from http://www.truss-frame.com/truss-mfg.html.
7. Houston Fire Department Continuing Education. (2001). *Wood trusses, Part I CE0013*. Retrieved September 13, 2005, from the Houston Fire Department website: http://www.houstontx.gov/fire/firefighterinfo/ce/2001/January/Jan01CE.htm.
8. Fisette, Paul. (2000). The evolution of engineered wood I-joists. *Building materials and wood technology*. Retrieved September 13, 2005, from the University of Massachusetts website: http://www.umass.edu/bmatwt/publications/articles/i_joist.html.
9. Davis, Richard. (2003). *Fire protection handbook*. (19th ed.). Quincy, MA: National Fire Protection Association.
10. Prendergast, Edward. (1999). *Building construction related to the fire service*. (2nd ed.). Stillwater, OK: International Fire Service Training Association.
11. Davis, Richard. (2003). *Fire protection handbook*. (19th ed.). Quincy, MA: National Fire Protection Association.
12. Ibid.
13. Tide, R. H. R. (First Quarter 1998). Integrity of structural steel after exposure to fire. *Engineering Journal*. Retrieved September 12, 2005, from http://www.steelstructures.com/IIW%20files/35_1_026.pdf.
14. Brannigan, Francis. (1999). *Building construction for the fire service*. (3rd ed.). Quincy, MA: National Fire

Protection Association. Retrieved from http://www.wpi.edu/Pubs/ETD/ Available/etd-0821102-115014/ unrestricted/dparkinson.pdf.

15. Prendergast, Edward. (1999). *Building construction related to the fire service.* (2nd ed.). Stillwater, OK: International Fire Service Training Association.

16. Brannigan, Francis. (1999). *Building construction for the fire service.* (3rd ed.). Quincy, MA: National Fire Protection Association.

17. Kodur, V. R., & Sultan, M .A. (September 1998). *Structural behaviour of high strength concrete columns exposed to fire.* (NRCC-41736). National Research Council Canada. Institute for Research in Construction. International Symposium on High Performance and Reactive Powder Concrete. Sherbrooke, Quebec. Retrieved from http://irc. nrc-cnrc.gc.ca/ircpubs.

18. Davis, Richard. (2003). *Fire protection handbook.* (19th ed.). Quincy, MA: National Fire Protection Association.

19. Brannigan, Francis. (1999). *Building construction for the fire service.* (3rd ed.). Quincy, MA: National Fire Protection Association.

20. Davis, Richard. (2003). *Fire protection handbook.* (19th ed.). Quincy, MA: National Fire Protection Association.

21. White, Robert. (December 1985). *Reporting of fire incidents in heavy timber structures.* (Research Paper FPL 464). United States Department of Agriculture Forest Service. Retrieved from http://www.fpl.fs.fed.us/documnts/ fplrp/fplrp464.pdf.

22. Brannigan, Francis. (1999). *Building construction for the fire service.* (3rd ed.). Quincy, MA: National Fire Protection Association.

23. Buchanan, Andy. (September 13, 2005). *Fire resistance of solid timber structures.* Christchurch, New Zealand: University of Canterbury. Retrieved from http:// www.branz.co.nz/branzltd/pdfs/ AndyBuchanan.pdf.

Suggested Reading

Brannigan, Francis. (1999). *Building construction for the fire service.* (3rd ed.). Quincy, MA: National Fire Protection Association.

Buchanan, Andy. (September 13, 2005). *Fire resistance of solid timber structures.* Christchurch, New Zealand: University of Canterbury. Retrieved from http:// www.branz.co.nz/branzltd/pdfs/Andy-Buchanan.pdf.

Davis, Richard. (2003). *Fire protection handbook.* (19th ed.). Quincy, MA: National Fire Protection Association.

Determine the construction classification number. *Firewise.* National Wildland/Urban Interface Fire Program. Retrieved September 13, 2005, from http://www.firewise.org/resources/ operationWater/step1.htm

Feirer, John, & Hutchings, Gilbert. (1986). *Carpentry and building construction.* (3rd ed.). New York: Popular Science Books.

Fisette, Paul. (2000). The evolution of engineered wood I-joists. *Building materials and wood technology.* Retrieved September 13, 2005, from the University of Massachusetts website: http:// www.umass.edu/bmatwt/publications/ articles/i_joist.html.

Guralnik, David. (1982). *Webster's new world dictionary.* (2nd College ed.). New York: Simon and Schuster.

The history of wood trusses. (2003). *Wood truss design and manufacturing.* Weyerhaeuser. Retrieved September 13, 2005, from Truss-Frame.com website: http:/ /www.truss-frame.com/truss-mfg.html.

Houston Fire Department Continuing Education. (2001). *Wood trusses, Part I CE0013.* Retrieved September 13, 2005, from Houston Fire Department website: http://www.houstontx.gov/fire/ firefighterinfo/ce/2001/January/ Jan01CE.htm.

Kodur, V. R., & Sultan, M .A. (September 1998). *Structural behaviour of high strength concrete columns exposed to fire.* (NRCC-41736). National Research Council Canada. Institute for Research in Construction. International Symposium on High Performance and Reactive Powder Concrete. Sherbrooke, Quebec.

Olson, Craig, & Laura, Smith. (May 9, 1997). Post-tensioned concrete for today's market. *The Seattle Daily Journal of Commerce.* Retrieved from http://www.djc.com/special/concrete97/10024302.

Parkinson, David. (2002). *Performance based design of structural steel for fire conditions.* Paper submitted to the Faculty of the Worchester Polytechnic Institute. Retrieved from http://www.wpi.edu/Pubs/ETD/Available/etd-0821102-115014/unrestricted/dparkinson.pdf.

Prendergast, Edward. (1999). *Building construction related to the fire service.* (2nd ed.). Stillwater, OK: International Fire Service Training Association.

Super bridge. (October 2000). *Nova.* Austin, TX: Public Television. Retrieved September 12, 2005, from http://www.pbs.org/wgbh/nova/bridge/concrete.html.

Tide, R. H. R. (First Quarter 1998). Integrity of structural steel after exposure to fire. *Engineering Journal.* Retrieved September 12, 2005, from http://www.steelstructures.com/IIW%20files/35_1_026.pdf.

White, Robert. (December 1985). *Reporting of fire incidents in heavy timber structures.* (Research Paper FPL 464). United States Department of Agriculture Forest Service. Retrieved from http://fire-service.blessedhopelongdistance.com/reporting-of-fire-incidents-in-heavy-timber-structures-114579

Fire Prevention Codes and Ordinances

8 CHAPTER

Key Terms

building codes, p. 148
citation, p. 156
discretionary
 authority p. 156
engineering
 corrections, p. 149
fire prevention codes, p. 148
injunction, p. 156

mandatory
 authority p. 156
mini-maxi code, p. 154
minimum codes, p. 154
model codes, p. 149
National Board of Fire
 Underwriters
 (NBFU), p. 149

negligence, p. 156
occupation
 classification, p. 154
ordinance, p. 148
permit model, p. 153
statute, p. 154
summons, p. 156
warrants, p. 153

Objectives

After reading this chapter, you should be able to:

- Describe the relationship between fire and building codes.
- Identify the different model code organizations.
- Determine the legal basis for fire code enforcement.
- Identify the types of fire codes and explain how fire codes are adopted.
- Explain the legal right that a fire official has to issue warrants, summonses, citations, and injunctions.

◆ INTRODUCTION

In Chapter 7, we looked at the resources that the fire service uses to protect the public from fires and, increasingly, from other kinds of emergencies as well. In this chapter, we will broaden our scope to look at ways in which the fire service cooperates with government, at the state and local level, to extend its protective reach. Specifically, we will look at codes and ordinances, the legal backbone of public safety. A *code* is a set of mandated standards set forth by a professional organization. When a

ordinance

A law that is established by a local jurisdiction, such as a county, city, or town.

standard is given legal status by a municipality or other local jurisdiction, it becomes an **ordinance**. We will define the function of codes and ordinances and examine how fire departments and other government agencies work together to administer them. We will consider some model codes and the organizations that create them. We will briefly trace the history of codes in the United States, explore some of the jurisdictional and civil liberties issues surrounding their implementation, detail how fire codes are adopted, and describe the enforcement process.

Major fires have played an all-too-frequent role in U.S. history. Almost every sizable city has suffered at least one major conflagration. It is important to remember that behind the rather dry language of codes and ordinances are real-life calamities involving tragic losses of life. For example, at 3:15 PM on December 30, 1903, a fire erupted in the Iroquois theater in Chicago. About 1,900 people filled the theater that had a seating capacity of 1,724. Although there were 27 exits, most were locked, covered, or blocked with iron gates. A total of 602 perished in the fire. As a result of the tragedy, the fire codes were changed, requiring all theater doors to open outward and to have clearly marked exits. The new code also required that theater management conduct fire drills with theater employees.[1] In Boston on November 28, 1942, a fire started in the Cocoanut Grove nightclub. About 1,000 people filled the structure that had a licensed capacity of 500. A total of 492 people died as a result of locked exits and a revolving door. After the fire, new fire codes were written to require outward-swinging exit doors with push-down "panic hardware" on each side of a revolving door. Codes were also written limiting the use of flammable materials for interior decorations in nightclubs.[2] Codes and ordinances are written, and just as frequently revised, to prevent such disasters from recurring.

The writing of a code is only the beginning. In order to function, codes must be adopted as ordinances and rigorously enforced. Enforcement means more than policing, however. First and foremost, it means informing the public about the importance of fire codes and explaining the reasoning behind code provisions and their importance. Public education is the duty of all fire service organizations. Firefighters can step in when a fire alarm goes off, but the primary responsibility for preventing that fire in the first place rests squarely on the shoulders of the informed citizen.

#16

◆ BUILDING CODES AND FIRE PREVENTION CODES

building codes

Laws adopted by states, counties, cities, or other government sources to regulate all aspects of building construction.

fire prevention codes

Body of laws designed to enforce fire prevention and safety.

There are two categories of codes that relate to public fire safety: building codes and fire prevention codes.

Building codes address fire and other safety issues related to building construction. They establish standards for the fire resistance of building materials and interior finishes; for means of egress; for vertical openings, such as stairwells, elevator, and utility shafts; and for other structural features. Typically, a municipality's building department is responsible for administering and enforcing building codes.

Fire prevention codes focus on fire safety issues that relate to how a building is used and maintained. These codes set standards for fire protection systems and equipment. They also define fire hazards, such as faulty electric appliances and improperly stored hazardous materials. Fire codes are usually administered and enforced by the fire department.

Building codes and fire prevention codes are different but are closely related, however, and a good working relationship between the building department and the fire department is the best recipe for protecting the public.[3] The relationship should be based on understanding that three basic ingredients are necessary for success: quality engineering, strict enforcement, and comprehensive public education programs. Cooperation is essential in all three areas.

Both the building department and the fire department should review new construction plans and make **engineering corrections** that enhance fire safety. These corrections may include the installation of fixed fire protection systems, such as automatic sprinklers. These sprinklers are designed to extinguish a fire in its early stage of development and provide a protected path of travel to exits, which will then reduce the amount of time required to evacuate a building Coordinating efforts for code enforcement between the building and the fire departments will result in greater convenience for the public, reduce redundancy, and save time and money for both departments. Coordinated education programs will enhance public compliance with inspections and will foster better understanding on the part of both departments for changing community attitudes and needs.[4]

engineering corrections

Corrections that enhance fire safety. These corrections may include the installation of fixed fire protection systems, such as automatic sprinklers

#16

◆ HISTORY OF MODEL CODES IN THE UNITED STATES

When establishing legal standards on construction and fire safety, state authorities can choose from **model codes**, which they can adopt or adapt, as the law allows, to fit the needs of their state. These codes are written, published, and revised (generally every three to five years) by professional membership organizations, some of which will be discussed later.

The origin of codes in the United States goes back to the 19th century. As a result of industrialization and population growth in the cities, urban fires became increasingly frequent and increasingly costly. By the middle of the century, major insurance companies were suffering such huge losses that some were forced into bankruptcy. After a major fire in Portland, Maine, in 1866, the insurance companies formed the **National Board of Fire Underwriters (NBFU)**. This new organization, which later became the American Insurance Association (AIA), was created to standardize rates and find ways to deter arson. Although it was never successful in standardizing rates, the NBFU did develop guidelines for fire apparatus and for municipal fire supplies. These guidelines became the nation's first fire codes.

In 1893, insurance companies became alarmed about the proposed Palace of Electricity at the World's Columbian Exposition in Chicago. The array of untested electrical connections, in close proximity to the flammable decorative façade, was considered so hazardous that the insurance companies threatened to pull their backing, which nearly ended the Exposition. A young Boston electrician, William Henry Merrill, saved the day. He reviewed the wiring, made certain that all proper safeguards were in place, and got the insurance companies back on board. With the backing of the NBFU, Merrill went on to form the Underwriters Electrical Bureau, which later became Underwriters Laboratories (UL).

In 1904, in response to the great Baltimore, Maryland, fire, the National Board of Fire Underwriters issued its Standard Grading Schedule for all municipalities. This standard was used to establish a municipality's basic insurance rate, which was contingent on the municipality's ability to protect its residents from fire. It factored in the number and types of fire apparatus available to the city's fire department, its water supply, and its average response time. The board also published the first National

model codes

Building codes developed and maintained by a standards organization separate from the governing body for building codes.

National Board of Fire Underwriters (NBFU)

Organized in 1866, the NBFU developed standards for fire safety in building construction and control of fire hazards.

Building Code in 1905. This code set height and area restrictions and established standards for such safety features as means of egress, fire-resistive construction, and fire protection systems. It was offered free of charge to any state or local government that was willing to adopt the code. Fourteen editions of the National Building Code were published before it was discontinued in 1976.[5] The American Insurance Association also published fire and building codes until 1971.

BUILDING OFFICIALS AND CODE ADMINISTRATORS INTERNATIONAL (BOCA)

The Building Officials and Code Administrators International (BOCA) is a membership organization serving the northeastern and midwestern United States and Canada. It comprises of state and local governments and agencies that are responsible for code enforcement, as well as members of the building construction disciplines. It was established in 1915 and published its Basic Building Code in 1950, which later became known as the BOCA National Building Code.[6]

INTERNATIONAL CONFERENCE OF BUILDING OFFICIALS (ICBO)

The International Conference of Building Officials was established in 1921. By 1927, it had published a number of codes pertaining to the building industry. One of these—the Uniform Building Code (UBC)—mandated construction design and building material requirements. This code was primarily used in the Pacific Coast, Rocky Mountain, and Great Plains states until publication of the code ceased in 2003.[7]

SOUTHERN BUILDING CODE CONGRESS INTERNATIONAL (SBCCI)

The Southern Building Code Congress International (SBCCI) was established in 1940 and published its Standard Building Code in 1945. This code included special provision for dangers, such as hurricanes, that occur in the area that the SBCCI served—the southeastern and southwestern United States. It was similar to other model building codes in other respects.[8]

INTERNATIONAL CODE COUNCIL (ICC)

In 1994, BOCA, ICBO, and SBCCI joined forces to found the International Code Council (ICC). Its mission was to develop a set of model international building codes. By 2000, the ICC's International Building, Fire, and Property Maintenance Codes were available for adoption. Three years later, BOCA, ICBO, and SBCCI consolidated all code development services into ICC and ceased publishing their own codes. ICC's work made standardized fire protection possible in the United States and, as an additional benefit, made it possible for building and construction inspectors to work in different parts of the country.[9]

NATIONAL FIRE PROTECTION ASSOCIATION (NFPA)

The National Fire Protection Association (NFPA) was established by the National Board of Fire Underwriters in 1896. In previous chapters, we looked at NFPA's work in developing national standards for firefighting education and apparatus. The organization has also developed almost 300 fire prevention and building codes and standards that have been adopted by reference in many jurisdictions.

CHAPTER 8 *Fire Prevention Codes and Ordinances* **151** ◆

TABLE 8.1 ◆ Building and Fire Code Organizations	
The Building Officials and Code Administrators International (BOCA)	Serves the northeastern and midwestern United States and Canada. It comprises state and local governments and agencies that are responsible for code enforcement, as well as members of the building construction disciplines.
The International Conference of Building Officials (ICBO)	Published a number of codes pertaining to the building industry, including the Uniform Building Code.
Southern Building Code Congress International (SBCCI)	Published the Standard Building Code in 1945. This code was very similar to the other model building codes, except it included special provisions for events, such as hurricanes, that affect the region it served. Membership consisted of representatives from the southeastern and southwestern United States.
International Code Council (ICC)	In 1994, BOCA, ICBO, and SBCCI joined forces and formed the International Code Council (ICC) and published a set of model building codes called International Codes.
The National Fire Protection Association (NFPA)	Established by the National Board of Fire Underwriters in 1896. This organization has been instrumental in developing almost 300 fire and building codes and standards that are adopted by reference in many jurisdictions.

Among NFPA's landmark documents are its National Electric Code, first issued in 1911 and still in publication today (NFPA 70). NFPA 101, *Life-Safety Code,* divides building occupancy into 13 categories and establishes means of egress standards for new and existing structures Because NFPA 101 is so comprehensive in addressing life-safety requirements, many jurisdictions incorporate it into their fire and building codes.[10] In 2002, NFPA published an alternative to the ICC International Codes. This publication is known as NFPA 5000, *Building Construction and Safety Code.* The ICC International Codes and the NFPA 5000 are written by private organizations. The codes are recommendations, not mandates. Each state or province may choose (through legislation) which code to adopt when writing building and fire prevention ordinances. Table 8.1 describes the differences among the codes. There has been an attempt by the ICC to consolidate the codes into one International Code.

◆ HISTORICAL BASIS FOR THE LEGAL AUTHORITY OF CODE ENFORCEMENT

In the United States, ordinances governing such matters as building codes and fire safety are enacted at the local level. However, the local government's discretion in these matters is limited by the state government, which establishes minimum standards, usually selecting the code that is to be used.

There is a history of controversy over the relative role of municipalities and states in legislating local standards (analogous to the national controversy between federal and states' rights). During the 150 years of America's colonial period, town governments enjoyed considerable autonomy in managing their local affairs. This concept, known as *home rule,* was based on principles inherited by the colonists from English law (the Carta Civibus Londoniarum of 1100 AD and the Magna Carta of 1215 AD).[11]

The Constitution of the United States, by contrast, does not provide for local government autonomy, instead granting to the individual states the rights and privileges that had traditionally belonged to towns. The Tenth Amendment to the Constitution reserves power to the states or the people that are not delegated to the federal government or prohibited to the states. The wording of the amendment is open to interpretation. Local jurisdictions in several states have argued that the phrase "or to the people" gives local governments the authority to develop their own codes and ordinances without interference from the state. The issue was brought before several state supreme courts before it was eventually decided by the U.S. Supreme Court in 1903.

DILLON'S RULE, THE COOLEY DOCTRINE, AND *ATKINS V. KANSAS*

In 1868, Iowa State Supreme Court Justice John Dillon wrote a landmark decision in the case of *City of Clinton v. Cedar Rapids and Missouri Railroad Company*. His opinion, known as Dillon's Rule, stated that local jurisdictions in Iowa could exercise only those powers that are expressly granted to them by the state.[12]

This strict constructionist decision was not welcomed in many states. Michigan Supreme Court Justice Thomas Cooley challenged Dillon's opinion by arguing that the Tenth Amendment reserved the right of home rule because it delegates power to the states "or to the people." Cooley argued that "to the people" meant "to the local government." Local governments should have powers to initiate policy. The Cooley Doctrine found support in the states of Indiana, Kentucky, Texas, and in some courts in Iowa. The home rule movement continued into the 20th century.[13]

In 1903, the U.S. Supreme Court heard the case of *Atkins v. Kansas*. The Court rejected Cooley's Doctrine and upheld Dillon's Rule, declaring that local governments can have only those powers specifically given to them by the state. In 1923, the Court further bolstered Dillon's Rule with a decision declaring that a state has the right to grant, withhold, or withdraw powers from local governments as the state deems necessary.[14] These decisions established the framework on which today's fire codes and ordinances are enforced.

Dillon's principle became the law of the land. However, in actual practice, states do sometimes grant a degree of decision-making autonomy to local government officials, and this may be reflected in the local codes and ordinances they enact.[15] In all cases, however, this power is granted at the discretion of the state.

WARRANTLESS ENTRY

Having explained the constitutional background for the authority granted to local governments by the states, we will now examine their use of this authority in enforcing codes and ordinances. Unquestionably, the most controversial aspect of code enforcement is the right granted to fire prevention authorities to enter premises either for inspections or to take emergency measures. Even though the state grants powers of code enforcement to local governments, and the local government in turn confers these rights to fire authorities, all authorities are legally bound by the U.S. Constitution and are limited in their ability to act. In most states, fire prevention codes are applicable to all premises, including single-family dwellings. However, what right does a representative of the government have to enter private property? The Fourth Amendment to the U.S. Constitution clearly states, "The right of the people to be secure in their persons, houses, papers, and effects, against unreasonable searches and seizures, shall not be violated, and no **Warrants** shall issue, but upon probable cause, supported by Oath or affirmation, and particularly describing the place to be searched, and the persons or things to be seized."[16]

The language of the Fourth Amendment would seem to forbid warrantless entry. The U.S. Supreme Court, however, has ruled that warrantless entry is legal under limited emergency circumstances. In three separate decisions—*Mapp v. Ohio*, *Frank v. Maryland,* and *Camara v. Municipal Court*—the Court ruled that warrentless entry is unconstitutional without the consent of the occupant or owner. It is justified if entry is made in response to an emergency. In another case, the Court ruled that a burning building is considered an emergency; therefore, warrantless entry is justified In many states, the law recognizes that even if a building is burning, the right to privacy outweighs the police powers of the government if the burning building is so isolated that it poses no threat to other property or persons.[17]

RIGHT OF ENTRY

All model building codes contain clauses pertaining to the right of entry by fire officials, but these officials are bound to act within carefully delineated guidelines. Under normal circumstances, permission to enter a property must be granted by the owner or occupant, or a warrant must be issued. The fire official must present identification and the proper credentials. The purpose for entry must be related to code compliance issues, and the entry request should be scheduled during normal operating hours of the occupancy or at an hour that is considered reasonable. Fire officials may enter a property for the purpose of code compliance verification without a warrant or permission only if a life-safety condition exists or if the area being inspected is visible from a public way (meaning that the inspection can be conducted without actually setting foot on private property).

The right of entry to private property by a fire official often depends on the type of model used in establishing inspection priorities. The two models used for prioritizing inspections are the **permit model** and the *inspection model*.[18]

Permit Model

Under the permit model, businesses that have a high potential for loss of life, such as nightclubs, restaurants, and hazardous materials processing and storage facilities, require a permit to operate. They must prove that their facility meets code requirements before a permit is issued. The permit is usually site specific, which means that it identifies the hazard, location, time of operation, and purpose. It allows fire officials the opportunity to reference other codes, such as those that establish building and zoning requirements. The permit is considered the property of the issuing agency and does not belong to the individual receiving the permit. The issuance of a permit gives the jurisdiction the right of entry to the premises for the purpose of inspection. If after receiving a permit for operation, the occupant or owner of a business refuses to allow an inspection of the facility during a "reasonable hour," the permit can be revoked.[19]

The permit model does not give a fire official unlimited access to premises, however. In the U.S. Supreme Court case of *See v. The City of Seattle*, a Seattle

warrants

Orders from a judge allowing fire officials the right to search or seize private property or arrest an individual.

permit model

Businesses that have a high potential for loss of life, such as nightclubs, restaurants, and hazardous materials processing and storage facilities, require a permit to operate.

businessman refused to allow fire inspectors into an area of his business, claiming that the area was used for private business matters and was not accessible to the general public. The Court ruled against the city, stipulating that a business owner can deny government officials entry to areas of a business that are not normally open to the public. If permission to enter that area is denied, the inspector must obtain a warrant before entering.[20]

Inspection Model

occupation classification

Classification given to a structure by a model code used in that jurisdiction and pursuant to the given use of that structure.

When using the inspection model, fire officials schedule inspections in their jurisdiction based on the occupation category of the buildings. The **occupation classification** of a building is based on the purposes it serves: residential, educational, mercantile, storage, and so on. Each occupancy type is designated for inspection during a set time period. For example, the fire inspector may decide that all fuel storage areas will be inspected during the month of May. Inspection schedules and procedures must be published, and the right of entry by the fire official will be based on the schedule. Any deviation from the schedules is discouraged, unless a life-safety hazard is found to exist. A **statute** establishes that a particular type of occupancy must be inspected a certain number of times each year and that fire officials have a duty to carry out those inspections. This reduces the possibility that a business owner will claim harassment.[21]

statute

A law established by a state or federal legislative body.

> ### Stop and Think 8.1
>
> Under what circumstances are unscheduled inspections allowed under the jurisdiction in your area?

◆ TYPES OF CODES AND CODE ADOPTION

The process by which fire codes are adopted is determined by state governments, but state governments can allow local jurisdictions discretion to adapt codes to meet particular local needs.

MINIMUM CODES VERSUS MINI-MAXI CODES

minimum codes

Lowest acceptable standards.

In some states, **minimum codes** are adopted by the state, and local jurisdictions are not allowed to implement regulations that have less restrictive requirements. They are empowered to implement more stringent code requirements if they see fit.

mini-maxi code

Establishes minimum and maximum building code requirements.

Under the **mini-maxi code** concept, local jurisdictions are not allowed to adapt or modify any regulation. The code is mandated by the state and is enforced statewide. The rationale behind the mini-maxi code is that it allows for uniformity in the application of the code. Builders, contractors, and architects support the concept because it simplifies their work—regulations are the same in all jurisdictions. All lobbying efforts for code changes are made at the state level. Fire officials are opposed to mini-maxi codes because they do not allow local jurisdictions the flexibility to write and enforce regulations that meet needs unique to their community. A positive aspect of the mini-maxi code concept is that it forces the fire departments in the state to work together and become proactive in code development.[22]

MODEL CODES

The code adoption process is defined by state law. Before a code can be adopted, notice must be published scheduling a public meeting so that citizens can voice their support or objections to the code. An effective date for the code must be written into the ordinance, and the agency responsible for code enforcement must be clearly identified. Model codes can be adopted by reference or by transcription.

Adoption by Reference or Transcription

When adopting by reference, the jurisdiction simply refers to an edition of the model code in the adopting ordinance. If the adoption is by transcription, the entire model code must be published in the adopting ordinance. The life expectancy of most model code editions is three to five years.

Prescription- or Performance-Based Models

States have the option to base their fire prevention efforts on one of two models: the prescription model or the performance model. These are different ways of viewing how codes work to protect the public. The choice of one over the other is strictly a matter of perspective.

Prescriptive Codes

Prescriptive codes establish specific requirements for building and fire safety. In the prescriptive model, fire safety is achieved through the use of mandated requirements in building materials, construction components, means of egress, and fire protection systems. These mandates are based on past fire experiences in which large losses of life or property occurred; the mandates are generally not revised until new lessons are learned, often as a result of another major fire loss. Prescriptive codes do not identify fire safety objectives; rather, the assumption is that if all of the mandated building requirements are followed, fire safety will be achieved.[23]

Performance Codes

Performance codes begin from the perspective of goals and objectives. This type of code does not mandate specific materials and so on, but relies on quantitative measures to determine the fire safety level of building construction assemblies and of the building as a whole. Code compliance can be verified by one of three methods:

> *Laboratory or On-Site Testing.* The design or material is tested in a laboratory or on the building site to determine if it meets code requirements.
> *Verification that a Technical Solution Satisfies Performance Requirements.* A standard unit of measurement is agreed on, so that test results can be validated in repeat tests.
> *Verification by a Technical Expert or a Certifying Entity.* Performance is evaluated and certified by outside technical experts.[24]

◆ CODE ADMINISTRATION AND ENFORCEMENT

The administration section of a model code identifies the enforcing authority and outlines the legal framework for its enforcement activities. It contains legal language that, like all legal language, calls for careful reading. The wording of the document is critical because it establishes the legal foundation for right of entry as well as for the issuance of orders for evacuations and other emergency measures intended to preserve life. Wording in this section should not only give the fire officials full authority to do their job effectively but should also specify the responsibility of the property owner for building maintenance and fire safety.[25] Wording in the code will also determine if the actions taken by a fire official are mandatory or discretionary. For example, if the code uses the word "shall," the action of the fire official is **mandatory authority** under the law. If the word "may" is used, the fire official is given **discretionary authority**. The fire official must be familiar with the wording in the code and follow the "letter of the law" for all mandated requirements. If the code mandates the inspection of certain occupancies, fire officials have a duty to inspect those facilities. Failure to do so may result in charges of **negligence**, resulting in legal action against the official.

DOCUMENTATION

To protect themselves legally, fire officials should document all inspections or visits to a facility. All actions taken must be recorded. If a violation of the code is discovered, a legal notice of violation must be issued. The form this takes depends on the code that the jurisdiction is using, but there are some basic violation notice procedures that are common to all codes. First, the notice must be in writing, identifying the exact location (legal address) of the inspected facility, the date of inspection, and the name of the inspecting officer. Second, violations must be specifically enumerated, and the code sections supporting the notice must be cited. Third, any required corrective action must be clearly described. Finally, the notice must contain a date for a follow-up inspection to verify code compliance.[26]

> ### Stop and Think 8.2
>
> What are the legal ramifications if an inspector fails to perform a follow-up inspection when a code violation is documented?

SUMMONS AND CITATIONS

Inspecting officials may be able to use other enforcement tools to ensure compliance with fire and building codes. The availability of these tools varies from jurisdiction to jurisdiction. Some jurisdictions give inspecting officials the power to issue a **summons** or **citation** (the two terms are synonymous). These documents are legal notices that effectively charge an individual with a criminal misdemeanor violation of the fire codes and direct that person to appear in a court of law on a certain date to answer the charges. Some codes require the inspector to issue a citation if a violation is discovered, whereas others may give officials discretionary powers to issue warnings and give individuals a chance to correct the violation before they face a criminal charge. In either case, a thorough and accurate report of the inspection must be submitted. If the owner of a facility does not correct a violation after receiving a summons or citation, the fire official can request that a court issue an **injunction** requiring the facility to cease operation until code requirements are met. An injunction is a court order. Failure to comply with stipulations stated in the injunction is considered contempt of court, and the guilty party can be held in jail until compliance with the code is achieved.[27]

mandatory authority

Makes the actions mandatory under the law.

discretionary authority

Allows the fire official to use discretion when enforcing the law.

negligence

Failure of a fire official to perform in the same manner as a prudent person would under the same circumstances.

summons

A written document from a judge or a law enforcement officer ordering a code violator to appear in court on a certain date.

citation

Same as a summons except it is issued by a law enforcement officer instead of a judge.

injunction

A legal document from a judge that orders a person or entity to cease or perform certain activities.

◆ SUMMARY

When the public thinks of the fire service, they picture gleaming red engines, their sirens screaming, racing to the scene of a raging inferno. That picture is valid, certainly, but it is only the tip of the iceberg. The mission of today's fire service is not limited to emergency response. Fire prevention efforts are an equally vital part of that mission, and those efforts include providing public education and monitoring building design and construction. In addition to extinguishing fires, firefighters now have input at the grassroots level of community development and planning. If they have the enforcement authority and training to do their jobs effectively, firefighters can, in a sense, extinguish fires before they are ignited.

Codes and ordinances are the legal backbone of fire prevention. Considerable overlap exists between the fire safety principles that encourage intelligent building design and construction and the fire safety measures that prevent fires. Building codes, administered by a community's building department, and fire codes, administered by its fire department, are different but are closely related. Therefore, it is very important for municipal building and fire department personnel to work together to ensure compliance with all codes and to enhance overall fire safety.

Until 1994, there were a variety of model building codes and several building code organizations. In 1994, the Building Officials and Code Administrators International, the International Conference of Building Officials, and the Southern Building Code Congress International joined forces and formed the International Code Council. In 2000, ICC created the International Building, Fire, and Property Maintenance Codes and made them available for adoption. In 2002, the National Fire Protection Association published the NFPA 5000, *Building Construction and Safety Code*.

The legal authority for a fire prevention organization rests in its fire code. If the code mandates that a certain business be inspected, the fire prevention officer has a duty to inspect that business. Code violations can be enforced in several ways. The owner can be issued a citation or summons to appear in court to answer criminal (misdemeanor) charges. If the owner does not correct the violation after receiving a citation or summons, the inspector can seek a court-ordered injunction against the individual, requiring that all business operations cease until the violation is corrected. Failure to comply with the mandates in an injunction is considered contempt of court and can result in the individual's being incarcerated until code requirements are satisfied.

Full and accurate inspection reports must be kept on file. Fire officials are sometimes summoned to court several years after an inspection. A complete report will allow the inspector to "refresh" his or her memory before testifying. An inspection report is an official document and is considered public information. *Remember: If it is not documented, it never happened!*

■■■

On Scene

CASE 1

A fire department responds to a building fire in a rural area. The building is approximately 1,000 feet from the road and several hundred feet from all other property lines. No exposures are threatened by the fire. On arrival, the first responding fire company is met by the landowner, who has blocked the entrance to the property and refuses to allow the fire department to enter.

1. What would you do if you were the fire officer in charge?
2. Does the owner have the legal right to refuse entry to his or her property in this situation?

CASE 2

The Fourth Amendment to the U.S. Constitution protects citizens from unlawful search and seizure and limits the state's police powers. However, warrantless entry is allowed under certain circumstances.

1. What circumstances would allow nonconsensual entry?

Review Questions

1. What influence do building codes have on fire codes?
2. Explain the concepts of Dillon's Rule and the Cooley Doctrine.
3. How does the Fourth Amendment to the U.S. Constitution affect the duties of fire officials in enforcing codes and ordinances?
4. What is the difference between a minimum code and a mini-maxi code?
5. How are model codes adopted?
6. What is the difference between a prescription-based code and a performance-based code?

Notes

1. Eastland Memorial Society. (March 24, 2008). *The Iroquois Theater fire.* Retrieved March 24, 2008, from the Eastland Memorial Society website: http://www.inficad.com/~ksup/iroquois.html.
2. *Cocoanut Grove fire Boston.* Retrieved March 24, 2008, from the Celebrate-Boston website: http://www.celebrateboston.com/disasters/fires/cocoanutgrovefire.htm.
3. Robertson, James. (2000). *Introduction to fire prevention.* (5th ed.). Upper Saddle River, NJ: Prentice Hall Health.
4. Crawford, Jim. (2002). *Fire prevention: A comprehensive approach.* Upper Saddle River, NJ: Prentice Hall Health.
5. Diamantes, David. (1998). *Fire prevention inspection and code enforcement.* Clifton Park, NY: Thompson Delmar Learning.
6. Cote, A., & Grant, C. (1997). *Fire protection handbook.* (18th ed.).

Quincy, MA: National Fire Protection Association.
7. Ibid.
8. Ibid.
9. Ibid.
10. Diamantes, David. (1998). *Fire prevention inspection and code enforcement.* Clifton Park, NY: Thompson Delmar Learning.
11. Krane, Dale. (1998) *Local government autonomy and discretion in the USA.* National Academy of Public Administration. Retrieved December 20, 2005, from http://www.napawash.org/aa federal system/98 national local.html.
12. *Dillon's Rule: State primacy over local governments.* Pennsylvania General Assembly Local Government Commission. Retrieved December 20, 2005, from Basics 02_Dillon'sRule_StatePrimacyOverLocal Governments.pub.

13. Zimmerman, Joseph. Evolving state–local relations. *The book of the states*. Retrieved December 20, 2005, from The book of the states 2002 Volume 34 - Entire Soft Cover Edition at http://www.csg.org/programs/ncic/documents/Kincaid1–State–FederalRelations–ContinuingRegulatoryFederalism.pdf

14. Ibid.

15. Krane, Dale. *Local government autonomy and discretion in the USA*. National Academy of Public Administration. Retrieved December 20, 2005, from http://www.napawash.org/aa federal system/98 national local.html.

16. *U.S. Constitution*. Legal Information Institute. Retrieved December 20, 2005, from http://www.law.cornell.edu/constitution/constitution.billofrights.html.

17. Diamantes, David. (1998). *Fire prevention inspection and code enforcement*. Clifton Park, NY: Thompson Delmar Learning.

18. Ibid.

19. Ibid.

20. Farr, Ronald, & Sawyer, Steven. (2003).*The fire protection handbook*. (19th ed.). Quincy, MA: National Fire Protection Association.

21. Diamantes, David. (1998). *Fire prevention inspection and code enforcement*. Clifton Park, NY: Thomson Delmar Learning.

22. Ibid.

23. Ibid.

24. Farr, Ronald, & Sawyer, Steven. (2003).*The fire protection handbook*. (19th ed.). Quincy, MA: National Fire Protection Association.

25. Robertson, James. (2000). *Introduction to fire prevention*. (5th ed.). Upper Saddle River, NJ: Prentice Hall Health.

26. Diamantes, David. (1998). *Fire prevention inspection and code enforcement*. Clifton Park, NY: Thomson Delmar Learning.

27. Ibid.

■ ■

Suggested Reading

Crawford, Jim. (2002). *Fire prevention: A comprehensive approach*. Upper Saddle River, NJ: Prentice Hall Health.

Diamantes, David. (1998). *Fire prevention inspection and code enforcement*. Clifton Park, NY: Thomson Delmar Learning.

Diamantes, David. (2005). *Principles of fire prevention*. Clifton Park, NY: Thomson Delmar Learning.

Dillon's Rule: State primacy over local governments. Pennsylvania General Assembly Local Government Commission. Retrieved December 20, 2005, from Basics 02_Dillon's Rule_State Primacy Over Local Governments.pub.

Farr, Ronald, & Sawyer, Steven. (2003).*The fire protection handbook*. (19th ed.). Quincy, MA: National Fire Protection Association.

Krane, Dale. *Local government autonomy and discretion in the USA*. National Academy of Public Administration. Retrieved December 20, 2005, from http://www.napawash.org/aa federal system/98 national local.html.

Robertson, James. (2000). *Introduction to fire prevention*. (5th ed.). Upper Saddle River, NJ: Prentice Hall Health.

U.S. Constitution. Legal Information Institute. Retrieved December 20, 2005, from: http://www.law.cornell.edu/constitution/constitution.billofrights.html.

Zimmerman, Joseph. Evolving state–local relations. *The Book of the States*. Retrieved December 20, 2005, from The Book of the States 2002 Volume 34 - Entire Soft Cover Edition at http://www.csg.org/programs/ncic/documents/Kincaid1–State–FederalRelations–ContinuingRegulatoryFederalism.pdf

CHAPTER 9

Fire Protection Systems

Key Terms

branch lines, p. 183
combination flame
  detectors, p. 196
combination
  system, p. 163
crossmains, p. 183
direct pumping
  system, p. 163
distribution grid, p. 166
distributor pipes, p. 166
encrustation, p. 167
flow pressure, p. 175
friction loss, p. 176
gravity system, p. 163
groundwater, p. 161
head pressure, p. 163
high-pressure
  systems, p. 198
indicating valve, p. 179

infrared flame
  detectors, p. 196
ionization detectors, p. 194
light-obscuration
  systems, p. 194
light-scattering
  systems, p. 195
local application
  systems, p. 198
low-pressure systems, p. 198
multipurpose dry chemical
  systems, p. 200
nonindicating valve, p. 166
nonpressurized dry hydrant
  system, p. 170
ordinary dry chemical
  systems, p. 200
photoelectric
  detectors, p. 194

pressurized dry hydrant
  system, p. 170
primary feeder
  mains, p. 166
residual pressure, p. 167
riser, p. 173
secondary feeder
  mains, p. 166
sedimentation, p. 167
static pressure, p. 167
subsidence, p. 161
surface water, p. 161
target areas, p. 172
total flooding
  systems, p. 198
tuberculation, p. 167
ultraviolet flame
  detectors, p. 196
watersheds, p. 162

Objectives

After reading this chapter, you should be able to:

- Define the two sources of water supply.
- Describe the four components of a water system.
- List and describe the two categories of water supply systems and explain the advantages and disadvantages of each system.
- Explain the different types of pressures.
- List and describe the three basic types of water supply systems.
- Identify the three components of the water distribution grid.
- Identify and describe the three types of fire hydrants.

- Describe the various types of sprinkler systems.
- Understand the various types of detection and alarm systems.
- Describe the various types of special fire protection systems.
- List and describe the different types of extinguishing agents used in fire protection systems.

◆ INTRODUCTION

This chapter is designed to introduce firefighters to fire protection systems. Topics covered include water supply and water distribution systems, fire hydrants, water-based sprinkler systems, standpipe systems, special systems, extinguishing agents, and detection and alarm systems. Entire courses and textbooks can be dedicated to each of these topics. The intent of this chapter is to present only the basic design features and operating principles of these systems, with the understanding that students will go into greater depth on these subjects in more advanced study.

The first part of this chapter is dedicated to water supplies and water distribution systems. Because the majority of fire protection systems encountered by firefighters use water as the extinguishing agent, it is important that the firefighter understand how this extinguishing agent (water) is supplied. Inadequately designed or antiquated distribution systems can wreak havoc at emergency fire scenes and endanger firefighters' lives. The remainder of the chapter is dedicated to the various sprinkler systems, standpipe systems, and special systems found in private sector construction.

◆ WATER SUPPLIES AND DISTRIBUTION SYSTEMS

Why are water supplies being discussed in a chapter dedicated to fire protection systems? The answer is simple. When firefighters connect their hoses or apparatus to a fire hydrant, the system that supplies that hydrant becomes a fire protection system. The volume of water and the amount of pressure the hydrant provides is directly related to the size of the mains, the design of the distribution grid, the type of system, and the volume and availability of the water supply. It is the responsibility of all firefighters to acquire a working knowledge of the types of systems that serve their community.

SOURCES OF WATER SUPPLY

There are two basic sources of water supply. **Groundwater** is found in flowing springs or in underground geologic formations known as aquifers. **Surface water** comes from rivers, lakes, streams, reservoirs, or various kinds of storage tanks.

Groundwater

The amount of water that can be drawn from an aquifer depends on the stability of the aquifer formation and the amount of rainfall in any particular season. In most cases, aquifers in limestone or other rock strata can withstand fluctuations of the water table during times of high demand and low recharge without extensive damage to the aquifer formation. In some areas of the country, the aquifers are in strata of sand and clay. If water is drawn from these aquifers faster than the water is replenished, the formation has a tendency to collapse. In some of the coastal areas of the country, collapse of these formations has caused a phenomenon known as **subsidence**,

groundwater

Water located beneath the ground surface.

surface water

Water collected on the ground from a stream, lake, ocean, or wetland.

subsidence

A phenomenon in which the surface of the ground sinks as water is drawn from the underground aquifer faster that it can be replenished.

FIGURE 9.1 ◆ Aquifers are replenished by rainfall in the recharge zone.

in which the ground sinks as the aquifer collapses. Municipalities in some affected areas have had to convert from relying on groundwater to using surface water. Houston relied heavily on groundwater from large wells. In the 1970s, Houston experienced a very large increase in population. The aquifers under Houston and the surrounding areas could not support the increased flow demand. As the water was drawn from the ground, the aquifers were unable to replenish fast enough to meet demand and began to collapse. Low-lying areas around Houston began to sink. Areas near the Gulf Coast dropped below sea level and were flooded by Gulf waters. The city and county governments realized that Houston could no longer rely on groundwater as a primary supply source. Large lakes were constructed around the Houston area, and the use of groundwater ceased (see Figure 9.1).

Surface Water

Like groundwater supply, surface supply is dependent on rainfall or runoff during spring thaws in some areas. Surface supplies are more readily recharged than groundwater supplies, especially if they have large recharge zones, also known as **watersheds**. It is possible to funnel the water that drains into a watershed into reservoirs constructed to store the water to supplement supply needs during periods of drought. Prolonged droughts can make this source of water supply unreliable (see Figure 9.2).

WATER SYSTEM COMPONENTS

In addition to a water supply source, all water systems require three additional components: a way to transport the water from the source to its final destination, a water treatment facility, and a distribution grid.

watersheds

Geographic areas around a body of water that have a topography that allows all rainfall in that area to flow into and replenish that body of water.

Figure 9.2 ◆ Surface water is replenished by rainfall in the watershed.

Water Transport Systems

Water may be supplied by the gravity system, the direct pumping system, or by some combination of the two.

The **gravity system** is the most efficient. In this system, water is moved through pipes from an elevated water source to the treatment facility and distribution system. There is no need for pumping apparatus. The height of the surface of the water source determines the amount of pressure in the system. This is called **head pressure**. Head pressure increases by 0.434 pound per square inch (psi) with each foot of elevation. Therefore, the higher the elevation of the water source, the higher the pressure in the system. For example, if the surface of a water reservoir was 1,000 feet above sea level and the area being served was at sea level, theoretically, the pressure in the system at sea level would be 434 psi (see Figure 9.3).

A **direct pumping system** relies completely on mechanical apparatus to move the water. Pumps are used to draw water from groundwater supplies or from surface supplies that cannot supply elevation head pressure. The pumps are usually located at or near the water supply and direct the water to a treatment facility and then on into the distribution system (see Figure 9.4).

A **combination system** uses features of both the gravity and pumping systems. Pumps move the water from the source to the distribution system. Some of this water is directed into elevated storage tanks located along the distribution path. These storage tanks not only act as a supplementary water supply but also are used to temporarily maintain pressure should the system pumps fail (see Figure 9.5).

Water Treatment Systems

Regardless of the type of system used, all public water systems must have a means to filter and purify the water so that it is safe for human consumption. The complexity of the treatment system depends on the size of the distribution area and the

gravity system
A water supply system that uses only gravity to move water.

head pressure
The height of the surface of the water supply above the discharge orifice is equal to 0.434 pound per square inch (psi) per foot of elevation.

direct pumping system
A water supply system that relies on mechanical pumps to move water.

combination system
A water supply system that uses mechanical pumps and gravity to supply operating pressure for moving water.

FIGURE 9.3 ◆ The gravity system is the most reliable distribution system.
Courtesy of James D. Richardson.

amount of purification needed. Groundwater is often naturally filtered, to some degree, as it moves through the aquifer and may not require as much purification as surface water. Some communities use chlorine injection to treat against coliform bacteria contamination. Others inject both chlorine and fluoride to provide protection

FIGURE 9.4 ◆ Direct pumping relies on mechanical equipment to move the water.
Courtesy of James D. Richardson.

FIGURE 9.5 ◆ The combination system utilizes mechanical pumps and gravity to move water through the system.
Courtesy of James D. Richardson.

against dental decay. Firefighters responding to an emergency in a water treatment facility should be aware that these chemicals are toxic and pose a severe health hazard to anyone not wearing the proper personal protective equipment (see Figure 9.6).

FIGURE 9.6 ◆ Water must be treated to ensure that it is safe for human consumption.
Courtesy of James D. Richardson.

distribution grid

The system in which power travels from a power plant to individual customers.

primary feeder mains

Large-diameter pipes that form the outside perimeter of the water distribution grid and receive water from the water source.

secondary feeder mains

Piping that interconnects with the primary feeder main.

distributor pipes

Piping located inside a building.

nonindicating valve

A valve that has to be physically operated to determine if it is open or closed.

Distribution Grids

Water is distributed to the individual customer through a configuration of pipes, referred to as a **distribution grid**. The size of the pipe used for each section of the grid depends on water demand and the size of the area covered by the grid. The grid consists of three sections:

- **Primary feeder mains** are very large pipes that are capable of carrying large volumes of water into a given area. Primary feeders form the outside borders of the grid.
- **Secondary feeder mains** are slightly smaller in diameter than the primary feeder pipes. These secondary mains interconnect portions of the primary feeders so that the water supply will flow from two directions. These pipes supply the distributors.
- **Distributor pipes** supply water to street fire hydrants and to the individual customers up and down the block.

A common practice is to place control valves in the grid mains so that no more than 500 feet of main will be shut down at any one time. Because the valves are within the grid structure, they may be buried many feet underground, depending on the depth of the frost line. Concrete or metal shafts, leading from the valve to the surface of the ground, are constructed to allow access to the valves. The shafts have metal or polymer lids identifying the location of the valve. A special long shaft "key" tool is needed to operate the valve. Such a valve is classified as a **nonindicating valve**, which means that the valve's operating position (open or closed) cannot be determined by visual inspection. The valve has to be physically operated to determine if it is open or closed (see Figure 9.7).

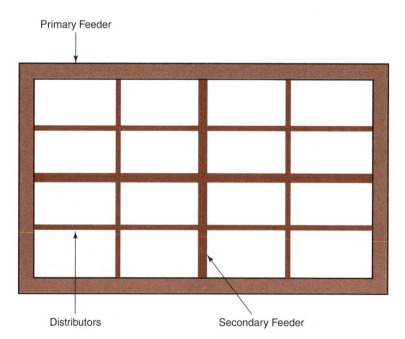

FIGURE 9.7 ◆ The distribution grid must be designed to provide adequate water supplies and flow to the community.
Courtesy of James D. Richardson.

PIPE SIZE REQUIREMENTS

The National Fire Protection Association (NFPA) recommends that pipes used to supply firefighting needs should be no less than six inches in diameter. Six-inch pipe, however, should be used only in a looped grid in which no leg is longer than 600 feet. A *looped grid* is a grid that is constructed in the configuration of a square or rectangle, which allows the water from the water source inlet to flow in two directions inside the loop. In congested areas, distributors should be no less than 8 inches in diameter—or no less than 12 inches in diameter if they form part of a long grid line. It is recommended that for fire protection purposes, pipe diameters should be increased one size over the required minimum. Increasing the pipe diameter by one size can dramatically increase water flow.

Flow ratio can be determined by using this simple formula:

$$\frac{LD^2}{SD^2} = \text{Flow ratio}$$

LD = Large-diameter pipe
SD = Small-diameter pipe

To compute the increased flow ratio, divide the square of the diameter of the SD into the square of the diameter of the LD.

Example: If the distributor size is increased from 8 inches to 10 inches, the flow increase can be determined as follows:

$$\frac{10^2 = 100}{8^2 = 64} = 1.56$$

By increasing the pipe by one size, the amount of flow is increased 1.56 times.

The condition of the interior lining of a pipe also determines the amount of flow that it can provide. Water pipes can be constructed of cast iron, steel, reinforced concrete, ductile iron, or polyvinyl chloride. Steel and cast iron pipes are subject to damage by various phenomena, such as tuberculation, encrustation, and sedimentation. **Tuberculation** is a buildup of rust on the interior lining of metal pipe. **Encrustation** is caused by chemical deposits or biological growth inside the pipe walls. **Sedimentation** is the accumulation of mud, sand, and other debris. All of these reduce the interior diameter of pipes, which reduces water flow. Concrete and polyvinyl chloride pipes are not subject to tuberculation because they do not rust, but they are affected by encrustation and sedimentation. Water distribution systems must be routinely flushed out and flow-tested to ensure that flow rates have not been reduced due to these problems. One indication of a restricted flow is when gauges show a good **static pressure** (pressure in the system without water flow), but **residual pressure** (the pressure that remains after a valve is opened and water is flowing) drops more than 25% when a flow valve is opened.

Stop and Think 9.1

While conducting flow tests in your district, you find a hydrant that has a static pressure of 85 psi, but when you open the hydrant valve, the residual pressure drops to only 40 psi. What information should you seek? How will this situation affect you and your fire crew?

tuberculation

The buildup of rust on the inside surface of a metal water pipe, which reduces the inside diameter of the pipe.

encrustation

The buildup of chemical deposits and biological growth on the inside walls of a water pipe, which reduces the inside diameter of the pipe.

sedimentation

The accumulation of mud, sand, and other debris inside a water pipe, which reduces the inside diameter of the pipe.

static pressure

Energy that is available to force water through pipes and fittings.

residual pressure

The pressure that remains in the system when water is being discharged.

TABLE 9.1 ◆ Advantages and Disadvantages of a Separate System	
Advantages	*Disadvantages*
The system is under complete control of the local fire department.	The community must pay for the construction of two distribution systems.
Cross-contamination of the local drinking water is avoided.	Firefighters must travel to a pump station to start and operate the pumps during an emergency. This reduces the available staffing for fire ground operations.
Nonpotable water can be used, and the water supply does not require treatment.	The pumps and other apparatus are in a nonoperational mode and may fail to start during an emergency if they have not been subjected to a regular schedule of maintenance and operation.
The water supply for firefighting is not impacted by population growth or by periods of heavy demand.	
The system can be custom-designed to meet all firefighting requirements.	

WATER DISTRIBUTION MODELS

Water distribution models can be categorized as either separate or dual. The separate model has two completely independent water supply systems. One system serves the everyday needs of the community's population and industry. The second system is used exclusively by the fire service and is independent of the community water distribution system, having its own water supply, pumps, and distribution mains. The dual model has one system that supplies water for all domestic, industrial, and fire protection needs. Both options have advantages and disadvantages.. These are outlined in Tables 9.1 and 9.2.

Standards for installation and maintenance of various components of these systems are provided by organizations such as the National Fire Protection Association and the American Water Works Association. These standards are supported through testing by Underwriters Laboratories and Underwriters' Laboratories of Canada.[1]

TABLE 9.2 ◆ Advantages and Disadvantages of a Dual System	
Advantages	*Disadvantages*
The pumps and other apparatus at the source of supply are in operation 24 hours a day, seven days a week.	There is a possibility of cross-contamination of the community's drinking water.
There are redundant systems that can be put into operation in case of equipment failure. Personnel are available for maintenance and repair at all times.	A common water source is used for domestic, industrial, and firefighting operations. Therefore, the water supply for firefighting is affected by population growth and periods of heavy demand.

The humble fire hydrant is the most visible component of the public fire protection system. As previously mentioned, the ability of a hydrant to support firefighting operations depends on the ability of the distribution system to support the hydrant. The first water distribution systems in the United States consisted of buried water mains constructed of hollowed wooden logs. To acquire water in an emergency, firefighters would dig a hole in the street to expose the main and then drill into the main. The hole would fill with water, becoming a cistern from which pumpers could be supplied. In 1803, a Philadelphia engineer named Frederick Graff designed a standpipe that could be permanently inserted into the wooden main. A control valve on the standpipe allowed the water to be turned off so that the pipe could be drained to protect the system during freezing temperatures. The visible part of this early hydrant was a wooden or metal box that protruded about 2 feet above the ground encasing the standpipe. It would be over 50 years before this design would be improved. (Ironically it is not possible to verify that Mr. Graff held the first U.S. patent for a fire hydrant, because the Patent Office burned to the ground in 1836, destroying all U.S. patent records!)

There are three classifications of fire hydrants: the base-valve dry-barrel hydrant (often referred to as a dry-barrel hydrant), the wet-barrel hydrant (also known as the California hydrant), and the dry hydrant. Base-valve dry-barrel and wet-barrel hydrants are linked to a distribution grid that provides a pressurized water supply. They are commonly found in towns and urban areas. The dry hydrant does not rely on a municipal distribution system and is adapted for use in rural areas.

THE BASE-VALVE DRY-BARREL HYDRANT

Samuel R. C. Mathews was a Detroit inventor who worked for a rotary pump fire engine manufacturing company. In 1858, he patented the design for a base-valve dry-barrel fire hydrant. This type of hydrant is constructed of cast iron and designed to be free standing (no wooden or metal box is needed for protection). The barrel of the hydrant is dry because the base valve is installed below the frost line. Water does not enter the barrel until the base valve is opened. To open this valve, it is necessary to turn a nut that is located on top of the hydrant bonnet for easy access. When the valve is closed, drain ports open, which allow water to drain from the hydrant barrel. This feature is of great benefit in areas where freezing temperatures occur (see Figure 9.8). When the barrel is filled, water is made available to all hydrant outlets. The main disadvantage of this design is that when more lines need to be connected to the hydrant, the hydrant must be shut down so that the outlet caps can be removed.

Base-valve dry-barrel hydrants are the most common hydrants used in the fire service today. They are manufactured of three basic materials. The barrel, caps, and bonnet are usually cast iron; all internal moving parts are bronze; and the compression valve face is neoprene or rubber.

Standards for base-valve dry-barrel hydrants are found in the American Water Works Association Standard C502-94, *Standard for Dry Barrel Fire Hydrants*, and Underwriters Laboratory Standard UL 246, *Hydrants for Fire Protection Service*.

FIGURE 9.8 ◆ The base-valve dry-barrel hydrant is designed to ensure proper operation during freezing weather.
Courtesy of James D. Richardson.

THE WET-BARREL (CALIFORNIA) HYDRANT

A young immigrant from France named Morris Greenburg invented the wet-barrel hydrant in San Francisco in 1851. It is constructed of the same materials as the dry-barrel hydrant, but pressurized water is supplied to the barrel of the hydrant at all times, hence, the name wet-barrel. Wet-barrel hydrants do not require a base valve. Individual valves control each outlet on the hydrant. This simple design allows additional hose lines to be attached without disrupting the water flow. Hydrants of this type are used in areas that are not subject to freezing temperatures.[2]

> Standards for wet-barrel hydrants are found in the American Water Works Association Standard C503-88, *Standard for Wet Barrel Fire Hydrants*, and Underwriters Laboratory Standard UL 246, *Hydrants for Fire Protection Service*.

DRY HYDRANTS

nonpressurized dry hydrant system

A hydrant system that is used to draft water from a static water source.

pressurized dry hydrant system

A piping grid that connects several dry hydrants. A fire department connection is placed at the base of the grid near a water source. During firefighting operations, a pumping apparatus connects to the water supply and supplies water to the dry hydrant system.

Dry hydrants are generally found in rural areas where there are no water distribution systems or where the existing systems cannot adequately support firefighting needs. They can be designed as either nonpressurized or pressurized. The **nonpressurized dry hydrant system** is installed at a static water source, such as a pond, lake, river, or bay. Construction consists of a pipe that extends into the water source on one end and has a threaded connection for attaching fire apparatus at the other end. The apparatus connects to the hydrant and drafts water from the water supply. The maximum vertical lift from the water surface to the pump should not exceed 15 feet, and an all-weather road should be provided to allow 24-hour access to the hydrant (see Figure 9.9).

When the area to be protected is at an elevation that a nonpressurized system cannot reach, a **pressurized dry hydrant system** must be devised. This consists of a piping grid that connects several dry hydrants. A fire department connection is placed at the base of the grid near a water source. During firefighting operations, a pumping

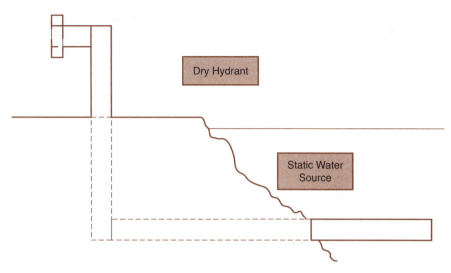

FIGURE 9.9 ◆ The dry hydrant is used to draft water from a static water supply.
Courtesy of James D. Richardson.

apparatus connects to the water supply and supplies water to the dry hydrant system through the fire department connection. In this type of system, the fire department connection is usually a base-valve dry-barrel fire hydrant that is labeled with the letters FDC (Fire Department Connection) to identify it as a water supply intake valve (see Figure 9.10). The system should be designed to provide a minimum fire flow of 1,000 GPM.[3]

The Natural Resource Conservation Service (NRCS), an office of the U.S. Department of Agriculture, provides recommendations to rural communities for dry hydrant operations. NRCS suggests that a dry hydrant water supply have a minimum capacity of 30,000 gallons.

FIGURE 9.10 ◆ Water is supplied through the 4 inch steamer connection of the hydrant.
Courtesy of James D. Richardson.

TABLE 9.3 ◆ NFPA 291, Fire Hydrant Classification and Color-Code Recommendation		
Class	**Flow Rate**	**Color of Bonnet and Caps**
AA	1,500 GPM or greater	Light Blue
A	1,000 to 1,499 GPM	Green
B	500 to 999 GPM	Orange
C	Less than 500 GPM	Red

HYDRANT MAINTENANCE AND FLOW TESTING

The National Fire Protection Association suggests that private fire hydrants should be visually inspected once each month, lubricated twice yearly, and opened and closed once yearly. Graphite is the preferred lubricant. Petroleum-based products can cause dirt and other foreign matter to collect on the outlet threads, which could impede proper hose connection. The NFPA has also published standards for flow-testing and marking fire hydrants. Flow-testing not only determines the amount of water that can be expected from a particular hydrant but also makes it possible to determine the flow capability of an entire area's distribution grid. After the flow rate of a fire hydrant is determined, NFPA 291 recommends that the hydrant be classified and color-coded so that firefighters can visually determine if the hydrant is capable of supplying their needs. NFPA standards are not laws; they are only recommendations. Many jurisdictions use their own color-code system. It is the responsibility of all firefighters to familiarize themselves with the color-code system used in their response areas (see Table 9.3).

> For NFPA standards for maintenance, inspection, and flow-testing of private fire hydrants, see NFPA 25, *Standard for the Inspection, Testing, and Maintenance of Water-Based Fire Protection Systems,* and NFPA 291, *Recommended Practice for Fire Flow Testing and Marking of Hydrants.*

target areas
Areas that are of high value or where there is a potential for a large loss of life.

The NFPA recommends that the distance between fire hydrants should not exceed 800 feet. In closely built or **target areas**, fire hydrants should be spaced at an interval of 500 feet or less.[4]

◆ STANDPIPE SYSTEMS

The NFPA defines standpipe systems as "fixed piping systems and associated equipment that provides a means to transport water from a reliable water supply to designated areas of buildings where hoses can be deployed for fire fighting."[5] These systems are found in multistory buildings and other areas, such as large warehouses and wharves, that cannot be reached by hose lines from hydrants or fire apparatus. They are intended for manual firefighting and allow the firefighter to bring the hydrant to the fire.

The insurance industry has taken the lead in lobbying for local or state legislation and for the passage of building codes that mandate standpipe systems. These can vary

FIGURE 9.11 ◆ Only trained and experience personnel should use a Class I standpipe system. Courtesy of Pickett's Video & Training.

from the very simple to the very complex. The simplest design consists of a water supply (feeder main), a check valve (backflow preventer), a **riser** (the piping), and a hose valve station. The NFPA recognizes three classes of systems: Classes I, II, and III.

> NFPA 14, *Standard for the Installation of Standpipe and Hose Systems.*

CLASS I SYSTEM

The Class I system is designed to be used only by trained firefighting professionals for full-scale firefighting. It is required to have one or more 2½-inch hose connections. The system is also required to have an outside fire department connection that allows water to be supplied by fire department apparatus. If the system is in a high-rise building with several fire zones, a fire department connection must be provided for each zone.

In buildings higher than three stories, a Class I standpipe system may be required even if the building is equipped with sprinklers (see Figure 9.11).

CLASS II SYSTEM

The Class II system is designed for use by building occupants. A 1½-inch outlet and preconnected 1½-inch hose and nozzle are provided. They are usually located in wall-mounted cabinets in the building hallways.

Class II systems were once required in large unsprinklered buildings, but today their use is discouraged for three principal reasons. First, the hose lines in these systems can carry as much as 100 GPM. This can create safety issues when such heavy lines are manned by individuals not trained in the strategy and tactics needed to use them effectively. Second, a delay in reporting the fire may result while occupants are trying to extinguish the fire. Third, energy spent on firefighting might be put to

riser

The largest pipe in a sprinkler system. It is connected directly to the main water supply line and contains the sprinkler alarm valve. It supplies water to the crossmain.

FIGURE 9.12 ◆ Class II systems delay evacuation and expose untrained and unprotected individuals to potentially deadly hazards.
Courtesy of Pickett's Video & Training.

better use in evacuating the building. Individuals not trained in the use of hose lines can do little against a fire once it has passed the incipient stage. Once it has entered the free-burning state, heat and smoke will rapidly increase in intensity, and unprotected occupants should abandon hose lines and evacuate the building immediately (see Figure 9.12).

> **Stop and Think 9.2**
>
> You and your crew advance to the fifth floor of an old hotel and find heavy smoke and one room heavily involved in fire. The building has no sprinkler system and only a Class II standpipe system. You see a charged hose line from the standpipe extending down the hallway to the involved room. What actions should you take to initiate extinguishment operations?

CLASS III SYSTEM

The Class III system is a combination of the Class I and II systems. A 2½-inch outlet is provided for full-scale firefighting and a 1½-inch outlet with preconnected hose and nozzle for occupant use. This system is declining in use and is discouraged for the same reasons as those for Class II systems.

TYPES OF STANDPIPE SYSTEMS

In 1993, NFPA 14 classified standpipe systems into five categories or types: automatic wet systems, automatic dry systems, semiautomatic dry systems, manual dry systems, and manual wet systems.

Automatic Wet System

The automatic wet system is the most common standpipe system. The piping is filled with pressurized water from the water supply. When the hose valve is opened at the standpipe station, water is immediately available for firefighting operations.

Automatic Dry System

The automatic dry system is used in areas that are subject to freezing temperatures. The system is filled with pressurized air that is controlled by a dry pipe valve. When the hose valve is opened at the standpipe station, air is expelled, and the dry pipe valve opens, allowing the water supply to enter the system.

Semiautomatic Dry System

The piping in the semiautomatic dry system is also filled with pressurized air. When a remote device is activated at the hose station, the air is exhausted from the system. A deluge valve opens and allows the water supply to fill the piping.

Manual Dry System

The manual dry system is not connected to a permanent water supply. The piping is filled with air at normal atmospheric pressure. Water supplies must be manually provided by the fire department through a fire department connection.

Manual Wet System

The manual wet system is not connected to a permanent water supply. The piping is filled with pressurized water from the domestic water supply. This supply is inadequate for firefighting. In the event of a fire, the fire department must supply water manually to the system through the fire department connection.

STANDPIPE FLOW RATE REQUIREMENTS

NFPA 14 requires that Class I and Class III systems be capable of delivering 500 GPM from the first standpipe and 250 GPM from each additional standpipe, up to a maximum total of 1,250 GPM. The water supply must be adequate to sustain this flow rate for a minimum of 30 minutes. These systems must also maintain a residual pressure of 100 psi while flowing 250 GPM at the two topmost outlets, plus 250 GPM from each additional standpipe, up to the maximum flow rate of 1,250 GPM. Class I and Class III systems must provide a minimum **flow pressure** (the pressure at which water is discharged from the outlet valve) of 100 psi to each outlet of the system. Class II systems must provide 100 GPM for 30 minutes and maintain a residual pressure of 65 psi at the topmost outlet while flowing 100 GPM.

flow pressure
The pressure created by the forward velocity of a fluid at the discharge orifice.

In Class I and Class III standpipe systems, the risers (pipes that carry water to ceilings and upper stories), must meet the following requirements:

- If the building's height is 100 feet or less, the standpipe riser diameter must be a minimum of 4 inches.
- If the building's height exceeds 100 feet, the standpipe riser diameter must be a minimum of 6 inches; however, the top 100 feet of the standpipe can be 4 inches in diameter.

Standpipe risers for Class II systems must meet the following requirements:

- If the building height is less than 50 feet, the riser must be 2 inches in diameter.
- If the building height exceeds 50 feet, the minimum riser diameter is 2½ inches.
- If the building height is over 275 feet, the system must be divided into zones.

PRESSURE-REGULATING DEVICES

As the height of the standpipe increases, the pressures at the lower outlets increase. This is caused by head pressure, the pressure of the column of water inside the riser. The fourth principle of pressure states that pressure in an open vessel is proportional to the depth of the vessel. To reduce the dangers of high pressure on lower outlets, standpipe systems must be divided into zones in buildings over 275 feet in height. At this height, the head pressure at the lowest outlet will be approximately 120 psi (275 × 0.434 = 119.35 psi). It is possible to install pressure-reducing devices that will lower outlet pressures to 100 psi or less. If such devices are installed in a standpipe system, the zone height requirement can be extended from 275 feet to 400 feet. A major disadvantage of working with a pressure-reducing device is that most fire departments use fog nozzles on 2½-inch hose lines. These nozzles are designed to be effective and reach maximum flow rates at 100 psi nozzle pressure. If the outlet is regulated to a pressure of 100 psi, there is no compensation for **friction loss** in the fire department hose line. Friction loss is the pressure that is lost in a fire hose or standpipe as a result of water turbulence and resistance caused by the water moving against the sides of the conduit. Friction loss reduces the flow rate and maximum effectiveness of the hose stream. Some fire departments extend maximum outlet pressures to 120 psi to compensate for friction loss.

The entire standpipe system should be tested annually to ensure proper installation and adjustment.[6]

friction loss

Pressure loss caused by turbulence as water flows through a conduit.

Stay Safe

◆ AUTOMATIC SPRINKLER SYSTEMS

John Carey invented the first automatic sprinkler system in England in 1806. The ingenious system consisted of perforated pipes suspended from the ceiling. Water supply was controlled by a set of valves held closed by counterweights. Strings suspended the counterweights. In the event of a fire, the strings would burn through, and the counterweights would drop, opening the valves and sending water through the perforated pipe. Water would be discharged over the entire area, not just on the fire. As this system evolved, solid pipes with open-style sprinkler heads replaced the perforated pipes.

Henry Parmelee, a Connecticut piano maker, is given credit for inventing the first practical automatic sprinkler in 1874. To prevent extensive water damage to the expensive wood that he used in his craft, he soldered brass caps over the open sprinkler heads. The system could be filled with water because the caps would hold the water back. In the event of fire, heat would melt the solder holding the caps, and water pressure would blow them away from the sprinkler heads, allowing water to be directed onto the fire. Areas not affected by the fire would remain dry.

Frederick Grinnell improved the Parmelee Sprinkler by designing a sprinkler head with a more sensitive, heat-actuated release mechanism that could be calibrated to open at a predetermined temperature and a valve seat that eliminated leakage due to water pressure fluctuations. The improved Parmelee Sprinkler later incorporated a

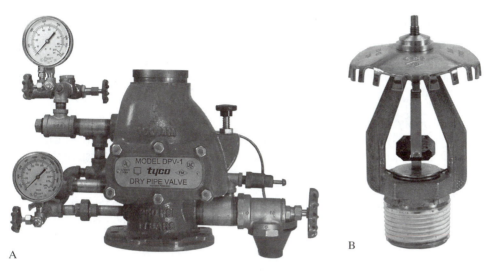

A   B

FIGURE 9.13 ◆ The basic operation of these devices has changed little since their inception.
(A) Sprinkler alarm valve; (B) sprinkler head.
Courtesy of Tyco Fire and Building Products.

variable-pressure alarm valve as well, a device invented by an Englishman named John Taylor. These innovations are still used in modern sprinkler systems (see Figure 9.13 A and B).

Today, the automatic sprinkler system is considered the most effective method of controlling fires in their incipient stage. Automatic sprinklers were primarily installed for property conservation purposes, but their life-safety benefits were soon realized. With the addition of John Taylor's alarm valve, the system not only attacked the fire in its early stages, it also sounded an alarm to warn occupants of fire danger. Sprinkler systems are so effective in controlling and extinguishing fires that the NFPA Life-Safety Code 101 allows increased travel distances to exits and reduced fire ratings on wall assemblies and interior finishes in building with complete sprinkler coverage. The exemplary performance of automatic sprinklers in supporting life safety was verified in 1995 when the NFPA's Fire Analysis and Research Division released a report entitled "U.S. Experience with Sprinklers: Who Has Them? How Well Do They Work?" It states, "NFPA has no record of a fire killing more than two people in a completely sprinklered building where the system was working properly."[7]

The National Fire Protection Association published the first codes for automatic sprinkler installation in 1896. Since then, it has established three standards and two recommendations dealing with these systems. It is important that all firefighters familiarize themselves with these standards and recommendations, especially with NFPA 13.[8]

NFPA 13, *Standard for the Installation of Sprinkler System;* NFPA 13D, *Standard for the Installation of Sprinkler Systems in One- and Two-Family Dwellings and Mobile Homes;* NFPA 13R, *Standard for the Installation of Sprinkler Systems in Residential Occupancies up to Four Stories in Height;* NFPA 13A, *Recommended Practice for the Inspection, Testing and Maintenance of Sprinkler Systems;* and NFPA 13E, *Recommendations for Fire Department Operations in Properties Protected by Sprinkler and Standpipe Systems.*

AUTOMATIC SPRINKLER SYSTEM COMPONENTS

An automatic sprinkler system consists of the following components: a water supply main, a backflow preventer (check valve), a water flow control valve (shut off valve), an alarm valve, a fire department connection, a local audible alarm, a riser, a cross-main, branch lines, and sprinkler heads.

Water Supply Main

The water supply main should be at least as large as the riser, a minimum of 6 inches in diameter. It must be able to deliver the required fire flow to the highest sprinkler with a residual pressure of 15 psi. If the system is designed by hydraulic calculation, the residual pressure to the highest sprinkler can be lowered to 7 psi. When the water supply is delivered from an elevated storage tank, the bottom of the tank must be a minimum of 35 feet above the highest sprinkler. This will maintain a minimum residual pressure of 15 psi.

Backflow Preventer

If the sprinkler system is being supplied from a public water distribution system or a private system that is also for public use, a backflow preventer must be installed between the public supply main and the water flow control valve. A backflow preventer, or check valve, is a one-way valve that allows water to flow toward the sprinkler system but closes when pressure in the system is reversed, thus preventing water from flowing water back into the public water supply. Fire protection systems may sit idle for many years. If they are not flow-tested regularly, the water in the pipes may become stagnant. If this water were allowed to contaminate the public water supply, the health quality of the system would be compromised. When inspecting the installation of a sprinkler system, the backflow preventer should be examined to ensure that it is properly installed. An arrow imprinted on the side of the valve housing indicates the direction of flow (see Figure 9.14).

FIGURE 9.14 ◆ The backflow preventer is installed to reduce the possibility of contamination of the public water supply.
Courtesy of James D. Richardson.

FIGURE 9.15 ◆ If the stem of the OS&Y valve is extended out from the operating wheel, the valve is open.
Courtesy of James D. Richardson.

Water Flow Control Valves

The water flow control valves, or shut off valves, are located in the riser between the water supply main and the fire department connection. There are several types of control valves.

Outside Stem and Yoke (OS&Y) Valve. This is the most common control valve used on sprinkler systems and is classified as a *gate valve*. The yoke controls a threaded stem that opens or closes the valve as the yoke is turned. The OS&Y is classified as an **indicating valve**, which means that the position of the valve can be determined by visual inspection. If the threaded stem protrudes from the yoke, the valve is open; if the stem is flush with the top of the yoke and the threads are not visible, the valve is closed (see Figure 9.15).

Post Indicator Valve. When sprinkler control valves are installed underground, a post indicator is provided so that the valve can be operated. The post indicator is a hollow post that houses a valve stem, which is connected to the operating yoke of the underground valve. That valve is opened and closed by turning a valve stem nut on top of the post. Indicators are attached to the stem. When the valve is closed, the words "Shut" or "Closed" will appear in a window opening on the side of the post. When the valve is opened, the word "Open" will appear in the window (see Figure 9.16).

Wall Post Indicator Valve. This valve is similar in operation to the post indicator valve, but it is located inside the building, next to an exterior wall. The valve stem extends through the exterior wall of the building and is connected to an operating wheel or lever mounted on an indicating post outside of the exterior wall . The post has an indicating window that shows the words "Shut" or "Open" to indicate the position of the valve gate (see Figure 9.17).

indicating valve
A valve that can be visually inspected to determine whether it is open or closed.

FIGURE 9.16 ♦ The post indicator valve is an operating device for a gate valve underground.
Courtesy of James D. Richardson.

Post Indicator Valve Assembly. This valve can be wall-mounted or placed in-line between the water supply and the alarm valve inside the alarm valve room. The post indicator valve assembly is the only one of the four valve types that operates as a butterfly valve. A crank handle or wheel is mounted on the side of the valve housing. When turned, a worm gear rotates the butterfly 90° inside the valve housing waterway. The indicator is a pointer mounted on top of the valve housing that points to the words "Open" or "Closed."

FIGURE 9.17 ♦ The wall post indicator valve is an operating device for a gate valve inside the building.
Courtesy of James D. Richardson.

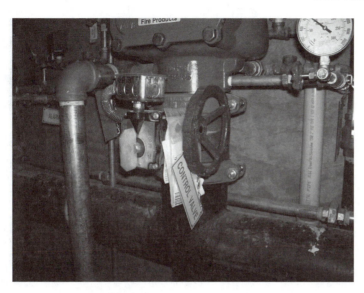

FIGURE 9.18 ◆ The post indicator valve assembly is a butterfly valve, the only indicator valve that is not a gate valve.
Courtesy of James D. Richardson.

Note: All water flow control valves on sprinkler systems are held in the "Open" position by a chain or lock, and most are electronically monitored to signal a valve shutdown (see Figure 9.18).

Fire Department Connection

Fire department connections are required on all sprinkler and standpipe systems. They allow fire department pumping apparatus to attach hose lines and supply additional pressure and water into the system. The connection can be found on the exterior wall of the building or mounted as a manifold outside of the building. There are several variations of fire connections. The most common consists of two 2½-inch inlets that "siamese" into a 4-inch pipe. For multiple or large systems, the connection may be set up as a manifold, having three or more 2½-inch or larger inlets to accommodate a large-diameter hose. All inlets have female threads. If the system has only one riser, the fire connection pipe will connect to the sprinkler side of the system, above the alarm valve. If more than one riser is attached to the supply main, the pipe will attach to the supply side of the alarm valve (see Figure 9.19).

Alarm Valves

The alarm valve is a clapper valve that can be designed to serve several functions. It can be operated as a wet valve, a dry valve, or a deluge valve. Different configurations enable the valve to be operated hydraulically, pneumatically, or electrically. The purpose of the valve is to protect against false alarms and direct the flow of water to the sprinklers and alarm signaling system (see Figure 9.20).

Often referred to as a "water gong," a local audible alarm is attached to the alarm valve by piping. As water flows through the valve, a portion of the water is diverted through the pipe and travels to the gong. The force of the water passing through a chamber in the gong turns a wheel that causes a clapper to strike a gong. The gong

FIGURE 9.19 ◆ The fire department is responsible for knowing which zones the fire department connections supply.
Courtesy of James D. Richardson.

will sound as long as water is flowing through the sprinkler system. After the water passes through the gong, it is discharged outside the building (see Figure 9.21).

Risers, Crossmains, and Branch Lines

The riser is the largest pipe in the sprinkler system. It is the portion of pipe that carries the water from the water supply to the ceiling and upper levels. The water supply main, alarm valve, crossmains, and fire department connection (on single-riser

FIGURE 9.20 ◆ An alarm valve may be designed for wet-pipe, dry-pipe, preaction, and deluge systems.
Courtesy of Tyco Fire and Building Products.

FIGURE 9.21 ◆ Bees have been known to nest in water gongs and make them inoperable.
Courtesy of James D. Richardson.

systems) are connected to the riser. **Crossmains** are slightly smaller pipes that extend from, and run perpendicular to, the riser. They carry water from the riser to other areas of a building. **Branch lines** are the smallest pipes in the system. They extend from the crossmain into specific areas or rooms. The sprinkler heads are attached to the branch lines (see Figure 9.22).

Sprinkler Heads

The sprinkler head serves a dual purpose. It is both a heat detector and a systems activation device. The heat detection device on the sprinkler can be designed as a fusible link, frangible bulb, chemical pellet, or fusible pellet. The fusible link, chemical pellet, and fusible pellet operate by melting after reaching a predetermined temperature. This releases the mechanism holding the nozzle cap and allows water pressure to blow the cap away from the sprinkler's discharge orifice. Temperature activation settings are determined by the amount of surface area, thickness, and type of material used for the release mechanism. Fusible links are generally constructed from an alloy of cadmium, bismuth, lead, and tin.

The frangible bulb sprinkler utilizes a liquid-filled glass bulb to hold the nozzle cap in place. Heat causes the liquid inside the bulb to expand until the glass bulb ruptures, releasing the nozzle cap and allowing water flow. The activation temperature is determined by the amount of liquid in the glass bulb (see Figure 9.23).[9]

Sprinkler heads are designed for installation in either a pendent, upright, or wall configuration, and they cannot be installed in any configuration except that for which they were designed (see Figure 9.24).

The sprinkler must be able to provide the required flow rate for the area being protected. The flow rate depends on the size of the discharge orifice and the amount of discharge pressure being provided. Sprinkler discharge orifice diameters vary from ⅜ inch to ¾ inch. The most common orifice diameter is ½ inch. The formula used to calculate sprinkler flow rate is:

$$Q = K\sqrt{P}$$

crossmains

Pipes that supply the branch lines from the main riser. They are slightly smaller than the riser but are larger than the branch lines.

branch lines

The smallest pipes of a sprinkler system on which the individual sprinkler heads are mounted.

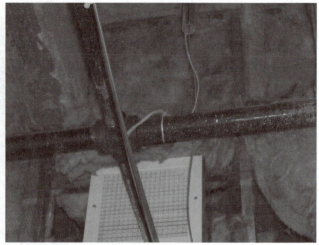

A

B

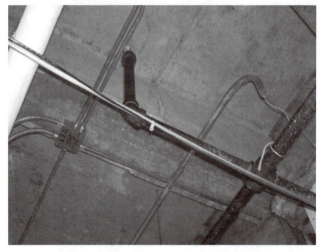

C

FIGURE 9.22 ◆ (A) The riser supplies water to the upper levels and the crossmain. (B) The crossmain delivers water from the riser to different areas of the building being protected. (C) The branch lines are supplied by the crossmain and are the mains on which the sprinkler heads are mounted.
Courtesy of James D. Richardson.

Q = Quantity of flow
K = Discharge coefficient based on the orifice size and surface characteristics
P = Discharge pressure at the sprinkler

The K factor coefficient for a ½-inch sprinkler ranges from 5.3 to 5.8, depending on the condition of the interior surface of the discharge orifice. The minimum discharge pressure for a nonhydraulically calculated sprinkler is 15 psi. If the discharge coefficient of a sprinkler is 5.5 and the discharge pressure is 15 psi, the flow rate can be calculated by multiplying 5.5 times the square root of the discharge pressure (15 psi).

$$5.5 \times \sqrt{15} = \text{GPM}$$
$$5.5 \times 3.87 = 21.28 \text{ GPM}$$

FIGURE 9.23 ◆ Sprinkler heads may be activated by a fusible link, bimetallic disc or strip, or frangible bulb.
Courtesy of James D. Richardson.

Flow rate is adjusted by varying the pressure or changing the diameter of the sprinkler orifice.[10]

TYPES OF AUTOMATIC SPRINKLER SYSTEMS

Four categories of automatic sprinkler systems are generally recognized: wet-pipe systems, dry-pipe systems, preaction systems, and deluge systems. The type of alarm

FIGURE 9.24 ◆ Sprinkler heads are designed to be mounted in only one configuration. For example, pendent sprinkler heads cannot be mounted in an upright configuration.
Courtesy of Pickett's Video & Training.

valve used in a system determines its category. These systems can be complicated in design, engineered to protect against very specific types of hazards, such as electrical transformer vaults or high-value building contents. No matter how sophisticated the system, they all share the same basic components.

Wet-Pipe System

The basic design of the wet-pipe system has changed little since its inception in the 1870s. Because of its simple design and steadfast reliability, it is the most common sprinkler system in use today. Wet-pipe systems are normally installed in areas that are not subjected to freezing temperatures. However, portions of a wet-pipe system can be used to protect cold areas. If such a system is connected to a public water supply, the water is removed from the sprinkler piping and replaced with U.S. Pharmacopeia grade glycerin or propylene glycol. Special designs and safety features prevent the antifreeze from entering other portions of the sprinkler system.

When a sprinkler head on a wet-pipe system is activated, the water pressure in the pipe above the alarm valve drops. The clapper (a one-way valve that covers the opening of the water supply pipe) in the alarm valve is thrown open by the force of the pressurized water supply. The water in the piping is immediately discharged from the sprinkler head. As the water flows through the alarm valve, a portion of the water is diverted through another pipe leading from the alarm valve. This pipe diverts water to the retard chamber; after the chamber is filled, the water travels through piping and activates an electronic signaling device and a water-flow-operated bell (water gong), which sounds a local alarm. The water is then discharged onto the ground under the water gong.

Water distribution systems are subject to a phenomenon known as *pressure surge* that is caused by the constant use of the water system (the opening and closing of water valves). During a pressure surge, the clapper in the wet-pipe alarm valve is lifted off its seat by the force of the surge. This allows water to enter the alarm valve piping. To prevent false alarms, the wet-pipe system is equipped with a retard chamber. This appliance is placed in the piping that connects the alarm valve to the water gong. The chamber has a capacity of approximately 2 gallons of water. Water must fill the chamber completely before it can reach the alarm devices. Water flow during a pressure surge is of short duration and is not adequate to fill the retard chamber, thus preventing the false alarm. The amount of water that does enter the chamber is released through a calibrated drain valve, mounted on the bottom of the device, after the alarm valve clapper reseats. When sprinklers are activated, and the clapper is locked in the open position, the volume of water is high enough to override the drain valve, fill the retard chamber, and activate the alarm devices (see Figure 9.25).

Dry-Pipe System

The dry-pipe system, more complicated than the wet-pipe system, is found in areas where temperatures can fall to 40°F or below. The riser, crossmains, and branch lines are filled with pressurized air instead of water. Air pressure in the piping is maintained by an air compressor or compressed air system. The clapper in the dry-pipe valve is held closed by utilizing the pressure differential principle: a small amount of air pressure above the valve holds back a larger amount of water pressure on the water supply side of the valve.

There are two basic designs that utilize this differential principle: the double-seated clapper and the single clapper. The double-seated clapper valve has two clappers: the upper air seat clapper and the lower water seat clapper. The upper air seat

FIGURE 9.25 ◆ The retard chamber is mounted only on wet-pipe systems.
Courtesy of Tyco Fire and Building Products.

clapper is substantially larger than the water seat clapper and is held closed by the air pressure in the piping. This creates the pressure differential needed to hold the lower water clapper closed. The single clapper valve simply incorporates the features of the double clapper valve into a single clapper. The top of the clapper has a much larger surface area than the water side of the clapper, thus creating the pressure differential. The normal pressure differential in a standard dry-pipe system is approximately 6 to 1. With this design, the dry-pipe system is not affected by pressure surges and does not require the installation of a retard chamber. The pressure of the water supply will determine the amount of air pressure required in the system. For example, if the water supply pressure is 100 psi, the dry-pipe system will require a minimum air pressure of approximately 17 psi to hold the dry-pipe valve closed: 100/6 = 16.66 psi. This is referred to as the *trip pressure*. The air pressure in the system should be 15 to 20 psi above the trip pressure. Therefore, the air pressure in our system would actually be set at 32 to 37 psi.

One disadvantage of a dry-pipe system is the time delay between sprinkler activation and water discharge. It takes time for the pressurized air to be purged from the system to allow water to flow into the piping. NFPA 13 requires a quick-opening device on any dry-pipe system that has a pipe capacity of 500 gallons or more. The two quick-opening devices currently used are accelerators and exhausters.

The accelerator is used in conjunction with the double-seated dry-pipe valve. It is designed to sense the first 1 or 2 psi of pressure drop when a sprinkler activates. It opens a valve that diverts the system's air pressure into a chamber under the upper air clapper. When the air pressure equalizes on both sides of the air clapper, the valve is tripped open by water pressure. This process takes about 10 to 15 seconds.

The exhauster can be utilized with either double-seated or single clapper valves. When it senses a drop in air pressure in the system, it opens a large valve to evacuate the pressurized air rapidly to the outside, thus allowing the dry-pipe valve to trip. When trip pressure is reached, the exhauster closes to prevent the loss of water (see Figure 9.26).

Preaction System

A preaction system is a dry-pipe system with a preaction valve and a supplemental detection system. The detection system is the central operating component. When the detection system senses a problem, it opens the preaction valve and fills the piping

FIGURE 9.26 ◆ The accelerator and exhauster is used only on the pressurized dry-pipe system. Courtesy of James D. Richardson.

with water. Therefore, when the sprinkler head fuses, the water is ready for discharge. What happens, essentially, is that a dry-pipe system is converted into a wet-pipe system prior to sprinkler operation. A positive feature of this system is that the preaction valve will not open until the detection system activates. A loss of air pressure, sprinkler activation, or pipe failure will not open the preaction valve. The detection system must sense a problem and send a signal to open the valve. This type of system is practical in situations where inadvertent water damage is a major concern. If the preaction system is configured as a double-interlock system, both the sprinkler and the detection system must activate before the preaction valve will open (see Figure 9.27).

Deluge System

The deluge system is a nonpressurized dry-pipe system with a detection system. It is used to protect an area that requires rapid application of large quantities of water, such as electrical transformer rooms or outside transformer pads. High-speed, rapid-response deluge systems may be installed in facilities where processes create explosive atmospheres, such as flammable gas cylinder filling facilities. The risers, crossmains, and branch lines in a deluge system contain air at atmospheric pressure. Deluge systems utilize open nozzles or sprinkler heads that do not contain operating elements. When the deluge valve is opened, water fills the piping and discharges from all of the nozzles or sprinkler heads simultaneously. The detection system controls the operation of the deluge valve. The detectors can be of the smoke-, heat-, or flame-detection variety, and the deluge valve can be configured to operate hydraulically, electronically, or pneumatically. A hydraulically operated system can use any kind of detection system. An electronically operated deluge valve requires a detection system capable of transmitting an electrical signal to the valve. A pneumatically operated deluge valve is designed to use detectors that operate by sensing the rate of increase in air pressure inside the detector that is caused by exposure to heat. Fire pumps are used to supply the large volumes of water that are required by the deluge system (see Figure 9.28).

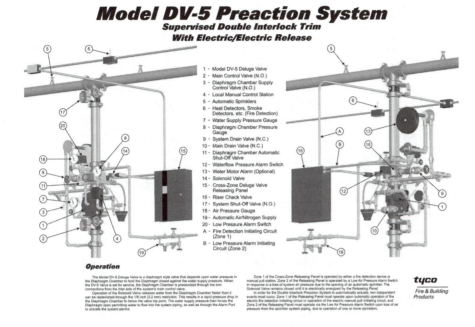

FIGURE 9.27 ◆ Operation of the alarm valve will occur only if a fire detection device activates the preaction system.
Courtesy of Tyco Fire and Building Products.

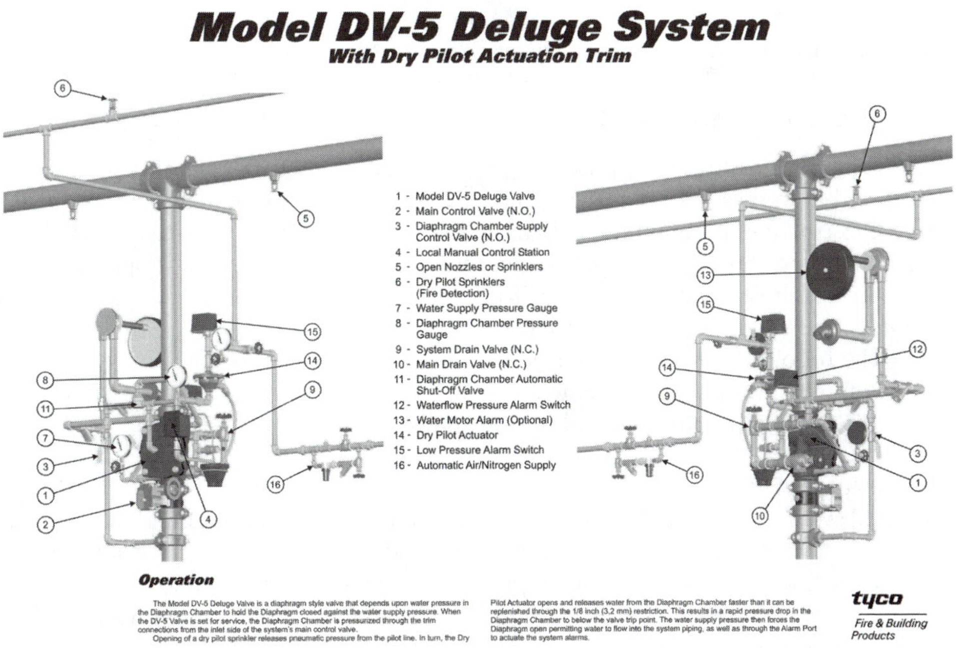

FIGURE 9.28 ◆ Fire pumps provide large quantities of water for the deluge system.
Courtesy of Tyco Fire and Building Products.

A signaling system is classified according to the function the system is expected to perform. All signaling systems, regardless of function, have six basic features.

Control Unit. This unit acts as the "brains" of the system. Its primary function is to receive a signal from the detection device and route the signal to a visual and audible alarm and, if designed to do so, to send an alarm to a monitoring agency.

Primary Power Supply. This is usually supplied by the public utility that serves the building.

Secondary Backup Power Supply. This can be a system of batteries or engine-driven generators. The secondary power supply is required to provide full signaling and detection system operational power within 30 seconds of primary power supply failure.

Trouble Signal Power Supply. This power supply can be from the building's public utility connection or from a separate source.

One or More Initiating Device Circuits. The initiating device may be activated by manual operation, heat, smoke, or flame (light).

One or More Alarm-Indicating Device Circuits. When the control unit receives a signal from an indicating device, it must activate audible and visual alarm devices.

Note: Some control units are able activate other life-safety systems in the building, such as automatic ventilation or extinguishing systems. They may also send signals that close fire doors and return all elevators to the ground floor.

To ensure that operational standards are met, all components of a signaling and detection system should be recognized by a testing laboratory, such as Underwriters Laboratory or Factory Mutual Research Corporation. Installation, maintenance, and use of the system should conform to the standards set forth in NFPA 70, *The National Electrical Code*, and NFPA 72, *The National Fire Alarm Code*.

TYPES OF SIGNALING SYSTEMS

The type of signaling system used depends on several factors, including the kinds of hazard that the system expects to handle, jurisdictional requirements, and the design and construction of the building. During a preincident survey, the firefighters should be able to recognize the type of system in use and understand how it operates. They will then be able to judge whether it is providing adequate protection to the occupants of the building and their property.

The following are basic descriptions of the most common signaling systems encountered by fire service personnel. We will look at local alarm systems, auxiliary systems, remote station systems, and central station systems.

Local Alarm System

A local system is designed to activate a visual and audible alarm and alert occupants on the immediate premises to evacuate the building. The system does not transmit a signal outside of the property being protected. The four most common types of local systems are the noncoded system, the zone-noncoded system, the master-coded system, and the zone-coded system.

Noncoded System. This is the simplest of all local alarm designs. When the control unit receives a detection signal, it initiates all audible and visual alarms

throughout the protected property. The unit is not designed to indicate the location of the problem. Firefighters must search the entire property to locate the cause of the alarm. Audible and visual alarms usually remain activated until the system is reset at the control unit.

Zone-Noncoded System. In this system, the protected property is divided into zones. Each zone is wired as a separate circuit to the control unit and is identified by a corresponding indication lamp on the unit. When an initiating device is activated in a zone, all visual and audible alarms on the premises will activate, and the indication lamp on the control unit assigned to that zone will illuminate. By viewing the control unit, firefighters can determine the general location of the problem.

Master-Coded System. This system utilizes audible and visual notification devices that are a normal part of a building's everyday operations, such as the bells that signal a lunch break. When the control unit receives a signal from a fire detection device, it initiates a different sound pattern from that normally used in the everyday operation. For example, if the signaling device normally sounds one long tone to indicate a break for lunch, the fire alarm control unit will cause the same device to sound several short tones. This system does not indicate the location of the trouble because all signaling devices throughout the premises are activated at once.

Zone-Coded System. This system utilizes the same zone circuitry and indicating lamp features as the zone-noncoded system. However, a coded audible circuit is added to each zone's circuitry at the control unit. When the control unit receives a signal from an indicating device, the zone indicator lamp is illuminated on the control unit, and all visual and audible alarms sound throughout the entire premises. The signals are coded, however, to indicate where the trouble is located. The pattern of the sound signal directs fire personnel to the affected zone.

Auxiliary System

This auxiliary system enables a building's alarm system to interface with a public alarm system. It can be used only in areas that utilize fire alarm boxes. When the building's control unit receives indication of an activated detector, it sends a signal to a municipal fire alarm box located on a nearby street. The signal activates the fire alarm box, which, in turn, sends a coded signal via dedicated phone line or radio signal to the fire department dispatch office. The code identifies the location of the fire alarm box. A fire-location indicator panel may be located inside the fire alarm box if it services more than one alarm system or if the system is zoned. The auxiliary system can be activated manually, by automatic fire detectors, or by water flow indicating devices.

Remote Station System

The remote station system operates in a similar manner to the auxiliary system in that it sends an alarm notification directly to the fire department dispatch office. This system does not, however, utilize a fire alarm box for signal transmission. Instead, the signal is sent directly from the protected premises through leased telephone lines to the telephone exchange, and then on to the dispatch officer. If multiple premises are being protected, the system must be coded. There are alternatives for communities that do not have a fire dispatch office. NFPA states, "If the public agency is unwilling to receive the remote station fire alarm signals, or if that agency is willing to allow another organization to receive those signals, then the signals may be received at a location acceptable to the authority having jurisdiction that is attended by trained personnel 24 hours a day."[11]

Proprietary System

The proprietary system is designed to protect large commercial and industrial complexes. It can also be found in high-rise buildings and in multiple buildings that are concentrated in one location. Each building or protected area has its own system. When the control unit receives a signal from an activated detector, it transmits that signal to an alarm panel that is located in a separate building on the premises or in a separate room within the structure that is protected from hazardous operations. NFPA 72 requires that the building or room housing the alarm panel, which is also called the *supervising station*, be staffed 24 hours a day by employees or representatives of the occupant. These individuals must be trained in the system's operations and in response procedures if the system is activated. They must also be able to notify the fire department through an automatic or manual notification system.

The control units of proprietary systems are required to transmit alarm, supervisory, and trouble signals to the alarm panel in the supervising station. NFPA 72 also requires that a runner be dispatched from the supervising station to investigate any such signal. If a fire alarm signal is received, supervising personnel must retransmit that signal to the fire department communication center. It must be emphasized that all personnel staffing the supervising station are employees or representatives of the protected entity.

Central Station System

A central station system is very similar to a proprietary system. The main difference is that the alarm panel and supervising station is not located on the protected property and is not supervised by employees or representatives of the occupant. Instead, the alarm panel and supervising station is located off premises and constantly staffed by employees of a contract service company whose sole purpose is providing central station service. Underwriters Laboratory or Factory Mutual Research Corporation must list the contract service company. The installation of a central station system must be certified by UL or FMRC and should meet the standards set forth by NFPA 71, *Standard for the Installation, Maintenance, and Use of Proprietary Central Station Signaling System.*

When a fire alarm signal is received, the central station operators notify both the fire service communications center and designated personnel, either at the protected location or in their homes. They must also dispatch a runner or technician to the location to reset the system. Runners must be able to reach the location within one hour. Runners must also be dispatched when supervisory or trouble alarms are transmitted.[12]

◆ AUTOMATIC DETECTION SYSTEMS

We described the different kinds of signaling systems that are used to alert building occupants and firefighters to danger. We will now look at the automatic detection devices that activate those signaling systems.

Automatic fire detectors are designed to sense one or more of the products of combustion: smoke, heat, or light (flames). The operations of these devices range from the very simple to the very complex.

HEAT DETECTORS

Heat detectors are the oldest automatic fire detection devices, dating back to the 1860s, about the time that the first automatic sprinkler heads came into use. They were used with early automatic sprinkler heads. Heat detectors that are designed to operate when a predetermined temperature has been reached are called *fixed-temperature* detectors. Others, called *rate-of-rise* detectors, respond to rapid temperature increases. Detectors that provide both fixed temperature and rate-of-rise capability are referred to as *combination* heat detectors. Heat detection systems are relatively inexpensive compared with more technologically sophisticated systems. They have the lowest false alarm rate of all automatic fire detection devices, but they are the slowest to activate under fire conditions.

Heat detectors should be located on or near the ceiling. To ensure optimum performance, they should be installed in areas where heat is expected to build rapidly, such as a small confined space, where environmental conditions would not allow the use of other types of fire detection systems, and where speed of detection is not a primary concern. Heat detectors may be used in environments that have dust or other suspended particles in the air. Smoke and light detectors would easily be activated in this type of atmosphere.

Fixed-Temperature Heat Detectors

The operating element of a fixed-temperature heat detector can be a fusible link, frangible bulb, bimetallic, or continuous line type. Fusible link, frangible bulb, and bimetallic detectors are referred to as *spot types*. They respond to environmental changes in the specific area where they are located. Fusible link and frangible bulb components are most often associated with automatic sprinklers.

When the fusible link is used in a heat detector, the eutectic (easily melted) metal alloy of cadmium, bismuth, lead, and tin acts as a solder to hold open a spring-loaded contact. When the temperature reaches the melting point of the solder, the contact is released, closing the circuit and initiating a signal.

The function of the frangible glass bulb in a heat detector is to hold two circuit contacts apart. When heat causes the liquid in the bulb to expand and rupture the glass, the contacts close and complete the circuit, thus initiating a signal. Fusible link and frangible bulb detectors do not reset when normal environmental conditions return and must be replaced after they have been activated.

The contacts in a bimetallic detector are held apart by a bimetallic strip or bimetallic snap disc. These elements are constructed by bonding two metals that have different heat expansion coefficients. When heat is applied to the bimetallic strip or disc, the bimetallic operating element will bend in the direction of the contact point, closing the circuit and initiating a signal. This type of detector automatically resets itself when the ambient temperature drops below its activation point.

Unlike spot-type detectors, *continuous line* detectors are designed to sense temperature changes over a large area. Two basic versions are available. In one design, two steel wires are separated by heat-sensitive insulation. When the ambient temperature reaches a predetermined level, the insulation melts from around the wires and allows them to make contact with each other. This closes a circuit and initiates an alarm signal. The fused segment of the line must be replaced in order to restore the system to normal operation. The other continuous line detector has a coaxial design. It consists of a conductive metal wire housed inside a stainless steel tube. A ceramic semiconductor separates the conductive metal wire and the stainless steel tubing. A minute electrical current flows through the circuit. Under normal conditions, this current is not strong enough to trip the alarm signal. However, when the stainless steel exterior tubing is heated, the electrical resistance of the ceramic semiconductor is lowered, allowing a stronger current to pass through the system and activate the alarm.

A fixed-temperature device activates when the ambient temperature at the ceiling exceeds its predetermined setting. At the time of activation, the actual ambient temperature at the ceiling will, however, be much higher than the activation setting of the device. This delay, called a *thermal lag*, is caused by the time needed for the operating element to absorb the heat necessary for activation. Thermal lags are an inevitable shortcoming of fixed-temperature heat detection systems.

Rate-of-Rise Heat Detectors

The rate-of-rise system eliminates thermal lag. When a fire occurs, the temperature of the air directly above the fire increases rapidly. The rate-of-rise detector is designed to sense this rapid increase in temperature and signal an alarm if the rate exceeds 12°F to 15°F per minute. These systems may be of the line or spot type, but both operate under the same principles.

The line-type rate-of-rise detector consists of metal tubing mounted on the ceiling in a loop configuration. This tubing is connected to a chamber that contains a diaphragm. There is an air chamber on one side of the diaphragm and a contact point on the other. When the air inside the tubing is heated, its pressure increases. This, in turn, brings increased pressure to bear on the air chamber, which forces the diaphragm outward and closes the contact points to initiate a signal. The detector is equipped with a vent hole in the air chamber. If the rate of temperature rise is less than 12°F to 15°F per minute, the vent is capable of maintaining pressure equilibrium in the chamber to prevent the diaphragm from moving. This prevents false alarms. The NFPA recommends that each circuit be limited to no more than 1,000 feet of tubing.

The spot-type rate-of-rise detector also works on a pneumatic principle. The main difference between the two is that the spot-type detector has a single air container instead of tubing.

Combination Heat Detectors

The combination heat detector employs both rate-of-rise and fixed-temperature principles in its design. It has an air chamber and diaphragm, as well as a contact that is held open by a fusible link. The rate-of-rise element gives it the capability to respond swiftly to rapidly developing fires, whereas the fixed-temperature element is more responsive to fires that develop slowly.

SMOKE DETECTORS

The smoke detectors that firefighters most commonly encounter operate on either the ionization principle or the photoelectric principle. **Ionization detectors** respond faster to high-energy or open-flame fires, whereas **photoelectric detectors** respond faster to low-energy or smoldering fires.

The ionization detector contains a small amount of radioactive material that is used to ionize air in a sensing chamber. When this occurs, the air becomes conductive, allowing a small electric current to flow between two charged electrodes. When smoke particles enter the chamber, they reduce the conductance of the air and disrupt the flow of current. This closes a circuit and activates the alarm (see Figure 9.29).

Photoelectric detectors work on either the light-obscuration or light-scattering principle. **Light-obscuration systems** consist of a light source, a light transmitter, and a light-sensitive device. A light beam is directed onto a light-sensitive receiver. When

ionization detectors

Smoke detectors that use a finite amount of radioactive material to make the air within a sensing chamber conduct electricity.

photoelectric detectors

Smoke detectors that use a small light source to detect smoke.

light-obscuration systems

Smoke detection systems that are activated when smoke particles block a light beam from a photoelectric light-sensitive cell.

FIGURE 9.29 ◆ Ionization smoke detector.
Courtesy of James D. Richardson.

smoke particles enter the beam, the light is blocked (or obscured) from reaching the sensing device. This reduction in light causes the initiation of an alarm. This type of system is used to provide protection for large open areas (see Figure 9.30).

Light-scattering systems are usually of the spot type. They consist of a chamber that houses a light source and a light-sensitive receiver. When smoke particles enter the chamber, the light beam becomes scattered. Some of this light is refracted onto the light-sensitive receiving device, and the alarm is activated (see Figure 9.31).

FLAME (LIGHT) DETECTORS

Flame detectors are designed to respond to either ultraviolet or infrared light waves. Because of their rapid detection capabilities, they are used in locations with a high

light-scattering systems

Smoke detection systems that are activated when light beams are scattered by smoke particles and strike a light-sensitive photoelectric cell.

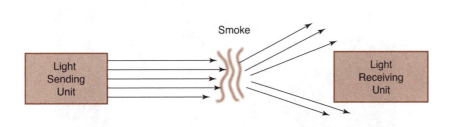

FIGURE 9.30 ◆ Light-obscuration smoke detection systems are used to protect large open areas.
Courtesy of James D. Richardson.

Light Scattering Smoke Detector

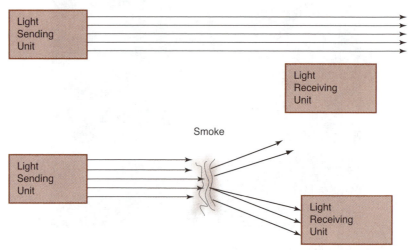

FIGURE 9.31 ◆ Light-scattering smoke detection systems are also used to protect large open areas.
Courtesy of James D. Richardson.

infrared flame detectors

Flame detectors that operate in extreme environmental conditions.

ultraviolet flame detectors

Fire detection devices that are activated when ultraviolet light waves are detected.

combination flame detectors

Alarm devices that can detect abnormal conditions by several means.

hazard potential, such as areas where explosive gases are present or where rapid fire spread might occur. **Infrared flame detectors** are best suited for protecting areas housing hydrocarbons. Because of their sensitivity to solar radiation, they should be sheltered from the sun's rays or provided with infrared filters. **Ultraviolet flame detectors** are sensitive to fires involving hydrocarbons, metals, sulfur, hydrogen, and ammonia. They do not respond to sunlight. **Combination flame detectors** are designed to sense both ultraviolet and infrared light waves (see Figure 9.32).[13]

FIGURE 9.32 ◆ Flame detectors are used in areas where very rapid detection is required.
Courtesy of James D. Richardson.

◆ SPECIAL EXTINGUISHING SYSTEMS

For some kinds of fires, water is not the most effective extinguishing agent. In areas where fires of this kind are apt to break out, special systems must be installed. A great many such systems are in use today and discussing all of them is beyond the scope of this chapter. It is important, however, that firefighters be able to identify the most common of these systems and understand their function and hazards. We will look briefly at systems using halons, carbon dioxide gas, various dry chemicals, and protein- or synthetic-based foams.

Unlike most automatic water-based extinguishing systems, special systems contain a limited amount of extinguishing agent. Because of this limited supply, these systems must be able to extinguish a fire effectively in a short duration of time. Automatic water-based extinguishing systems are rated as successful if they control the fire; special systems are rated successful only if they extinguish the fire.

HALON SYSTEMS

Halogenated hydrocarbon extinguishing agents, or halons, date back to the 19th century. They are derived from a process by which one or more of the hydrogen atoms in a hydrocarbon are replaced by atoms of the halogens fluorine, chlorine, bromine, or iodine. Halons are used to protect areas containing high-valued contents, such as electronic or computer rooms, records and document rooms, museum storerooms, or other materials that would be damaged by other extinguishing agents that leave extinguishing residues. The two most likely halon agents that a firefighter may encounter today are Halon 1211 and Halon 1301. The numbers indicate the number of carbon and halogen atoms in each molecule. The first digit is the number of carbon atoms; the second digit, fluorine; the third digit, chlorine; and the fourth digit, bromine. Halon 1211 is best suited for use in portable extinguishers because it is normally expelled as a liquid. It will, however, rapidly vaporize when it comes in contact with the fire. Halon 1301 is best suited for use in fixed extinguishing systems because it is normally discharged as a vapor. These two agents are classified as clean extinguishing agents because they do not leave any residue after extinguishing a fire, but they are dangerous. Firefighters working with these extinguishers should wear full protective clothing and use SCBA (self-contained breathing apparatus) systems in unoccupied areas or areas that are only temporarily occupied. The system should be equipped with a predischarge alarm to allow occupants time to evacuate before the extinguishing agent is released.

When Halon 1301 is decomposed by heat, the halogens combine with hydrogen formed by the combustion process to create toxic byproducts: hydrogen fluoride, hydrogen bromide, and bromine. Decomposition by-products of Halon 1211 include all of the above, plus hydrogen chloride and chlorine. Halon 1211 is, therefore, the more toxic of the two agents. Toxicity is not the only drawback of halons, however. Because of their harmful effects on the ozone layer, the production of halons was discontinued at the end of 1993. The Montreal Protocol, an international treaty, mandated that all existing halon systems be decommissioned by December 31, 2003. These systems are, however, still being used by the military and could still be encountered by firefighters.

Recommendations for installation, use, and maintenance of these systems are published in NFPA 12A, *Standard on Halon 1301 Fire Extinguishing Systems*, and NFPA 12B, *Standard on Halon 1211 Fire Extinguishing Systems*.

Alternative agents have been developed to replace halons. They are listed under several commercial names. These agents are ozone friendly but pose various levels of toxicity hazards to exposed personnel. Because many of these agents are compatible for use in existing Halon 1301 systems, it is very important that firefighters know which agent is being used when they perform prefire planning.[14]

CARBON DIOXIDE SYSTEMS

Carbon dioxide (CO_2) is an inert gas that is produced naturally in our atmosphere. It is colorless, odorless, nonconductive, and nontoxic. It is also nonflammable and 1½ times heavier than air. These last two characteristics allow CO_2 to displace oxygen-rich air on a burning surface, creating an atmosphere that inhibits combustion. Carbon dioxide has long been used in the commercial arena for refrigeration and in other industrial processes. It was first used as a fire-extinguishing agent in 1914, when the Bell Telephone Company developed the portable CO_2 fire extinguisher to protect their electrical wiring and equipment. Automatic fixed carbon dioxide extinguishing systems were developed in the 1920s, and the National Fire Protection Association published a standard on their proper installation, operation, and maintenance in 1929 (NFPA 12, *Standard on Carbon Dioxide Extinguishing Systems*).

Carbon dioxide systems are classified according to their method of application: total flooding or local application.

Total flooding systems are designed to fill the protected area with carbon dioxide, creating a CO_2-rich atmosphere in which combustion cannot continue. The system consists of a storage tank or cylinder, piping, and nozzles (referred to as *horns*). It can be activated by an automatic fire detection system, by normal manual operation, or by emergency manual operation. If the system is activated by an automatic fire detection system or by normal manual operation, it must be equipped with a predischarge alarm that will alert occupants and give them time to evacuate the area before the system activates. Systems with an emergency manual operation feature must be equipped with a discharge alarm but do not require a predischarge alarm. When this operation system is used, the occupants must evacuate while the system is discharging because no predischarge warning is sounded. Therefore, the emergency manual operation feature should be activated only when other initiating devices fail.

Local application systems are designed to direct the carbon dioxide onto the object being protected instead of flooding the entire area. The system components are the same as in a total flooding system: a storage tank, piping, discharge nozzles, and a fire detection system. When carbon dioxide is discharged in a local application system, dry-ice particles are produced. These particles have the potential to carry a charge of static electricity. To eliminate shock hazards to personnel operating the equipment and to prevent static discharge in a volatile atmosphere, the discharge nozzles must be properly grounded (see Figure 9.33).[15]

There are two ways in which the carbon dioxide can be stored in extinguishing systems. In **high-pressure systems**, the carbon dioxide is stored at room temperature in standard Department of Transportation–approved cylinders at a pressure of approximately 850 pounds per square inch. The maximum capacity of a high-pressure cylinder is 100 pounds of carbon dioxide. If more agent is needed, additional cylinders must be connected to the system through a manifold. In **low-pressure systems**, liquefied carbon dioxide at 0°F is contained in large refrigerated storage tanks pressurized to 300 pounds per square inch. Low-pressure storage tank capacities range from 1¼ to 60 tons. The effectiveness of a CO_2 system depends on the structural integrity of the area being protected. If the area is tight, the CO_2 that is discharged will be retained longer.

total flooding systems

Extinguishing systems that are designed to completely fill the protected area with extinguishing agent when activated.

local application systems

Permanent suppression systems that are required to cover a protected area with 2 feet of foam depth within two minutes of system activation.

high-pressure systems

Carbon dioxide is stored at room temperature in standard Department of Transportation–approved cylinders at a pressure of approximately 850 pounds per square inch.

low-pressure systems

Liquefied carbon dioxide at 0°F is contained in large refrigerated storage tanks pressurized to 300 pounds per square inch.

A

B

FIGURE 9.33 ◆ (A) Total flooding system; (B) local application system. These systems require a preactivation alarm to allow occupants time to evacuate the area.
Courtesy of Ansul Incorporated.

If the area has openings, especially on the sides and bottom, the heavier-than-air carbon dioxide may leak out before effectively extinguishing the fire.

Stop and Think 9.3

Your fire crew has been dispatched to an activation alarm at the local newspaper printing room. What hazards would you anticipate facing on arrival? How can you protect yourself and your crew?

DRY CHEMICAL SYSTEMS

ordinary dry chemical systems

Dry chemical systems designed primarily to extinguish flammable liquid fires (Class B fires).

multipurpose dry chemical systems

Dry chemical systems that use extinguishing agents that are rated to be effective on more than one type of fuel.

Dry chemical systems are characterized as either ordinary or multipurpose. **Ordinary dry chemical systems** are designed primarily to extinguish flammable liquid fires (Class B fires). They may also be used on electrical equipment that is subject to flammable-liquid fires, such as transformers, and on energized electrical fires (Class B fires). **Multipurpose dry chemical systems** are rated for Class A (ordinary combustibles), Class B, and Class C fires. Neither one of these systems should be used in areas such as computer rooms and switchboards where there is electronic or electrical equipment, because dry chemical residue may be harmful to sensitive circuitry.

The basic chemicals used to produce dry chemical agents are sodium bicarbonate, potassium bicarbonate, potassium chloride, urea–potassium bicarbonate, and monoammonium phosphate. Additives, such as metallic stearates, tricalcium phosphate, and silicone, are used to increase flow characteristics, shelf life, and water resistance. These chemicals in various combinations give dry chemical systems four fire-extinguishing properties:

- **Smothering.** When monoammonium phosphate decomposes, it leaves a residue (metaphosphoric acid) on the surface of the fuel. This residue smothers the fire by sealing off oxygen from the fuel.
- **Cooling.** Although this is not a primary extinguishing property of dry chemical agents, it has been proven that these compounds absorb large amounts of heat when they decompose. This adds to the extinguishing ability of the agents, albeit only in a minor way.
- **Radiation Shielding.** When dry chemical agents are discharged, a large dense cloud of powder is generated between the fuel and the flame. This prevents heat from being radiated back to the fuel from the flame, thus aiding in the extinguishing process.
- **Breaking the Chain Reaction Process.** This is the most powerful of the four extinguishing properties of dry chemical agents. When dry chemical agents are exposed to heat, they decompose. Particles from the decomposition process interfere with the chain reaction of combustion, reducing the number of free radicals in the flame. Preventing these reactive particles from coming together breaks the chain reaction needed to continue the combustion process.

Dry chemical agents are nontoxic and nonconductive, but they can irritate the eyes, mucous membranes, and the respiratory system. Some agents may cause chemical burns on moist skin. Personnel must always wear full protective equipment, including SCBA, when entering areas that have been extinguished with dry chemical agents (see Figure 9.34).[16]

Dry chemical systems can be either fixed systems or handheld hose line systems and must meet the requirements of NFPA 17, *Standard for Dry Chemical Extinguishing Systems.*

It should be noted that although dry chemical agents are effective on most flammable liquid fires, they have a low performance rating on cooking oil fires (Class K fires). Wet chemical agents with a high saponification value are the extinguishing agents of choice for these fires. *Saponification* is a process in which the fatty acids in the cooking oil is converted into soap, or foam, which covers the surface area of the liquid. When dry chemical agents are used on cooking oil fires, the soap that is formed tends to break down rapidly. This stops the cooling process, increasing the possibility of a reflash.

FIGURE 9.34 ◆ Dry chemical agents are nontoxic but can be highly irritating.
Courtesy of Ansul Incorporated.

FOAM SYSTEMS

Foam extinguishing systems are especially effective in fighting fires fueled by hydro-carbons (methane, propane, benzene, naphthalene, and various polymers such as polyethylene and polypropylene). The foams work by forming a blanket over the burning fuel, separating it from combustion vapors and oxygen.

Firefighting foams are classified as low-expansion, medium-expansion, and high-expansion. Low-expansion foam has a 20:1 expansion ratio. This means that 1 gallon of foam concentrate will expand to 20 gallons of foam when mixed with water and agitated with air. Medium-expansion foam has an expansion ratio of 20:1 to 200:1, and high-expansion foam expansion ratios range from 200:1 to 1000:1.

Foam concentrates can be synthetic- or protein-based. Synthetic foams are detergent-based and are very effective on liquid hydrocarbon fires. They can be premixed in water storage tanks. An effective foam blanket, with a 20:1 expansion ratio, can be applied with a standard fog nozzle; an aerating nozzle is not required. Synthetic alcohol-resistant foams are also available. Protein foams are made from plant or animal matter. These foams are not often used today.

There are six basic types of foam systems:

- **Self-Contained Apparatus.** Found on fire apparatus, this system has all of the components needed to develop foam: a water storage tank, a foam storage tank, piping, a foam inductor, and the equipment needed to agitate the foam and water mixture.
- **Fixed.** This system is a permanently installed unit. It has a central foam station that delivers foam through piping to a discharge nozzle in the protected area. This unit may be of the total flooding or local application type.
- **Semi-Fixed Type A.** This system has installed piping and nozzles but no permanent source of foam. An apparatus must be connected to the system to pump the foam solution into the piping. This arrangement is similar to the dry standpipe system.

- **Semi-Fixed Type B.** This system provides foam protection by piping the foam solution through piping to foam hydrants located throughout the protected property. This system does not discharge the foam but makes the foam available for use. This can be compared to a fire hydrant on a water distribution system.
- **High-Expansion Foam.** The high-expansion foam system consists of an activation device (automatic or manual), a foam generator, a water supply, a foam supply, and piping. It can be used for local application or total flooding application, but in either case, it must supply enough high-expansion foam to cover the hazard with a minimum of 2 feet of extinguishing agent. High-expansion foam systems are used to protect aircraft hangers, basement areas, and confined spaces where flammable atmospheres or processes may exist. The high expansion ratio (200:1 to 1000:1) allows large areas to be filled with agent with very little water damage.
- **Foam and Water.** This system is a deluge system that discharges foam. The piping contains a foam induction system and aerating sprinkler heads, which are mounted at the end of the piping. This configuration allows the foam to expand five or six times when discharged.
- **Compressed Air Foam (CAFS).** These systems can be fixed or self-contained. They consist of a fire pump, a rotary air compressor, and a foam-proportioning system. The foam is generated by injecting compressed air into the foam solution as it travels through piping. The compressed air injection increases the momentum of the foam solution, allowing the stream to be projected farther than a water stream, which results in better fire penetration. A CAFS uses less foam solution and less water than other types of foam induction systems. This greatly reduces property damage caused by the extinguishing process. The expansion ratios for CAFS-generated foam range from 1:4 to 1:20. However, tests have shown that foam solutions with an expansion ratio exceeding 1:10 are too dry to be very effective in fire suppression.[17]

◆ SUMMARY

This chapter introduced much new terminology and presented a wealth of technical information. Read the text carefully, study the illustrations, and begin to use the proper vocabulary in talking about water systems, hydrants, sprinklers, and extinguishers.

Full mastery of the topics covered in this chapter will come with future in-depth study and with all-important on-the-job experience. If the student has gained an appreciation for the sheer volume of what there is to know and a healthy respect for the importance of this information, the chapter will have served its purpose. Two main points can stand repeating:

- Firefighters must know the firefighting needs of their jurisdiction and be able to identify and understand the equipment (water systems, hydrants, standpipes, alarms, fire detectors, and extinguishing equipment) used in buildings they serve. They must be aware not only of the abilities of each system but also of the dangers these systems may pose for responding personnel.
- Communities and their needs change. When the population of a rural district grows or new industry comes into a community, the fire service needs of the area change. Firefighters must keep abreast of these developments and be vigorously proactive in ensuring that the local water distribution system can handle the increased demand and that updated and expanded fire protection systems are in place before they are needed.

On Scene

The fire chief of a small city bordering San Antonio, Texas, was informed that a developer had been given a permit to construct over 1,000 residential structures on a large parcel of land inside the city limits. It was suggested that this growth would be good for the city's economic growth. The fire chief is responsible for approving any new water distribution systems. Groundwater is the major water source for the city of San Antonio and the 28 municipalities bordering it. Groundwater also supplies several large military installations in the county.

1. What issues face this fire chief?
2. What construction requirements must he or she demand to ensure that an adequate water supply will be maintained now and in the future?
3. What questions must he or she ask the developer and city administration?

Review Questions

1. List and describe the four components of a water distribution system.
2. What methods are used to move water in a distribution system?
3. How are water distribution systems categorized?
4. Describe the differences between a base-valve dry-barrel hydrant and a wet-barrel hydrant.
5. The NFPA requires that Class I and Class III standpipe systems be capable of delivering _____ GPM from the first standpipe and _____ GPM from each additional standpipe.
6. _____ is the standard for the installation of sprinkler systems.

7. List the four types of sprinkler control valves.
8. Which of the four sprinkler control valves operates as a butterfly valve?
9. How are heat detectors categorized?
10. Describe the two methods used to initiate alarms with photoelectric smoke detectors.
11. Describe the hazards involved with the following special extinguishing agents:
    a. Halon 1301
    b. Halon 1211
    c. Carbon dioxide
    d. Dry chemicals

Notes

1. Wieder, M., Smith, Carol, & Brakhage, Cynthia. (1998). *Private fire protection systems.* (2nd ed.). Stillwater, OK: Fire Protection Publications, Oklahoma State University.
2. Ibid.
3. Lamb, Willis. (2001). *Pressurized dry hydrants*. Retrieved April 16, 2008, from Pressurized Dry Hydrants website: http://www.firehydrant.org/info/dryhsys1.html.

4. Schultz, Gerald R. (1997). Water distribution systems. *NFPA.* (18th ed.). Quincy, MA: National Fire Protection Association.
5. Shapiro, Jeffrey M. (1997). Standpipe and hose systems. *NFPA.* (18th ed.). Quincy, MA: National Fire Protection Association. Pp. 6–249
6. Wieder M., Smith, Carol, & Brakhage, Cynthia. (1998). *Private fire protection systems.* (2nd ed.). Stillwater, OK: Fire

Protection Publications, Oklahoma State University.

7. Solomon, Robert E. (1997). Automatic sprinkler systems. *NFPA*. (18th ed.). Quincy, MA: National Fire Protection Association. P. 6-138

8. Isman, Kenneth E. (1997). Automatic sprinklers. *NFPA*. (18th ed.). Quincy, MA: National Fire Protection Association.

9. Wieder, M., Smith, Carol, & Brakhage, Cynthia. (1998). *Private fire protection systems.* (2nd ed.). Stillwater, OK: Fire Protection Publications, Oklahoma State University.

10. Isman, Kenneth E. (1997). Automatic sprinklers. *NFPA*. (18th ed.). Quincy, MA: National Fire Protection Association.

11. Wilson, Dean K. (1997). Fire alarm systems. *NFPA*. (18th ed.). Quincy, MA: National Fire Protection Association. P. 5-10

12. Wieder, M., Smith, Carol, & Brakhage, Cynthia. (1998). *Private fire protection*

systems. (2nd ed.). Stillwater, OK: Fire Protection Publications, Oklahoma State University.

13. Ibid.

14. DiNenno, Philip J. (1997). Direct halon replacement agents and systems. *NFPA*. (18th ed.). Quincy, MA: National Fire Protection Association.

15. Makowka, Norb. (2005). *Why carbon dioxide (CO_2) in fire suppression systems?* National Association of Fire Equipment Distributors. Retrieved from http://www.nafed.org/library/whyco2.cfm.

16. Bell, K., & Zastrow, K. W. (2004). *Pre-engineered chemical systems.* Retrieved from http://www.nafed.org/library/prechem.cfm.

17. Kim, A. K., & Crampton, G. E. (2000). *A new compressed-air-foam technology.* National Research Council of Canada, Halon Options Technical Working Conference, Albuquerque, NM. Washington, DC: National Institute of Standards and Technology, .

■ ■

Suggested Reading

Bachtler, R., & Brennan, T. *The fire chief's handbook.* (5th ed.). Saddle Brook, NJ: Fire Engineering Books and Videos.

Bell, K., & Zastrow, K. W. (2004). *Pre-engineered chemical systems.* Retrieved from http://www.nafed.org/library/prechem.cfm.

Berg, Stephen. (2004). *FM 200 MSDS.* Retrieved from http://www.h3r.com/halon/FM200.pdf.

A brief history of the hydrant. (2003). Retrieved from http://www.firehydrant.org/pictures/hydrant_history.html.

Browmann, Mark. (2004). Fire protection: Usage of common standpipes. *PM Engineer.* Issue 4/04. Retrieved from http://www.pmengineer.com/CDA/ArticleInformation/features/BNP_Features_Item/0,273.

Bush, Loren S., & McLaughlin, James H. (1979). *Introduction to fire science.*

(2nd ed.). Encino, CA: Glencoe Publishing.

Casey, James F. (1970). *Fire service hydraulics.* (2nd ed.). New York: Reuben H. Donnelley Corp.

Colletti, Dominic. (1998). *Class A foam—Best practice for structure firefighters.* (1st ed.). Royersford, PA: Lyon's Publishing.

DiNenno, Philip J. (1997). Direct halon replacement agents and systems. *NFPA*. (18th ed.). Quincy, MA: National Fire Protection Association.

Dry fire hydrants. (2004). Ashtabula Soil and Water Conservation District. Retrieved from http://www2.suite224.net/~ashtswed/dry_dry_fire_hydrants.htm.

The dry hydrant program. (2004). Retrieved from http://www.rtis.com/reg/big8/DryFireHydrant.htm.

Grant, Casey C. (2005). The birth of NFPA. *NFPA Journal.* Retrieved from http://www.nfpa.org/itemDetail. asp?categoryID = 500&itemID = 18020&url = About%20Us/.

Hague, David R. (1997). Dry chemical agents and application systems. *NFPA*. (18th ed.). Quincy, MA: National Fire Protection Association.

History of fire sprinkler systems. (2004). Fire Protection Group. Retrieved from http://www.apifiregroup.com/ firesprinkler/sprinkler-history.html.

Isman, Kenneth E. (1997). *Automatic sprinklers.* (18th ed.). Quincy, MA: National Fire Protection Association.

Kim, A. K., & Crampton, G. P. (2000). *A new compressed-air-foam technology*. National Research Council of Canada, Halon Options Technical Working Conference, Albuquerque, NM. Washington, DC: National Institute of Standards and Technology.

Lamm, Willis. (2001). *History of California fire hydrants*. Retrieved from http://www. firehydrant.org/info/history1.html.

Lamm, Willis. (2001). *Pressurized dry hydrant systems.* Retrieved from http://www.firehydrant.org/info/ dryhsys1.html.

Makowka, Norb. (2005). *Why carbon dioxide (CO$_2$) in fire suppression systems?* National Association of Fire Equipment Distributors. Retrieved from http:// www.nafed.org/library/whyco2.cfm.

Scheffey, Joseph L. (1997) Foam extinguishing agents and systems. *NFPA*.(18th ed.). Quincy, MA: National Fire Protection Association.

Schultz, Gerald R. (1997). Water distribution systems. *NFPA*.(18th ed.). Quincy, MA: National Fire Protection Association.

Shapiro, Jeffrey M. (1997). Standpipe and hose systems. *NFPA*. (18th ed.). Quincy, MA: National Fire Protection Association.

Taylor, Gary M. (1997). Halogenated agents and systems. *NFPA*. (18th ed.). Quincy, MA: National Fire Protection Association.

Weldon, Corey. (May 2004). Commercially available clean agents in the U.S.A. *International Fire Protection Magazine*. Retrieved from http://www.ifpmag.com.

Wickham, Robert T. *Development of halon 1301 & 1211 status of industry efforts to replace halon fire extinguishing agents*. Retrieved from http://rtwickham.home. comcast.net/images/wickham-halon-status.

Wysocki, Thomas J. (1997). Carbon dioxide and application systems. *NFPA*. (18th ed.). Quincy, MA: National Fire Protection Association.

Wieder, M., Smith, Carol, & Brakhage, Cynthia. (1998). *Private fire protection systems.* (2nd ed.). Stillwater, OK: Fire Protection Publications, Oklahoma State University.

CHAPTER 10 # Organizational Structure and Emergency Incident Management Systems

Key Terms

compensation and claims
   unit, p. 214
cost unit, p. 214
Federal Emergency
   Management Agency
   (FEMA), p. 214
fire ground command
   system (FGC), p. 211
FIRESCOPE, p. 211
incident command
   system (ICS), p. 209

incident
   effectiveness, p. 207
Incident Management
   System (IMS), p. 212
line operation, p. 210
National Fire Service
   Incident Management
   System (NFSIMS), p. 215
National Incident
   Management System
   (NIMS), p. 211

National Interagency
   Incident Management
   System (NIIMS), p. 215
paramilitary
   organization, p. 207
procurement unit, p. 214
public safety
   department, p. 208
span of control, p. 219
staff operation, p. 210
time unit, p. 214
unity of command, p. 219

Objectives

After reading this chapter, you should be able to:

- Describe the principles of command used in fire and emergency services.
- Describe a fire department's chain of command.
- Identify the different ranks and their general responsibilities.
- List and describe the four types of fire departments.
- Explain the history of the incident command system.
- Describe the differences among the FIRESCOPE, FGC, IMS, NIIMS, and NIMS ICS structures.
- Identify the components of the incident command structure.
- Describe the areas of responsibility of the ICS components.
- Explain why a national incident management system is needed.

To work effectively in a fire department, personnel need to understand how the department is organized. Who reports to whom? Who is responsible for what? How does each member fit in? This chapter describes the different kinds of fire departments, how they are structured, and how structure affects function. We look at the day-to-day operation of the department as well as how it operates in an emergency. We will also trace the development of the Incident Management System (IMS), which culminated in the Department of Homeland Security's National Incident Management System (NIMS).

◆ **FIRE DEPARTMENT BASICS**

In a certain sense, a fire department is like any other organization of people working together to achieve a common purpose. A critical difference, however, is the urgency and importance of the fire department's mission. When human lives are on the line, there can be zero tolerance for mistakes and misunderstandings. In an emergency situation, priorities must be set, objectives established, and an action plan initiated without debate and without delay. Lines of command need to be clear, and communications among all parties must be kept open at all times. To achieve this, fire departments and emergency services typically adopt a structure similar to that of a **paramilitary organization**. This organizational style is hierarchical. Personnel usually wear uniforms and carry insignia denoting their rank. It is very clear who makes the decisions and who gives the orders. Keep in mind, however, the importance of rank does not outweigh the importance of teamwork. When teamwork is compromised, injuries and death take a tragic toll.

paramilitary organization
A group of civilians organized in a military fashion.

TYPES OF FIRE DEPARTMENTS

There are four basic types of fire departments in the United States: volunteer, combination, public safety, and full-time paid. Industrial fire brigades are a fifth specialized type. We will devote a brief discussion to the features that differentiate these department types. All, however, have certain features in common. Almost without exception, they adhere to the one-department model. Standard equipment is supplied by the department to all employees, and all employees follow the same guidelines. All personnel have in common the desire for **incident effectiveness**. This refers to the ability of the department's personnel to perform their functions safely and effectively during emergency situations. It demands that physical fitness, skills training, and equipment training are department priorities. Finally, all departments share a commitment to that most important ingredient in the formula for success: team building.

incident effectiveness
The ability of the department's personnel to perform their functions safely and effectively during an emergency. This requires physical fitness, training, and equipment.

Volunteer Fire Departments

In 2004, the U.S. Fire Administration (USFA) began counting the number of fire departments in the United States. At that time, there were over 30,400—probably more by the time you are reading this. Of these departments, 90% are volunteer or mostly volunteer, and they protect 39% of the nation's population.[1]

Volunteer fire departments tend to be small, ranging from 15 to 50 members, and generally serve either small towns or rural areas. There are exceptions to this rule. For example, Pasadena, Texas, has a population of 125,000 residents. The volunteer fire department has a force of 200 personnel. In the United States, the members of the first fire departments were all volunteers, and volunteer departments clearly are still a major factor in public safety and a focus of life in the communities they serve. Because the firefighters are not paid, these departments are able to control costs and can use all of the money they bring in from donations and fundraising events to purchase equipment and fund training opportunities. The National Fire Academy (NFA) has numerous training programs designed for volunteer firefighters.

Combination Fire Departments

The combination department has some paid employees and some volunteers. The demographics of rural communities are changing. The quaint, once-quiet townships are becoming busy satellite communities to the urban areas they surround. With increased population and business growth, they find themselves unable to provide adequate fire protection with an all-volunteer department. However, they also do not have the adequate revenues to fund a fully paid fire department. Many of these communities have chosen to form combination fire departments. Frequently, in combination departments, the officers, drivers, and apparatus operators are paid, whereas the line firefighters are volunteers. This arrangement obviously results in cost savings. Personnel costs are kept low, but trained professionals are on hand when a situation calls for experienced leadership or some special kind of expertise. The volunteers in these departments are sometimes referred to as reserves. The reserves receive the same training and experience as the fully paid personnel.

Public Safety Departments

public safety department

A department that employs a person who is triple-certified as a police officer, emergency medical technician (Basic or Advanced level), and firefighter.

In some jurisdictions, a **public safety department** takes the place of separate fire, police, and emergency medical technician (EMT) departments. A public safety department officer, often referred to as a *public safety officer* (PSO), must be triple-certified as a police officer, EMT (Basic or Advanced level), and firefighter. Public safety departments are most often found in communities that experience seasonal population fluctuations, such as winter ski and summer lake recreational areas. Public safety departments save money in personnel costs, but they also have their disadvantages. The same employee can handle all three types of calls, but he or she cannot be expected to have the same level of skills as the professional who specializes in only one of these highly demanding jobs.

Fully Paid Fire Departments

Fully paid fire departments are the rule in large metropolitan areas where service demands require full-time professionals. The jurisdiction normally has more control over these employees than in other kinds of fire departments. Expert management is a must, and prerequisites for employment at all levels tend to be high. Generally, firefighting experience is not enough. College hours, advanced certifications, and even officer experience or advanced degrees may be required.

Specialized Fire Brigades

A specialized fire brigade is a private fire department organized by a business, usually an industrial concern, such as an oil refinery or a manufacturing plant. These fire brigades consist of personnel hired and paid by the company.

ORGANIZATION AND RANK STRUCTURE

Fire departments are often divided into subunits. The number and type of units depend on the size of the fire department and the potential hazards in the community.

Fire Company

The fire company is the smallest unit in the organizational structure. It consists of a fire apparatus and crew and is named for the type of apparatus, such as engine company, ladder company, or rescue company. The company is usually composed of four personnel: an officer with a rank of lieutenant or captain, an engineer/ operator, and two firefighters. This unit is responsible for responding to emergencies and performing initial operations as directed by the incident commander. A fire station may house several fire companies, such as an engine company and a ladder company. These companies respond together in an emergency. In areas where fire departments are small, companies from several departments may join forces to respond to a multiple-alarm incident.

Battalion or District

The battalion or district organizational unit is found in large cities and communities that have many fire stations. This type of unit is usually comprises multiple fire stations and is supervised by a battalion or district chief. Line battalion chiefs are responsible for personnel administration in their respective district and assume the role of incident commander at emergency scenes within the battalion's area of responsibility.[2] Staff battalion or district chiefs are strictly administrative officers, who concentrate on planning, logistics, finance, and other support functions.

RANK STRUCTURE

To further break down these fire department units, a rank structure has been established. Each rank has assigned duties and areas of responsibility. In an emergency, their roles are coordinated by an overarching organizational structure referred to as the **incident command system (ICS)**. Incident command systems will be described later in this chapter. For now, let's look at the functions of the different ranks, from line firefighter up to district chief.

incident command system (ICS)

A blueprint for organizing the response to an emergency situation.

Firefighter

The firefighter is responsible for performing all physical tasks associated with emergency scene operations. This is the entry-level fire service position. Some fire departments have a probationary firefighter rank for new members. Probationers must achieve certain levels of proficiency during a designated time period that ranges from three months to two years. On successful completion of the probationary period, they are granted the rank of firefighter.

Driver

A fire company's drivers are responsible for delivering personnel and equipment to the emergency scene and returning them safely to the fire station. They must be proficient in apparatus operations, such as running pumps and aerial equipment. They are also responsible for ensuring the operational readiness of the apparatus, for inventory control, and for keeping track of tools and equipment taken off apparatus during emergency operations. Drivers may be referred to as fire apparatus operators (FAOs), engineer/operators, or chauffeurs. In some fire departments, this position is a promotional rank from firefighter. In other departments, being a driver is considered part of the firefighter's normal duties.

Lieutenant

The lieutenant is the lowest of the supervisory ranks. He or she is usually in charge of an engine or rescue company, although specific duties may vary based on departmental policy. If a lieutenant is the first officer arriving at the emergency scene, the lieutenant assumes command and implements the tactics and strategies deemed necessary to conduct a successful operation. The lieutenant will remain in command until relieved by an officer of higher rank.

Captain

The captain is the second level in the supervisory hierarchy. When a fire station houses several fire companies (each headed by a lieutenant), the captain is considered the station or shift officer in overall charge. The captain is responsible for supervising all officers and firefighters under his or her command during emergency and nonemergency operations. Captains must also be experienced and knowledgeable in tactical and strategic planning. They, like lieutenants, assume command of an emergency scene until a higher-ranking officer arrives.

Battalion or District Chief

line operation

Those activities that involve the direct physical tasks necessary for the completion of the mission.

The position of battalion or district chief may be in either the **line operation** or **staff operation** of a fire service department. The line battalion or district chief supervises multiple fire stations and their assigned companies. Line battalion or district chiefs are responsible for ensuring that the personnel under their command are adequately trained and equipped to respond to any emergency. Staff battalion or district chiefs are strictly administrative officers, who concentrate on planning, logistics, and finance.

Deputy or Assistant Chief

staff operation

The administrative activities necessary to accomplish day-to-day operations.

In larger departments, the rank of deputy or assistant chief is predominantly a staff position. However, some fire departments have line deputy chiefs, who supervise several battalions and respond to large, multiple-alarm fires. The deputy chief usually has the discretion either to assume command from the battalion chief or simply to observe the operation.[3] The organizational structures of fire departments vary. In some geographic areas, the deputy chief and assistant chief are separate ranks. In the Houston Fire Department, the deputy chief is the line chief over the districts (battalions) and district chiefs (battalion chiefs). The assistant chief is one rank below the fire chief. In North Carolina, the assistant chief supervises the battalions and battalion chiefs, and the deputy chief is subordinate to the fire chief. Individuals seeking a profession in the fire service should become acquainted with the organizational structures of the departments in their geographic area.

◆ INCIDENT COMMAND SYSTEMS

An incident command system (ICS) is a kind of blueprint for organizing the response to an emergency situation. It takes into consideration all aspects of the response and attempts to integrate them into one smoothly operating, efficient, and effective whole. Facilities, equipment, personnel, communications, and standard operating procedures are all part of this well-oiled machine. Over the years, ICS development has evolved until it has become a field of study in its own right.

We will look first at the events that led to the development of the incident command system. Then we will consider the later stages of development, which culminated in the **National Incident Management System (NIMS)**, a federally mandated program that is crucial to our nation's homeland security strategy. We will also consider aspects of incident command system theory, its development, and its effect on firefighting incident management.

FIRESCOPE

The concept of an incident command system was born in the early 1970s in the wake of a disastrous wildfire in California that resulted in the loss of 16 lives and 700 structures. The major problems identified by postfire analysis were ineffective communications, lack of a common command structure, lack of accountability, and inability to coordinate available resources. Clearly, better and more centralized leadership was needed during large-scale wildfire operations.

Firefighting Resources of California Organized for Potential Emergencies (FIRESCOPE) was designed to answer this need. Under the **FIRESCOPE** model, a single incident commander would direct and coordinate multiple agency responses during wildland firefighting operations. FIRESCOPE pinpointed four requirements for the success of such a plan:

- ◆ The organizational structure must be flexible to adapt to any size or type of incident.
- ◆ It must be able to be used for routine events as well as major incidents.
- ◆ The incident command system must be standardized to allow the addition of personnel from different agencies and geographic locations into the operation without disruption to the organizational structure.
- ◆ The system must be economically feasible.[4]

This system was soon found to be effective in an all-hazard environment. FIRESCOPE's flexibility allowed it to be utilized in any emergency situation, regardless of the size and complexity of the operation or hazardous situation. FIRESCOPE established common terminology and procedures, defined job responsibilities, and developed an organizational structure. Although the system was created for wildland firefighting, it was applied by many fire departments throughout Southern California for structure fires and other emergencies.

In 1982, the FIRESCOPE incident command system was revised and renamed the National Interagency Incident Management System (NIIMS) and adopted by the National Fire Academy for its incident command training programs.[5]

One of the most enduring features of the FIRESCOPE incident command system was its division of incident management into five areas of responsibility: command, operations, planning, logistics, and finance/administration. With only one addition (information and intelligence), these remain the basic command functions in today's incident management systems. We will look at this command structure later in this chapter.

FIRE GROUND COMMAND SYSTEM (FGC)

During the same time that FIRESCOPE was being developed, Chief Alan Brunacini of the Phoenix Fire Department was working on the **fire ground command (FGC) system**. FGC focused on urban fire issues: structural firefighting, hazardous materials emergencies, and mass casualty incidents. It was a simpler organizational model than

National Incident Management System (NIMS)

A reconstruction of NIIMS to accommodate additional needs arising from the nature of the new threat—terrorism—that the nation now faces.

FIRESCOPE

Firefighting Resources of California Organized for Potential Emergencies

fire ground command system (FGC)

An incident management system developed for fire ground and emergency medical operations.

FIRESCOPE, but it had the same central goal: allowing one incident commander to maintain control of fire ground operations. FGC was adopted by the National Fire Protection Association as its model for fire ground organization and control.

FIRESCOPE VERSUS THE FGC

FIRESCOPE, adopted by the NFA, and the FGC, adopted by the NFPA, had much in common, but there also were differences. The most critical of these concerned the command structure. The FGC model was less formal and less rigid than that of FIRESCOPE. These differences created problems when jurisdictions working together on an incident were using different models. A common incident command structure was needed that would be accepted—and used—by all emergency responders. The pros and cons of the two systems were debated throughout the 1980s. In 1982, the Incident Management System was introduced for national acceptance and was widely adopted.

◆ INCIDENT MANAGEMENT SYSTEM (IMS)

Incident Management System (IMS)

A merger of FIRESCOPE'S organizational design and command structure with the tactical and procedural components of the Fire Ground Command system.

The **Incident Management System (IMS)** merged FIRESCOPE'S organizational design and command structure with the tactical and procedural components of the FGC. A set of operational protocols was established for structural firefighting, emergency medical incidents, structural collapses, hazardous materials emergencies, wildland interface fires (wildland fires that spread into urban areas), high-rise fires, and highway incidents. Central to the effectiveness of the system were three basic features:

- ◆ It established a common vocabulary to ensure effective lines of communication.
- ◆ It established clear lines of authority and span-of-control standards that would keep supervisors from being overwhelmed. A line officer could be in charge of no more than three to seven subordinates, and a staff supervisor could command three to seven companies.
- ◆ As mentioned previously, the IMS adopted FIRESCOPE's command structure, clearly outlining five functional sectors. This command structure is the bedrock on which all fire and rescue incident management systems rest.[6]

FUNCTIONAL SECTORS OF THE IMS COMMAND SYSTEM

The IMS protocol for managing any incident calls for responsibility to be divided into five functional sectors: command, operations, planning, logistics, and administration/finance.

Command Sector

The command sector consists of the incident commander and his or her command staff. The commander manages the incident at the strategic level and has the authority to control resources, establish functional areas, and ensure the safety of all responding personnel. Command is an executive function and should remain an executive function, no matter how small the incident may be. The commander is responsible for organizing the incident response to ensure that the mission is accomplished successfully. This includes establishing incident control objectives and work priorities, approving action plan development and resource allocation, coordinating with interagency and public officials, and approving all information given to the media.[7]

The command staff consists of several positions. A public information officer (PIO) is responsible for providing accurate information to the news media, outside agencies, and individuals who have been affected by the incident. The PIO is also responsible for providing information updates to the incident commander, incoming crews, and other organizations that may be involved in the operation.[8] A safety officer is responsible for the safety of all responding personnel. The safety officer has the authority to stop operations if responders are exposed to an immediate danger. The liaison officer maintains communication and contact among responding agencies.[9] The liaison officer also meets and greets any individual or agency representative reporting to the command post; gives directions to individuals and agencies, forwards messages from the commander, and assigns outside agencies to duties that will best meet the needs of the situation.[10]

Operations Sector

The purpose of the operations functional area is to ensure that tactical assignments are carried out and to accomplish the strategic goals established by the incident commander. This functional area can become very complex, especially if the incident is large or multifaceted. The operations chief may be responsible for supervising simultaneous branch operations in fire, emergency medical services (EMS), hazardous materials, search and rescue, recovery, and so on. Each branch supervisor reports directly to the operations chief, who reports directly to the incident commander. The operations chief usually comes from the agency or organization with the most jurisdictional involvement in the incident.[11]

Planning Sector

The planning sector is responsible for evaluating incident information and developing the incident action plan. Its function is to provide the incident commander with up-to-date status reports on resource requirements and allocation, situation development, incident control, responder injuries, intelligence, and suggestions for stabilizing the incident.[12] This sector consists of four units: resources, situation, demobilization, and documentation. Technical specialists are utilized to interpret data, evaluate situational development, and forecast resource needs. Table 10.1 defines typical planning responsibilities.

TABLE 10.1 ◆ Planning Responsibilities	
Incident Planning	◆ Works with command and operations to update the strategic plan ◆ Assesses previous actions and strategies that have been used ◆ Makes adjustments to current and future plans and makes recommendations to command and operations ◆ Predicts the outcomes of the plans ◆ Estimates and evaluates possible resource needs with operations ◆ Reviews projected tasks established by command and operations
Resource Assessment	◆ Is responsible for personnel ◆ Is accountable for responding units and assignments ◆ Keeps track of special equipment
Situation Status	◆ Continually monitors events to maintain an up-to-date status of incident situations
Documentation	◆ Maintains a master sequence of events timeline and resource requests ◆ Monitors operations to maintain span of control
Demobilization	◆ Develops a plan to dismantle the organizational structure and return all apparatus, equipment, personnel, and other resources to preincident status after the event is over[13]

Logistics Sector

The logistics sector provides all of the outside resources and materials required for incident operations. This includes facilities (mass care, rehabilitation, family assistance centers, warehouses, and so on), transportation, supplies, equipment maintenance, fuel, food, communications and technological support, and EMS support services. The logistics officer usually assigns these requisition duties to a support and service section, collecting data from them, and then providing information to the incident commander.[14] The specific responsibilities of the logistics officer are defined in the following box.

Responsibilities of Logistics Officer

- Locates and provides supplies and equipment
- Inventories all arriving equipment and notifies command of available resources
- Provides all support activities to all units in operations
- Provides critical incident stress debriefings
- Provides adequate and continual communications throughout incident operations[15]

Federal Emergency Management Agency (FEMA)

The federal agency responsible for supporting state and local municipalities during a disaster.

compensation and claims unit

Part of the finance branch of ICS that documents property loss by citizens during a disaster.

procurement unit

Part of the finance branch that establishes contracts for equipment that may be needed during an emergency.

cost unit

Part of the finance unit that tracks the cost of equipment and supplies.

time unit

Department that tracks all time sheets for every individual working on the emergency scene throughout the duration of the incident.

ADMINISTRATION AND FINANCE SECTOR

The expense of a managing a large, long-term incident can be staggering. For example, the cost of the 1970 California wildfire that led to the creation of the FIRESCOPE project ran to $18 million per day. Today, a large-scale operation can carry an even higher price tag. The finance and administration officer is responsible for documenting the cost of materials and personnel. This is very important, especially if the incident is declared a national disaster. If a national disaster declaration is issued, the **Federal Emergency Management Agency (FEMA)** will reimburse up to 75% of the incident's operational cost to the community. This reimbursement will occur only if all of the operational expenses are properly documented and verified.

The finance and administration sector includes four subunits. The **compensation and claims unit** documents all citizens' claims of damage associated with conducting the operation. The **procurement unit** contracts for equipment that may be needed during the operation. This is usually accomplished during the development of the strategic plan prior to the incident. The **cost unit** keeps a record of expenses for equipment and supplies used during the incident. The **time unit** keeps time sheets for all workers (including volunteers) employed throughout the operation (see Figure 10.1).[16]

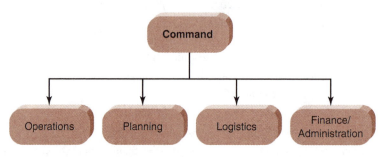

FIGURE 10.1 ◆ The basic incident command structure is divided into five areas of responsibility.
Courtesy of the U.S. Department of Homeland Security.

Stop and Think 10.1

You are assigned to be sector officer for finance and administration during a major disaster in your community. The incident commander has informed you that there will be a thorough audit after the incident to ensure accountability.

1. What do you do to ensure that all expenditures are properly documented?
2. Why do these records have to be maintained so precisely?
3. What repercussions would result if the financial records were not kept?

NATIONAL INTERAGENCY INCIDENT MANAGEMENT SYSTEM (NIIMS)

The **National Interagency Incident Management System (NIIMS)**, introduced in 1982, expanded on the IMS concept. It is a total systems approach to risk management that integrates all levels of national emergency incident planning—federal, state, and local. The NIIMS established five subsectors, each of which would tackle an area where standardization was vital. The goal was to develop and implement national standards in these five areas:

- The incident command system
- Training
- Qualifications and certification systems
- Publications
- Support technologies (satellite imagery, communications, geographic information systems [GIS], and so on.[17]

National Interagency Incident Management System (NIIMS)

A total systems approach to risk management that integrates all levels of national emergency incident planning—federal, state, and local.

NATIONAL FIRE SERVICE INCIDENT MANAGEMENT SYSTEM CONSORTIUM (NFSIMSC)

In 1990, the **National Fire Service Incident Management System Consortium (NFSIMSC)**—a working group of federal, state, and local agencies—was formed. It is a nonprofit corporation with the mission of sorting out issues and ironing out difficulties that arise as the national incident management effort unfolds.[18]

National Fire Service Incident Management System Consortium (NFSIMSC)

A nonprofit corporation with the mission of sorting out issues and ironing out difficulties that arise as the national incident management effort unfolds.

INCIDENT COMMAND SYSTEMS—A REPORT CARD

In February 2000, Assistant Chief Dana Cole of the California Department of Forestry and Fire Protection submitted a research paper to the National Fire Academy. In his paper, entitled *The Incident Command System: A 25-Year Evaluation By California Practitioners,* Chief Cole listed 16 attributes of ICS. Former and present command and general staff officers of 17 state and federal major incident management teams in California were asked to rate the attributes based on their experience of how effective they were in actual practice. Based on these ratings, Chief Cole divided the attributes into three tiers, as follows:

Tier One. These are the most effective attributes, which is not surprising because they are predetermined. The participants know the system and understand the rules that must be followed. Everyone has received training on how the system works. These are as follows:

- Predefined hierarchy, including chain of command and delineated responsibilities for every position

- Uniform terminology for identifying resources and organizational functions
- Modular organizational structure that is expanded and contracted as needed
- Incident action plans that are updated for each operational period
- Manageable span of control

Tier Two. The next six attributes were rated lower. They contribute to the effectiveness of ICS but create some complications as well. These are as follows:

- Standardized forms used for all incidents
- Ample flexibility and authority given to staff for accomplishing objectives
- Cross-jurisdictional and cross-functional working relationships when ICS is used
- Communications plan that is coordinated among responding agencies
- Clear decision-making process
- Process for transitioning command authority from one level of government to another as incident complexity changes

Tier Three. The last five attributes were rated lowest by respondents. Although still perceived positively, they were also viewed as potential problem areas. These are as follows:

- Resource mobilization effectiveness
- Effectiveness of integrating nonfire government agencies into the ICS structure
- Consistency of implementation among various agencies
- Effectiveness of integrating nongovernment organizations into the ICS structure
- Agreement among agencies about who has authority to modify the ICS "rules of the game"[19]

Chief Cole emphasized that strengthening these last five attributes should be the focus of efforts to develop national incident command systems further. The effectiveness of ICS when implemented during large-scale operations cannot be denied, but work is still to be done.

NATIONAL INCIDENT MANAGEMENT SYSTEM

If there was ever any question about the need for a nationwide incident management system, the disastrous events of September 11, 2001, removed all doubt. Emergency operations demand multiagency or multijurisdictional responses, common terminology, and common communications. A system needed to be developed and implemented with the full weight of law to back it. This task was given to the newly created Department of Homeland Security.

Fortunately, as we have seen, much of the framework for such a system was already in place. NIIMS was restructured to accommodate additional needs arising from the nature of the new threat—terrorism—that the nation now faced. The new system was christened the National Incident Management System (NIMS). On March 1, 2004, Secretary of Homeland Security Tom Ridge mandated that all agencies must adopt NIMS in order to receive federal preparedness funds.[20]

The organizational structure of NIMS comprises six components:

- Command and management
- Preparedness
- Resource management
- Communications and information management
- Supporting technologies
- Management and maintenance

The operating principles of NIIMS and NIMS are similar, but with terrorism in the forefront of its concerns, NIMS understandably places heavier emphasis on prevention and preparedness. NIMS retains the five functional areas of older IMS systems (command, operations, planning, logistics, and finance and administration) but adds information and intelligence management as a sixth function.

As we review the principles behind NIMS, it will become clear how much this system owes to earlier incident command systems.

Management by Objectives

The incident commander establishes specific, measurable objectives for each functional area. All measurable performance is documented to correct deficiencies.

Reliance on an Incident Action Plan

The incident action plan (strategic action plan) guides the objectives for both operational and support activities. It is the foundation that supports tactical decisions and operations.

Manageable Span of Control

An incident management supervisor can supervise three to seven subordinates effectively. It has been suggested that five subordinates are the ideal number to maintain a manageable span of control.

Predesignated Incident Mobilization Center Locations and Facilities

Certain types of incidents require specialized support facilities. The requirements of these facilities should be designated in the incident action plan. The incident commander is responsible for the identification and location of these facilities, based on the situation. Examples of predesignated facilities are the incident command post, rehabilitation area, apparatus staging area, hazardous materials decontamination area, and so on.

Comprehensive Resource Management

Accurate and continual accountability must be maintained for all personnel, equipment, and resources. Resource management is responsible for organizing, requesting, delivering, tracking, and recovering all resources used during the emergency. This should include paid and volunteer personnel, equipment that has been contracted for use, and volunteer organizations.[21]

Integrated Communication

Interoperable communications and a common communications plan are an absolute necessity to maintain operational continuity during multiagency or multijurisdiction response incidents. This is accomplished by developing an emergency operations plan, with cooperation from all of the agencies that may participate during an emergency. Field exercises focusing on interagency communications can be conducted to determine weaknesses in the plan.

Establishment and Transfer of Command

Development of the incident command structure during strategic planning must include an established procedure for the transfer of command. If command is transferred, the new commander must be thoroughly briefed on past and present actions for the command process to continue safely and effectively.

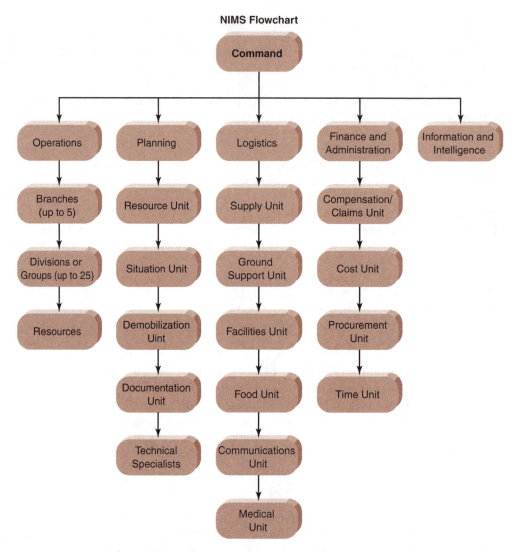

FIGURE 10.2 ◆ The basic NIMS command structure is divided into six areas of responsibility. Courtesy of the U.S. Department of Homeland Security.

Chain of Command and Unity of Command

The purpose of the incident command structure is to establish a line of authority to ensure accurate and efficient communications and information transfer. All participants must have designated supervisors to whom they report. These supervisors, in turn, are to report to their supervisor or directly to the incident commander.

Unified Command

Incidents involving multijurisdictions or multiagency response may require implementation of a unified command. This type of command allows different agencies to work together without compromising individual agency authority, responsibility, or accountability. The unified command structure is dependent on the incident's location and type.

Accountability of Resources and Personnel Deployment

Each level of the incident command structure requires implementation of an accountability system. All responding personnel must report in and be given a specific work assignment. Once assigned, they will report to only one supervisor **(unity of command)**, and that supervisor will be assigned a limited number of subordinates to supervise (**span of control**). The arrival and distribution of resources must be recorded, and updated status reports must be provided to the incident commander

Information and Intelligence Management

The effective and efficient management of an incident requires the gathering of accurate and timely information and intelligence. When the information is received, it must be shared with the appropriate agencies or authority. This function has been added to the command structure as a result of NIMS. The information and intelligence sector analyzes and then shares the information with others on a need-to-know basis (see Figure 10.2).[22]

unity of command

The concept that every member of the organization answers to just one supervisor.

span of control

The number of units or personnel that a supervisor can effectively manage.

Stop and Think 10.2

Consider the worst-case scenario in your area of response.

1. If you were responsible for developing the command structure, what part of NIMS would you consider using?
2. How will NIMS help you in this scenario?

◆ **SUMMARY**

The need for an incident command system (ICS) was inspired by forward-thinking emergency response officers who had witnessed the tragic consequences of chaotic large-scale operations. The evolution of ICS was slow and difficult. FIRESCOPE's ICS was developed for the management of large wildland fires; FGC was a less formal ICS that focused on structure fires, hazardous materials, and mass casualty incidents. Each concept was supported by different national organizations: FIRESCOPE by the National Fire Academy, and FGC by the National Fire Protection Association. It soon became evident that a national incident management system could not be developed without the full cooperation of all agencies responsible for public protection. After years of negotiations, all of the participants were able to resolve their differences and create the incident management system (IMS) that became the foundation for NIIMS. After 9/11, components of NIIMS were modified to create NIMS.

NIMS has now been mandated by the Department of Homeland Security to be used for all incidents by all agencies receiving federal grant funding from the Department of Homeland Security. Of course, funding should not be the only reason for using NIMS. NIMS provides for unity of command, span of control, effective communications, and a whole host of other benefits. It is the best possible approach for emergency workers to bring their mission to a successful completion in a timely, cost-effective manner. For the incident commander, the chief value of NIMS is not the time or money savings, however. It is the flexibility and control that NIMS provides to him or her to protect the lives of emergency responders and civilians.

■■■

On Scene

On the morning of June 24, 1995, companies of the Houston Fire Department responded to a warehouse complex on the east side of the city. The 175,000-square-foot burning warehouse contained 1,000-pound boxes of plastics, hundreds of oil drums, and 250 different types of chemicals. One of the exposed warehouses contained hydrazine, pesticides, and organic peroxides. A fully loaded liquid petroleum gas tank car, in close proximity to the burning warehouse, was parked on the railroad track separating a residential subdivision from the warehouse complex. The only access road to the complex and the initial water supply were inadequate. Additional alarms were requested to relay water supplies and provide additional personnel and handlines. By 12:30 PM, seven alarms had been requested. During the operation, oil-laden runoff water flowed across a parking lot and ignited, destroying an aerial ladder truck.

1. If you were incident commander, what parts of NIMS would you utilize?
2. Would this be an interagency operation?
3. If so, which agencies would participate?

■■■

Review Questions

1. Create a simple fire department organizational chart with at least five levels of rank represented.
2. Give an example of division of labor in the structure of the fire department.
3. Why was the incident command system developed?

4. What is the main difference among FIRESCOPE, ICS, and FGC?
5. What are the main components of ICS?
6. What are the areas of responsibility of the ICS components?
7. Why is a national incident management system needed?

■■■

Notes

1. U.S. Fire Administration. (December 28, 2006). *Fire statistics*. Retrieved February 7, 2007, from http://www.dhs.gov/statistics/departments/.
2. Calfee, Mica. (2004). *Fire department history, terminology, and tactics*. Retrieved February 7, 2007, from Captainmica.com website: http://captainmica.com.
3. Ibid.
4. U.S. Coast Guard. ICS introduction. *History of ICS development*. Retrieved March 4, 2006, from http://www.auxetrain.org/icsintro100.htm.

5. Coleman, John. (1997). *Incident management for the street-smart fire officer*. Saddle Brook, NJ: PennWell Publishing Company.
6. Wieder, Mike. (1999) National fire service incident management system consortium. *NFSIMS*. *SFCAV*. Retrieved March 11, 2006, from http://www.ims-consortium.org/backinfo.htm.
7. Irwin, Robert. Chapter 7: The incident command system (ICS). *Disaster response: Principles of preparation and coordination*. Disaster Response. Retrieved March 11, 2006, from http://orgmail2.coe-dmha.org/dr/flash.htm.

8. Coleman, John. (1997). *Incident management for the street-smart fire officer*. Saddle Brook, NJ: PennWell Publishing Company.

9. Wurtz, Thomas. (2004). *Firefighter's handbook: Essentials of firefighting and emergency response*. (2nd ed.). Clifton Park, NY: Delmar Publishing.

10. Coleman, John. (1997). *Incident management for the street-smart fire officer*. Saddle Brook, NJ: PennWell Publishing Company.

11. Federal Emergency Management Agency (FEMA). NIMS, March 1, 2004. *NIMS—Introduction and overview*. Retrieved February 14, 2006, from http://www.fema.gov/txt/nims/nims doc1.txt.

12. Wurtz, Thomas. (2004). *Firefighter's handbook: Essentials of firefighting and emergency response*. (2nd ed.). Clifton Park, NY: Delmar Publishing.

13. Coleman, John. (1997). *Incident management for the street-smart fire officer*. Saddle Brook, NJ: PennWell Publishing Company.

14. Wurtz, Thomas. (2004). *Firefighter's handbook: Essentials of firefighting and emergency response*. (2nd ed.). Clifton Park, NY: Delmar Publishing.

15. Coleman, John. (1997). *Incident management for the street-smart fire officer*. Saddle Brook, NJ: PennWell Publishing Company.

16. Wurtz, Thomas. (2004). *Firefighter's handbook: Essentials of firefighting and emergency response*. (2nd ed.). Clifton Park, NY: Delmar Publishing.

17. Cole, Dana., (2007), The incident command system: A 25-year evaluation by California practitioners. *National Fire Academy*. Retrieved April 23, 2006, from http://www.usfa.fema.gov/pdf/ efop/efo31023.pdf.

18. Wieder, Mike. (1999), National fire service incident management system consortium. *NFSIMS. SFCAV.* Retrieved March 11, 2006, from http://www. ims-consortium.org/backinfo.htm.

19. Coleman, John. (1997). *Incident management for the street-smart fire officer*. Saddle Brook, NJ: PennWell Publishing Company.

20. State of Virginia. National inter-agency incident management system (NIIMS). *Operational system description*. Retrieved April 23, 2006, from http://www1.va.gov/emshg/apps/kml/docs/ NIIMS_ICS_OperationalSysDesc.pdf.

21. Federal Emergency Management Agency (FEMA). Crosswalk/ comparison of NIIMS/NIMS. *NIMS*. Retrieved April 23, 2006, from http:// www.fema.gov/txt/nims/NIIMS-NIMSComparison.txt.

22. Ibid.

Suggested Reading

Cole, Dana. (2007), The incident command system: A 25-year evaluation by California practitioners. *National Fire Academy*. Retrieved April 23, 2006, from http://www.usfa.fema.gov/pdf/efop/ efo31023.pdf.

Coleman, John. (1997). *Incident management for the street-smart fire officer*. Saddle Brook, NJ: PennWell Publishing Company.

Federal Emergency Management Agency (FEMA). Crosswalk/comparison of NIIMS/NIMS. *NIMS*. Retrieved April 23, 2006, from http://www.fema.gov/txt/ nims/NIIMS-NIMSComparison.txt.

Federal Emergency Management Agency (FEMA). NIMS.(2006) *NIMS— Introduction and overview*. Retrieved February 14, 2006, from http://www. fema.gov/txt/nims/nimsdoc1.txt.

Federal Emergency Management Agency (FEMA).(2004), *NIMS and the incident command system*. NIMS. Retrieved February 11, 2006, from http://www. fema.gov/txt/nims/ nims_ics_position_paper.txt.

Irwin, Robert. Chapter 7: The incident command system (ICS). *Disaster response: Principles of preparation and coordination*. Disaster Response. Retrieved

March 11, 2006, from http://orgmail2.
coe-dmha.org/dr/flash.htm.

State of Virginia. National inter-agency incident management system (NIIMS). *Operational system description.* Retrieved April 23, 2006, from http://www1.va.gov/emshg/apps/kml/docs/NIIMS_ICS_OperationalSysDesc.pdf.

Wieder, Mike. (1999), National fire service incident management system consortium. *NFSIMS. SFCAV.* Retrieved March 11, 2006, from http://www.ims-consortium.org/backinfo.htm.

Wurtz, Thomas. (2004). *Firefighter's handbook: Essentials of firefighting and emergency response.* (2nd ed.). Clifton Park, NY: Delmar Publishing.

Preincident Planning, Fire Strategy, and Tactics

11 CHAPTER

Key terms

aspect, p. 230
barrier, p. 230
elevation, p. 230
external exposures, p. 234

internal exposures, p. 234
occupancy, p. 224
preincident planning, p. 224
shape of the area, p. 230

slope, p. 230
strategy, p. 224
tactics, p. 224
ventilation, p. 238

Objectives

After reading this chapter, you should be able to:

- Explain the difference between strategy and tactics.
- Describe the purpose for strategic planning.
- Identify the resources that must be considered in developing a strategic plan.

◆ **INTRODUCTION**

To the untrained eye, firefighting operations may seem chaotic. This perception may be perfectly accurate if firefighters have arrived on the scene with no preplanning. It is a grave and sometimes fatal mistake to neglect the work that must be done before an emergency arises. Strategy and tactics are not developed on the fire ground; these should have been established and documented in writing, weeks, months, or even years prior to the event.

The terms *strategy* and *tactics* are often used interchangeably; however, they have quite different meanings. Strategy outlines what needs to be done to achieve a positive outcome in a given set of circumstances. Tactics have to do with how a task is accomplished, how personnel are deployed, and how resources are brought to the scene. Although they are very dependent on each other, strategy and tactics are separate functions, and they should be distinguished. The purpose of this chapter is to introduce the basic principles of fire ground strategy and tactics.

◆ STRATEGY

strategy

Planning and directing the maneuvering of personnel and equipment into the most advantageous positions prior to actual operations.

tactics

The actual maneuvering of personnel and equipment in an operation to achieve short-term objectives.

preincident planning

Assessing risks and prioritizing potential hazards, then developing a strategic plan to mitigate the hazards.

occupancy

General term for how a building structure or home will be used and what it will contain.

Strategy is a broad-based plan for controlling an incident and achieving a positive outcome. It is based on the fire department's standard operating procedures and focuses on three priorities: life safety (firefighter and civilian), incident stabilization, and property preservation.[1] The strategic plan, often referred to as the *action plan*, identifies the tasks to be accomplished, identifies the material resources that will be needed, and specifies personnel requirements.[2] Strategy provides the foundation from which the fire ground **tactics** are implemented.

Several factors need to be considered when developing a fire ground strategy. The main concerns are available resources and environmental conditions. A strategy is well planned if there are enough resources to implement the plan and the plan is flexible enough to allow for unexpected or environmental changes. If there are not enough resources to execute the plan and it does not allow for flexibility, it does not matter how good it was in theory; it is virtually useless as a strategy.

PREINCIDENT PLAN

The most accurate way to determine the amount and availability of resources is through **preincident planning**. Preincident planning usually involves an inspection. Identification of fire dangers is not only an important tool in the fire prevention program; it also provides valuable information on resource needs and availability that can be used to develop a strategic action plan. This information enables the first arriving officer to make sound tactical decisions and accomplish the mission effectively and safely.

Preincident planning is usually done at the fire company level and involves on-site inspections. An effective preincident planning initiative is a sign of a strong departmental public relations program. Fire company personnel go into their areas of response, meet with local business owners, and ask permission to walk through their property. These company inspections help build a feeling of trust between the fire department and the business community, and they allow the company officer and personnel to gain first-hand knowledge of the target area or **occupancy**. The term occupancy refers to building use, such as residential, mercantile, and so on. By relating this new information to their past experiences and training, firefighters will have a foundation on which to build a strategic plan and to begin anticipating tactics. Company officers will also be able to make informed decisions about the kinds of training programs that will best prepare their companies for the challenges ahead (see Figure 11.1).

In his book *Fireground Strategies*, Deputy Chief Anthony Avillo, who serves with the North Hudson (NJ) Regional Fire and Rescue, identified 13 factors that should be considered during the preincident planning process. He formed these factors into the acronym COAL WAS WEALTH:[3]

- ◆ Construction
- ◆ Occupancy
- ◆ Apparatus and manpower
- ◆ Life hazards
- ◆ Water supply

UNIVERSAL CITY FIRE DEPARTMENT INSPECTION REPORT
DATE OF INSPECTION _____

PRE-C OF O( ) FINAL ( ) REGULAR ( ) RE-INSPECTION ( )

BUSINESS NAME: **PHONE NUMBER:**

ADDRESS:

NOTED DISCREPANCIES ( )   **NO DISCREPANCIES NOTED ( )**

BUILDING CONDITIONS
1. Building address not visible from street. Letters and numbers must be 4″.
2. Outside needs cleaning- High grass, weeds and accumulation of trash, etc.
3. Obstructed access to utilities: Electrical breakers   Inside   Outside   Gas shutoffs.

STORAGE
4. Storage closer than 24″ to ceiling.
5. Provide metal containers with metal covers for oily rags.
6. Accumulation of waste or other unusable materials.
7. Improper storage of flammable or combustible liquids.
8. Water heater closet or equipment room used for storage.
9. Compressed gas cylinders not secured to wall.

FIRE PROTECTION
10. Fire extinguishers need servicing (inspection), (recharge). Date of last inspection:
11. Fire extinguishers missing.
12. Fire extinguishers not accessible or visible.
13. Improper fire extinguishers.
14. Storage closer than 18″ to sprinkler heads.
15. Sprinkler head impaired or obstructed.   Date fire alarm last inspected:
16. Sprinkler control valve obstructed or closed.   Test fire alarm:
17. Date of last inspection of sprinkler system: Hydro test new sprinkler system:
18. Fire protection system for grills, fryers needs inspection.   Date of last inspection:
19. Fire department connections blocked and/or caps missing.
20. Fire lanes: need to be painted___ signs missing___ signs need to be replaced___
21. Fire system for paint booth (s) needs inspection. Date of last inspection:

EXITS
22. Exits locked___ blocked___ obstructed___
23. Exits not properly marked.
24. Exit lights not in operating condition.
25. Aisles are blocked with storage.
26. Emergency lights not in working condition.

HEATING AND AIR CONDITIONING
27. Heater not properly vented.
28. Soot showing in air chamber or vent.
29. Filters not cleaned regularly (monthly).
30. Combustibles kept too close to heater.
31. System not serviced regularly.

ELECTRICAL
32. Improper use of extension cord.   Cord in poor condition.
33. Electrical outlets, junction boxes, or panels not covered or broken or missing covers.

KITCHEN
34. Accumulation of grease on hood ducts, filters, equipment.
35. Combustibles too close to cooking unit.

KNOX-BOX
36. Have locks on doors been changed? If so, a new key is needed for the KNOX-BOX.
37. Need KNOX-BOX (Key Box) as required by the fire code.

This inspection is intended for your safety and your neighbor's safety. You are required by ordinance to complete the above Fire Code Requirements or climinate hazards by the date indicated.
Re-inspection date: _____

_____   _____
Signature of Owner/Manager/Employee   Fire Marshal
U.C. Form No. 405   659-0333 EXT.251

FIGURE 11.1 ◆ The inspection form eliminates oversights during the fire inspection.
Courtesy of the Universal City Fire Department, Universal City, Texas.

- Auxiliary appliances
- Street conditions
- Weather
- Exposures
- Area and height
- Location and extent
- Time
- Hazardous materials

Two more items should be added to this list. When sizing up a potential incident, the following should also be considered:

- Communications
- Command structure

WATER SUPPLY

Water is the primary extinguishing agent in the majority of firefighting operations. Most strategic plans begin with water supply and water requirement considerations. This is a concern whether the plan is being developed for an urban area with an established water distribution system or for a rural, wildland area where there are no water distribution systems. In urban areas that have established water distribution systems, the main concern in strategic planning will be to determine if the water system can provide adequate fire flow. Fire flow requirements can then be used to determine personnel and equipment needs on the fire ground.

Another strategic problem faced by the fire service is domestic development in rural areas. These areas do not have readily available water distribution systems. Water is usually supplied to residences by private wells drawing from a groundwater supply. These wells rarely have adequate flow capacity to support the needs of firefighting apparatus. Therefore, the fire department must rely heavily on tank water carried in by the fire apparatus or on surface water supplies from lakes or ponds (see Figure 11.2). A strategic plan for rural firefighting must include provisions for transporting, relaying, or drafting water operations.

Warning: It is critical to keep strategic plans up to date. If demographic changes have affected fire flow requirements in an area—urban or rural—and the strategic plan has not been altered to accommodate these changes, any future firefighting operations based on the outdated plan could be doomed to failure.

APPARATUS AND PERSONNEL

In 1954, Keith Royer and Bill Nelson of the Fire Service Institute at Iowa State University developed a formula for calculating flow rate. The formula states that the volume of the structure divided by 100 will equal the gallon-per-minute (GPM) flow needed to extinguish the fire in a structure.

$$V/100 = GPM$$

Example: A warehouse that is 200 feet long, 100 feet wide, and 10 feet high, with no attic space, would have a volume of $200 \times 100 \times 10 = 200{,}000$ cubic feet. By dividing this amount by 100, the total GPM flow can be determined.

$$200{,}000/100 = 2{,}000 \text{ GPM}$$

FIGURE 11.2 ◆ Many rural areas rely on water tenders (mobile water supplies) for their fire operations. Courtesy of James D. Richardson.

This formula assumes total fire involvement of the building.

Royer and Nelson also developed a method for determining personnel and apparatus needs based on required flow rates. In a study of 250 large fires during the late 1970s, it was determined that total personnel needs averaged one firefighter for every 50 GPM of required fire flow and one pumping apparatus for every 500 GPM of required flow. Going back to the example of the warehouse, a 2,000 GPM ideal flow rate would require:

$$2,000/50 = 40 \text{ firefighters}$$
$$2,000/500 = 4 \text{ pumping apparatus}$$

It should be stated that this formula provides only the minimum requirement. Personnel and equipment should be based on accurate prefire planning and worst-case scenarios.[4]

Rural and wildland areas pose a problem for strategic planners. As mentioned earlier, rural and wildland firefighting strategy should be based not only on the availability and amount of water but also on the method of its delivery. Water supplies may be provided by tanker trucks, relay operations from static water sources, direct drafting operations, or aircraft tanker operations. Moreover, wildland fires cover vast areas, and they produce extremely large amounts of heat. In the confined area of a structure fire, the conversion of water into steam displaces oxygen and aids in fire extinguishment. This phenomenon does not occur in open, wildland fires. Consequently, high volumes of water are needed to absorb the heat and cool the fuel for extinguishment. GPM flow rate computations will not be as important as total volume flow. The personnel and equipment requirements for wildland fires are also difficult to calculate because they vary widely depending on variables such as the size of the geographic area involved, the type of terrain, the amount of fuel, the water supply, and weather conditions.[5]

OCCUPANCY

The occupancy of a structure has a great influence on strategic planning. The term occupancy refers to the type of process being conducted on the premises, not the number of people in a structure. Occupancy affects decisions involving life safety and fire extinguishment methods.[6]

Regularly scheduled prefire planning should be conducted to maintain up-to-date information on occupancy changes in a protected area, especially if the area is experiencing demographic changes, such as urban renewal or population growth. An occupancy change may undermine the safety of a building by changing load impositions or by increasing fire loads to levels the structure was not designed to withstand (see Figure 11.3).

Stop and Think 11.1

In Houston, Fire Station Number One, a four-story structure, was sold and converted into a nightclub restaurant and aquarium. How does this change of occupancy change the load impositions on this structure and the occupancy load?

WEATHER

Weather is a major consideration when developing a strategic action plan. Although certain seasonal weather patterns can be assumed, day-to-day weather conditions such as wind, temperature, and humidity are harder to predict. They must, however, be taken into account.

FIGURE 11.3 ◆ Changes in occupancy may increase the fire load and hazards during fire operations. Courtesy of James D. Richardson.

Wind

Wind is one of the most influential forces affecting the behavior of fire. It is common knowledge that wind intensifies burning by supplying additional oxygen to the fire. One pound of fuel uses 200 cubic feet of air during the combustion process. By increasing available oxygen, wind increases the rate of fuel consumption.

Wind also brings additional problems to the fire ground strategist because it greatly influences the direction of fire travel. This is no surprise to those who are experienced in wildland and rural firefighting. However, studies have shown that air currents inside buildings move in the same direction as the wind outside the building. Even light winds can overcome the air pressure created by the combustion process and affect the direction of the fire.[7] Wind can cause hot embers to be dispersed to unburned areas and move heated fire gases into contact with dry fuels. It is a very dangerous variable in any fire situation. In conjunction with high temperature and low relative humidity, wind can create a fire scenario that will tax any fire organization's resources.

Temperature

Strategic plans should consider both seasonal changes in temperature and daily weather fluctuations. The temperature will affect the type of firefighting operation planned. During structural firefighting operations, temperature extremes greatly affect personnel. Any strategic plan must provide for personnel rehabilitation and relief. In wildland situations, other considerations come into play. Wildland fuels heated by the sun ignite more readily than fuel in areas of dense shade, and temperature (as well as wind and relative humidity) greatly influence the speed and intensity of a wildland fire.

Relative Humidity

Relative humidity is the relationship between the current amount of moisture in the air and the total amount of moisture the air can hold at a given temperature. Air picks up moisture when it moves across surfaces. If the relative humidity drops below 30%, the air can absorb more moisture from such fuels as light vegetation. This lowers the ignition temperature of the fuel and increases the danger of rapid fire spread.[8]

TOPOGRAPHY

Topography refers to the characteristics of terrain that include elevation, slope, and orientation. Because these features are fairly stable, topography tends to be the most predictable environmental element of a strategic plan.

Topography has always been considered an important factor when planning wildland firefighting operations. Some strategic planners, however, fail to consider its influence on structural firefighting. Urban prefire planning should identify conditions that could slow response times, prevent effective apparatus placement, or limit access to the interior of a structure. These may include bridges, road construction, and utility excavations. In fact, the urban landscape is actually less predictable than rural topography. Through an active prefire program, the strategic planner can stay abreast of changes that may affect firefighting response and operations. Cooperation and notification procedures should be established with other city departments and the state Department of Transportation so that responding companies will be aware of street or road closures due to construction or utility work.

The wildland fire strategist must take several topographic factors into consideration when developing an action plan. Each of these factors will affect ease of ignition and fire movement.[9]

slope

The angle of a hillside.

aspect

The slope of land that is exposed to the sun.

shape of the area

Topographic shapes, such as canyons, ridges, and mountains.

elevation

The height of the land above sea level.

barriers

Natural or artificial firebreaks for controlling fire spread.

- *Slope.* The angle of a hillside (its slope) affects the speed of fire traveling up the slope. The steeper the slope, the faster the fire will travel from the bottom to the top.
- *Aspect.* This refers to the side of a slope of land that is exposed to the sun. In the Northern Hemisphere, slopes have a southern aspect. Slopes that are exposed to the sun usually have lighter and drier fuels. The ambient temperature is higher, and the relative humidity is lower. These are areas where fire ignition is more likely and spread tends to be more rapid.
- *Shape of the Area.* Topographic shapes, such as canyons, ridges, mountains, and ravines, greatly influence fire behavior. Their effects on wind speed and direction influence the direction of fire travel, the intensity of burning, and the speed at which a fire will spread.
- *Elevation.* This affects the condition of the fuel and the amount of fuel available. Because temperatures are normally higher at lower elevations, fuels at lower elevations become drier earlier in the summer than fuels at higher elevations.

  As a rule, fuels are also less abundant at higher elevations; at extremely higher elevations, there may be no fuel at all. Precipitation and wind are also influenced by elevation.
- *Barriers.* Lakes, rivers, roads, highways, and constructed firebreaks are effective tools for controlling fire spread. These natural and artificial barriers should be major considerations in strategic plan development.

Whether the emergency response is in an urban or rural area, topography should be considered in the strategic plan. Contingency plans must be developed in both settings. It makes no difference whether a road is blocked by a mudslide or by a water main repair project, the road is still blocked. A good preincident plan will take this into consideration and establish alternative routes for emergency responders (see Figure 11.4).[10]

FIGURE 11.4 ◆ Topography can create many hazards for wildland firefighters. Courtesy of James D. Richardson.

BUILDING CONSTRUCTION

The advantages and disadvantages of various types of building construction were discussed in Chapter 8. We cannot overemphasize how important construction is in developing a strategic plan.

Strategic planners must take into account the building's size and configuration, its construction materials, and its interior finishes. They must also consider environmental influences, such as weather and topography, on fire behavior in different types of structures. The terrain on which the structure is built will determine personnel and apparatus needs. Seasonal changes may bring about drastic environmental conditions that could seriously impact firefighting strategies. Response times may be altered, or apparatus and personnel requirements may need revision. Strategic planners must include all of these variables.

Consider this incident: On March 9, 1995, San Francisco firefighters responded to a multistory residential structure fire. The residence was constructed on the side of a steep slope to take advantage of the panoramic view of the landscape. Entry to the structure was limited to the garage area. A fire crew made entry through the garage, which gave access to a living area on the second level of the structure. They were unaware that the main seat of the fire was on the first level below them and that it was being fed by 70-mile-per-hour winds. The intense heat caused the automatic garage door opener to activate and close the door, blocking the firefighters' escape route. One firefighter died, and two were seriously injured.[11] Weather, topography, and building construction added up to a deadly combination.

Stop and Think 11.2

1. How does topography influence firefighting operations in your area of responsibility?
2. What can you do to obtain information on residential structures?
3. How does weather affect your strategic and tactical planning?

AUXILIARY APPLIANCES

Saving lives is the first priority in any emergency operation. Preincident inspections will identify the life-safety devices that are in already in place, such as sprinklers, standpipes, or special extinguishing systems. Inspections will also expose the need for additional appliances that could increase the safety of occupants in the event of a fire.

Identifying in advance the location of these appliances in the building is vital. If a company knows where a standpipe is located, it can plan strategies for a rapid interior attack to attain atmosphere tenability and possibly save lives.

Knowing what types of appliances are in use is also critical. For example, some total flooding and deluge systems use extinguishing agents, such as halon and CO_2, that create toxic or hazardous atmospheres in the protected areas. Being forewarned of the presence of these systems allows firefighters to wear the right protective gear when responding to an emergency.

LIFE-SAFETY HAZARDS

In most cases, preincident planning will identify the life-safety hazards of an occupancy. However, this may not always be the case. The 2000 census reported that 12.4% (36,695,904) of the population of the United States were 65 years of age or older.[12] With escalating demand for elder-care facilities, unlicensed nursing homes and care facilities are becoming more common. Unlicensed nursing homes and elder day-care facilities are often located in residential neighborhoods. When viewed from the street, they appear to be single-family dwellings. In residential areas, preincident planning activities are conducted on request from the homeowner or occupant. Therefore, these illegal businesses can go unnoticed for months or years. Many of the residents of these facilities have limited or no mobility. This creates a major problem for the incident commander and responders, especially when they are not aware of the facility use until an emergency occurs.

TIME OF DAY

The time of day is a major factor in fire ground strategy. The prefire plan will identify when a building would be at or near normal occupancy. For example, most businesses operate between the hours of 8:00 AM and 5:00 PM. Firefighters may assume that an office building will be at normal occupancy levels during that time period. This does not mean, of course, that the building will be unoccupied after 5:00 PM. It means only that the life-safety hazard will be expected to be lower because there are fewer people in the building at night.

Remember: Incident commanders should always expect the unexpected. That "ordinary" house fire to which you respond in the middle of a workday may turn out to be an unlicensed child- or elder-care center.

STREET CONDITIONS

Emergency responders should be very familiar with the streets in their area of responsibility. This is called "knowing your territory." However, knowing where the location is and getting to it are different matters. Road construction, special events, traffic, or adverse weather can make streets impassable for emergency apparatus. All personnel must be able to identify and navigate alternative routes quickly to any location in their response area (see Figure 11.5).

FIGURE 11.5 ◆ Knowledge of area streets and roads reduce response time. Courtesy of James D. Richardson.

HAZARDOUS MATERIALS

Hazardous materials have become one of the primary concerns in firefighting response and operations. The location and quantity of these materials must be recorded during prefire planning and inspections. However, firefighters must be aware that almost every response will involve some type of hazardous material.

Fire personnel expect to encounter hazardous materials when responding to transport vehicle accidents but may not expect to encounter these during "routine" house or automobile fires. However, firefighters receive the heaviest exposure to hazardous materials during routine fires because they do not expect the exposure and are not properly protected. For example, when the insulation that protects an automobile's electronic components burns, the byproducts are carcinogenic. Many firefighters will not wear respiratory protection when combating an automobile fire. Firefighters also have a tendency to remove their SCBA when conducting overhaul operations at structure fires. Even though fire has been extinguished, the incident commander should not allow firefighters to enter an enclosed area without SCBA until the atmosphere is tested and declared safe (see Figure 11.6).

Stop and Think 11.3

1. What type of hazardous materials might fire personnel find in a rural area?
2. Would the hazardous materials found in a rural area be different from those found in an urban area?

FIGURE 11.6 ◆ The incident commander is responsible for protecting emergency responders from exposure to, or contamination from, hazardous materials during an operation. Courtesy of James D. Richardson.

The basic strategic plan for responding to any hazardous material incident is simple—it should be put into effect whenever the presence of hazardous materials is suspected. The steps to follow for hazardous material fires are as follows:

1. Approach the location from the upwind and uphill side.
2. Cordon off the area to isolate the incident and prevent entry.
3. Identify the product.
4. Seek help from qualified experts.[13]

Exposures

Protecting exposures is the next firefighting priority after life safety. An *exposure* is any building or area of building that is not yet on fire but could have possible fire damage. It can also refer to persons trapped in those areas. Exposures can be external or internal. **Internal exposures** include entrapped occupants, hazardous materials, or an uninvolved room. **External exposures** can be anything in close proximity to the fire, such as buildings, prairie grass, a forest, or any outdoor area. It is important that exposures be identified during prefire planning.

Many factors are considered when determining the priority of exposure protection. Naturally, life safety will come first, but building construction, private fire protections systems, hazardous materials, and access to the area are also important. In some circumstances the strategic plan makes exposure protection a priority over extinguishing the fire in an already involved structure.[14] For example, if a structure is totally involved with fire and rescue operations are not possible, it is better to focus on protecting exposed occupants and structures to prevent the fire from spreading.

internal exposures

Areas or persons in close proximity that can be threatened by a fire within the involved structure.

external exposures

External areas or persons in close proximity that can be threatened by a fire.

LOCATION AND EXTENT

The actual location and extent of a fire can, of course, be determined only after firefighters' arrival on the scene. Information gained from prefire planning and inspection reports will, however, give the incident commander clues to the probable location of the fire by identifying areas where a fire would most likely occur. Prefire plan information about a building's construction and design may also allow the incident commander to predict where the fire is apt to travel

The location and extent of a fire are primary factors in developing a strategic action plan and in deciding on tactics at the fire ground. These factors set the stage for all other activities, such as determining how and where the attack takes place, what exposures are to be protected, and which rescue operations may be necessary.[15]

COMMUNICATIONS

An incident command structure cannot be effective without an efficient communication network. Without a command structure, even the best-laid strategic plan is built on quicksand. The complexity of a communication system depends on the size of the emergency service organization and the area being served. A written standard operating procedure for the system's use must include rules that increase efficiency and reduce unnecessary radio traffic.

With the advent of the terrorist threats facing the United States, a multi-agency response to certain emergency incidents has become standard. These agencies must have access to a common communication system and be trained in the use of common terminology to communicate efficiently (see Figure 11.7). This is especially essential in smaller communities that rely heavily on mutual aid response from other

FIGURE 11.7 ◆ A common form of communication among the emergency responders is critical if the emergency operation is to succeed. Courtesy of James D. Richardson.

municipalities. A multi jurisdictional or multi agency response requires the use of an integrated communication system. A communications plan should contain the following elements:[16]

- *Systems.* Some jurisdictions are sharing the cost of setting up and maintaining a common communication system that can be used by all agencies responding to a multi jurisdictional incident.
- *Frequencies and Resource Use.* When a common system is used, each agency can be assigned a frequency to use to communicate with their resource personnel and additional tactical channels to be used during emergency responses to communicate with other agencies.
- *Information Transfer Procedures.* A communications plan must allow for the transfer of information not only within an organization but also between different agencies. All involved entities should work together and agree on a set of written procedures for transferring information. Commonly agreed-on procedures will help to keep misinformation at a minimum during an emergency.[17]

COMMAND STRUCTURE

Developing an incident command structure during an incident may be construed as a tactical operation. The actual implementation of the system is tactical, but the basis on which the commander develops that command structure is the strategic plan and a standard operating procedure for command development.

As stated earlier, fire ground operations must be strictly controlled and coordinated to ensure safety and effectiveness. All responders must know their assignments and know to whom they report. When an operation is well coordinated, the participants can focus on their individual assignments and accomplish their mission. The command structure establishes this well-defined line of authority.

◆ TACTICS

Tactics involves the deployment of personnel and other resources in a manner that resolves a situation effectively and efficiently. Tactical decisions should be based on a tactical plan of action. The strategic plan describes what should be accomplished, what resources are available to execute the plan, and who is responsible for which tasks. The tactical plan specifies how tasks are to be performed based on information from the strategic plan. Tactical planning is achieved at the operations level and is based on the availability of resources in relation to the situation at hand.[18]

TACTICAL DECISION MAKING

The tactical plan is heavily influenced by the decisions of the first arriving officer at a scene. He or she must note several factors in order to assess the situation.[16] First, the location of the fire should be ascertained. Regardless of the type of incident (structure, wildland, and so on), location is the most important factor. Determining the present extent of the fire and what life hazards may be involved depends on locating the fire.

The next step is to judge the probability that the fire will spread. This is affected by many factors, such as the amount of heat buildup, the type of fuel, the building's construction, weather conditions, and topography.

Finally, the first arriving officer must immediately decide whether enough resources are available to start tactical operations. If the fire is larger than the available resources can handle, the officer must request more resources. OSHA 1910.134 and NFPA 1500 stipulate that two firefighters are required to initiate an interior attack on a fire. Two additional firefighters should be fully equipped and ready outside the structure to monitor the interior attack team and serve as a rapid intervention team (RIT) should the attack team need help. This is referred to as the *two-in/two-out rule*. However, situations arise in which enough personnel may not initially be on the scene to initiate the two-in/two-out rule, and the building's occupants are in imminent danger. Remember, life safety is the number one priority, and life safety of firefighters comes first. However, if it appears that occupants may be saved, the fire officer in charge may determine that immediate rescue operations must be conducted. A combination of fire attack and rescue operation is then initiated, and the two-in/two-out rule is waived.

To put it simply, the first arriving officer must determine what is burning, in which direction the fire is likely to travel, who or what is in its path of destruction, and whether additional help is needed at the scene. Before a tactical fire officer can answer these questions, he or she must be familiar with the strategic plan, understand fire behavior, and know the priority of tactical objectives.[19]

There are seven tactical objectives that must be accomplished during firefighting operations. These priorities must be accomplished in the order listed as follows.[20]

Rescue

The primary objective in firefighting operations is saving lives. The safety of the operations personnel must be the first consideration when initiating a rescue operation. If a number of victims must be rescued, it is recommended that the sequence of priority be as follows. The first victims to consider for rescue are those closest to the fire. The largest number of victims who are threatened by the fire should be considered next. The third priority should be any other victims in the fire area. The last priority are victims who are in danger of being exposed to the fire.

Stop and Think 11.4

You arrive on the scene of a fire in an apartment complex. One building is heavily involved and has trapped victims. The fire is moving in a northerly direction. There is a report that five people are in the building on the north side of the involved structure. In a building on the south side of the fire, 20 people are on balconies yelling for help.

What is your priority of rescue in this situation?

Exposures

As we have explained, an exposure can be anything that is in the path of a fire: a person, a room in a building, another building, a field of grass or other vegetation, and so on. Exposures can be protected by either removing them from the path of the fire or by keeping them cool with water spray. Protecting exposures can serve the dual purpose of confinement if the fire is surrounded and prevented from spreading to other areas.

Confinement

Rapid confinement deals a death blow to a fire. Conversely, if a fire cannot be quickly confined, heavy fire losses can be expected. In fact, the majority of high-fatality fires are unconfined fires.

Concealed spaces in buildings, such as elevator and utility shafts, attics, or voids between trusses, allow fires to move rapidly from one area to another. Weather, wind, and topography affect wildland fire extension. The tactical officer should visualize a fire as having four sides, a top, and a bottom. Attention must be paid to all six dimensions if fire confinement is to be accomplished.

Extinguishment

The tactical aspect of fire extinguishment is a judgment call made by the first arriving fire officer. On arrival on the scene, the officer will estimate the type and amount of extinguishing agent that will be needed based on prefire plan information and observations made during scene evaluation.[19] An attack on the fire should not be attempted until an adequate amount of the proper kind of extinguishing agent is available at the incident scene. If on-hand supplies are inadequate, it may be prudent to use what is available to protect exposures and to contain the fire until enough agent is available for an all-out attempt at extinguishing the fire.

Overhaul

Overhaul is the process of ensuring that the fire is completely extinguished. There is nothing more embarrassing for a fire company than having to return to a fire scene because of a rekindle.

Overhaul operations are more complicated and considerably more dangerous than may be imagined. It is not simply a matter of just flushing everything out the back door and going back to the station. For example, care must be taken to prevent disturbance of possible crime scene evidence. Coordination between overhaul and fire scene investigation operations is an absolute necessity.

In many departments, overhaul operations are performed by the same fire companies that were assigned to the fire attack teams. These personnel may be suffering from fatigue and may lack the concentration necessary to identify hazardous conditions. The operations officer should consider bringing in fresh crews to perform the overhaul. Reducing exposure of overhaul operations personnel to toxic atmospheres should be a second major consideration for the operations officer. During firefighting operations, the firefighters wear respiratory protection. However, after the fire is extinguished, most firefighters remove their protective clothing. At this point, they are exposed to the highest concentrations of carbon monoxide and other products of combustion. The operations officer should ensure that all overhaul operations personnel continue to wear their breathing apparatus or appropriate respiratory protection until the atmosphere has been tested and deemed safe.[21]

Stay Safe

ventilation

The systematic removal of pressure, gases, and heat from a structure or confined area.

Ventilation

Ventilation is the process of removing the by-products of combustion (smoke, toxic gases, heat, and particulates of unburned fuel) from a confined area. The purpose of ventilation operations is to create an atmosphere that is as safe as possible, not only for firefighters entering the area but also for trapped victims. The tactical operations officer can utilize ventilation to direct the movement of the fire, reduce the heat, increase visibility for firefighters, and increase the oxygen content of the atmosphere. Ventilation can be natural, mechanical, or hydraulic. Natural ventilation takes advantage of air currents by opening windows and doors or cutting a hole in the roof of the building to allow the natural flow of heated gases to vent outside the structure and draw fresh air into the building. Mechanical ventilation involves the use of fans to either push air into the building (positive ventilation) or draw air out of the building (negative ventilation). Hydraulic ventilation is used by attack or overhaul teams to draw smoke outside the building. This is accomplished by directing a fog-pattern hose stream out of a window in the room being ventilated. When properly positioned, a negative air pressure is created behind the fog pattern. This is called a *venturi effect.* This negative pressure draws the products of combustion out of the window with the hose stream.

The choice of method depends on the tactical requirements of the situation. Ventilation operations should be continued until the fire is completely extinguished and overhaul operations are completed.[22]

Salvage

Although salvage is the last operation in the prescribed tactical sequence, it can actually be initiated at any stage of the operation. The purpose of salvage is to protect property and unaffected areas from the damaging effects of the products of combustion and from water damage. In structure fires, this may involve covering furniture and personal property with salvage covers to protect them from water and smoke damage, removing water from the structure during or after the fire, or recovering valuables from fire debris.

Salvage operations during wildland firefighting operations may be incorporated into the confinement stage of the operation and may involve removing debris to inhibit fire spread and to protect structures exposed by the fire. Another form of wildland salvage operation is planting new vegetation in burned-out areas to prevent erosion or mudslides.

Deciding when salvage operations should take place and what they entail is the duty of the tactical operations officer.

◆ **SUMMARY**

A strategic (or action) plan is a broad-based protocol that establishes the goals for incident management. Based on information gathered from preincident surveys, the strategic plan identifies resources and establishes standard operating procedures to be used as guidelines during tactical operations. The plan must be flexible and capable of modification should unexpected events arise during the incident.

Tactical plans describe how strategic plans will be put into effect. The operations officer is responsible for implementing the tactical plan in the best way according to the actual situation. The operations officer must also stay abreast of changes in the situation and be able to react to unexpected developments. This does not mean the officer abandons the plan; it does mean that the tactical officer must have the skills and ability to adjust the plan to meet all of the tactical objectives. Tactical objectives must be accomplished in the proper sequence to ensure a positive incident outcome.

Constant communication and coordination between the operations officer and the incident commander are absolute necessities. A well-defined command structure gives responders confidence and improves morale.

Finally, whether your goal is developing a strategic plan or a tactical plan, always remember Murphy's famous law: "Anything that can go wrong, will go wrong." Plan for the unexpected, and plan relentlessly.

#3

On Scene

Midday on July 31, 1979, the Houston Fire Department responded to an apartment fire. When they got to the scene, the first arriving companies found a structure housing several apartment units heavily involved in fire. The building was one of several on a 55-acre development. There were a total of 1,083 residential units in the development. The buildings were of wood-frame construction. The roofs and exterior walls were covered with cedar shingles. As the fire grew in intensity, a steady south wind sent burning embers downwind and ignited the shingle roofs of adjoining buildings. Before the fire was under control, seven alarms had been requested. A total of 35 fire apparatus and over 100 firefighters responded. An area over one-quarter of a mile long containing 437 apartment units was totally destroyed. Flying embers were reported in subdivisions 12 miles north of the fire.

1. What would you have done if you were the first arriving officer?
2. What do you think were the important factors to be considered on arrival at the scene?
3. What strategy and tactics would you have implemented?

Review Questions

1. Why is it incorrect to assume that tactics and strategy are synonymous?
2. What is the purpose of a strategic plan?
3. Why is the prefire plan so important in developing a strategic plan?
4. What factors must be considered when developing the strategic plan?
5. What regulations and standards govern the two-in/two-out rule?

■■■

Notes

1. Clark, William. (1981). *Fire fighting principles and practices*. New York: Technical Publishing.
2. Basic strategy and tactics. *Instructors guide*. Retrieved January 16, 2006, from firehouse.com: http://www.firehouse.com/training/drills/files/DM_0506.doc.
3. Avillo, Anthony. (2002). *Fireground strategies*. Tulsa, OK: PennWell Corporation.
4. Clark, William. (1981). *Fire fighting principles and practices*. New York: Technical Publishing.
5. Avillo, Anthony. (2002). *Fireground strategies*. Tulsa, OK: PennWell Corporation.
6. Clark, William. (1981). *Fire fighting principles and practices*. New York: Technical Publishing.
7. Schneider, Rachel, & Breedlove, Deborah. Fire management study unit. *National weather service program*. Retrieved December 2, 2005, from http://www.fs.fed.us/conf/ee/fire_mgt.pdf.
8. Ibid.
9. Fuhrman, Paul. (December 1995). *Fatal house fire San Francisco, CA*. St. Louis, MO: Working Fire Subscription Films. Retrieved February 19, 2006, from Video Archive.
10. Demographics of the United States. (May 1, 2006). *Wikipedia, The Free Encyclopedia*. Wikimedia Foundation.
11. Avillo, Anthony. (2002). *Fireground strategies*. Tulsa, OK: PennWell Corporation.
12. Ibid.
13. Ibid.
14. EMC012 review answers. *Emergency communications units—Information bulletin*. Auxiliary Communications Service of California Governor's Office of Emergency Services. Retrieved January 20, 2006, from http://acs.oes.ca.gov/EMCOMM/EMC001%20to%20060/EMC012.txt.
15. Coleman, John. (1997). *Incident management for the street-smart fire officer*. Saddle Brook, NJ: Fire Engineering.
16. Clark, William. (1981). *Fire fighting principles and practices*. New York: Technical Publishing.
17. Wurtz, Thomas. (2004). *Firefighter's handbook: Essentials of firefighting and emergency response.* (2nd ed.). Clifton Park, NY: Delmar Publishing.
18. Ibid.
19. Clark, William. (1981). *Fire fighting principles and practices*. New York: Technical Publishing.
20. Charlottesville Fire Department. (February 19, 2006). *Emergency operations standard operating procedures.* Charlottesville, VA: Charlottesville Fire Department.
21. Wurtz, Thomas. (2004). *Firefighter's handbook: Essentials of firefighting and emergency response.* (2nd ed.). Clifton Park, NY: Delmar Publishing.
22. Ibid.

■■■

Suggested Reading

Avillo, Anthony. (2002). *Fireground strategies*. Tulsa, OK: PennWell Corporation.

Basic strategy and tactics. *Instructors guide*. Retrieved January 16, 2006, from firehouse.com: http://www.firehouse.com/training/drills/files/DM_0506.doc.

Bush, Loren, & McLaughlin, James. (1970). *Introduction to fire science*. Beverly Hills, CA: Glencoe Press.

Charlottesville Fire Department. (February 19, 2006). *Emergency operations standard operating procedures.* Charlottesville, VA: Charlottesville Fire Department.

Clark, William. (1981). *Fire fighting principles and practices*. New York: Technical Publishing.

Coleman, John. (1997). *Incident management for the street-smart fire officer*. Saddle Brook, NJ: Fire Engineering.

Demographics of the United States. (May 1, 2006). *Wikipedia, The Free Encyclopedia*. Wikimedia Foundation.

EMC012 Review answers. *Emergency communications units—Information bulletin*. Auxiliary Communications Service of California Governor's Office of Emergency Services. Retrieved January 20, 2006, from http://acs.oes.ca.gov/EMCOMM/EMC001%20to%20060/EMC012.txt.

Fuhrman, Paul. (December 1995). *Fatal house fire San Francisco, CA*. St. Louis, MO: Working Fire Subscription Films. Retrieved February 19, 2006, from Video Archive.

Monkman, Chris. Murphy's laws origin. *Murphy's laws origin*. Retrieved January 15, 2006, from Murphy's Laws website: http://www.murphys-laws.com/murphy/murphy-true.html.

James, Page. (1973). *Effective company command*. Alhambra, CA: Borden Publishing Co.

Schneider, Rachel, & Breedlove, Deborah. Fire management study unit. *National weather service program*. Retrieved December 2, 2005, from http://www.fs.fed.us/conf/ee/fire_mgt.pdf.

Wurtz, Thomas. (2004). *Firefighter's handbook: Essentials of firefighting and emergency response*. (2nd ed.). Clifton Park, NY: Delmar Publishing.

CHAPTER 12

Support for Fire Emergency Services: Their Vital Functions

Key Terms

computer-aided dispatch
   systems (CAD), p. 250

enhanced 9-1-1, p. 249
NFPA 1221, p. 250

Objectives

After reading this chapter, you should be able to:

- Identify types of support organizations.
- Explain how these support organizations assist the fire service.
- Explain the difference between a managerial support function and a technical support function.
- Demonstrate the significance of partnerships and organizations to the fire service.
- Describe the duties and responsibilities of fire department support personnel.

◆ INTRODUCTION

This chapter is devoted to the people and organizations whose work supports the fire and emergency services. An organization as intricate as a modern U.S. fire department cannot stand alone. It requires support from international and national associations dedicated to the firefighting profession. It must interface with many other professional groups as well, drawing on their expertise. It must cooperate and coordinate with a long list of government agencies at the federal and state level. It draws community support from local civic groups. To do their jobs, firefighters need to know about these organizations—large and small, public and private—and about the kinds of help they have to offer.

Publications devoted to various aspects of the fire service are another source of support, helping firefighters keep up to date with new technology and new trends in the field.

Finally, the department's firefighting personnel rely heavily on support personnel to provide the kind of competent backup that makes it possible for them to perform their primary mission of fighting fires. By understanding these supporting positions and their functions, firefighters can better perform their own jobs within the framework of the department.

This chapter provides brief capsule descriptions of over 40 professional and governmental organizations and furnishes their web addresses. A list of helpful professional periodicals is provided. We also discuss the various roles of technical and administrative personnel.

◆ NATIONAL AND INTERNATIONAL ORGANIZATIONS

National and international membership organizations enhance the professionalism of the fire service and provide a highly visible platform where members can exchange views and voice concerns.

American Fire Sprinkler Association (AFSA). Organized in 1981, this nonprofit, international association representing open-shop fire sprinkler contractors is dedicated to the educational advancement of its members and promotion of the use of automatic fire sprinkler systems. http://www.sprinklernet.org

American National Standards Institute (ANSI). This organization serves as a clearinghouse for nationally coordinated voluntary safety, engineering, and industrial standards developed by industrial firms, trade associations, technical societies, consumer organizations, and government agencies. http://www.ansi.org

American Red Cross. By doing everything from joining forces to help a family devastated by fire to educating children about emergency preparedness and raising funds for disaster victims, American Red Cross chapters support fire departments across the country. Although the policies in different jurisdictions may vary, in most localities, fire personnel will call the Red Cross as soon as they realize that fire victims need help. That call, sometimes made from the scene of the fire, is often the first step in locating emergency housing and filling other immediate needs. Red Cross disaster relief focuses on meeting immediate emergency disaster-caused needs. When a disaster threatens or strikes, the Red Cross provides shelter, food, health, and mental health services. In addition to these services, the core of Red Cross disaster relief is the assistance given to individuals and families affected by disaster to enable them to resume their normal daily activities independently. The Red Cross also feeds emergency workers, handles inquiries from concerned family members outside the disaster area, provides blood and blood products to disaster victims, and helps those affected by disasters to access other available resources. http://www.RedCross.org

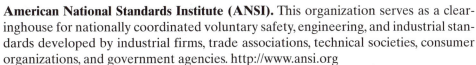

Stop and Think 12.1

List a few ways in which the American Red Cross can assist victims in emergency situations.

American Rescue Dog Association (ARDA). Founded in 1972, this is the nation's oldest air-scenting search dog organization. Its standards and training methods have served as the model for K-9 search and rescue units across the nation and around the world. http://www.ardainc.org

Board of Certified Safety Professionals (BCSP). This not-for-profit corporation was established in 1969. It operates solely as a peer certification board with the purpose of certifying practitioners in the safety professions. http://www.bcsp.org

Building and Fire Research Laboratory (BFRL). This research facility studies building materials, computer-integrated construction practices, fire science and fire safety engineering, as well as structural, mechanical, and environmental engineering. http://www.bfrl.nist.gov

Chemical Transportation Emergency Center (CHEMTREC). This center serves as a round-the-clock resource for obtaining immediate emergency response information for accidental chemical releases. CHEMTREC is linked to the largest network of chemical and hazardous material experts in the world, including chemicals and response specialists. http://www.chemtrec.org

Fire Apparatus Manufacturers' Association (FAMA). Organized in 1946, this nonprofit trade association is committed to enhancing quality in the fire apparatus industry and emergency service community through the manufacture and sale of safe, efficient fire apparatus and equipment. http://www.fama.org

International Association of Arson Investigators (IAAI). This organization works in cooperation with other associations and law enforcement agencies to prevent and suppress the crime of arson. Cooperating organizations include the U.S. Fire Administration, the Federal Emergency Management Agency, the National Fire Academy, the International Association of Fire Chiefs, the Insurance Committee for Arson Control, and many others. http://www.firearson.com

International Association of Fire Chiefs (IAFC). Established in 1873, this organization provides leadership to career and volunteer chiefs, chief fire officers, and managers of emergency service organizations throughout the international community through vision, information, education, services, and representation to enhance their professionalism and capabilities. http://www.iafc.org

International Association of Firefighters (IAFF). This is the largest union organization representing firefighters. http://www.iaff.org

International Fire Marshals Association (IFMA). This association serves as a forum for the exchange of information on fire prevention and arson [formerly known as the Fire Marshal's Association of North America (FMANA)]. http://www.nfpa.org

International Fire Service Training Association (IFSTA). Established in 1934, the mission of this association of fire service personnel is to identify areas of need for training materials and foster the development and validation of training materials for the fire service and related areas. http://www.ifsta.org

International Fire Service Accreditation Congress (IFSAC). The mission of this organization is to increase the level of professionalism of the fire service through accreditation of fire service training or education entities. http://www.ifsac.org

National Association of State Fire Marshals (NASFM). The members of this organization are the top fire officials in the 50 U.S. states. The organization is dedicated to improving the effectiveness of state fire marshals. http://www.firemarshals.org

National Fallen Firefighters Foundation. The U.S. Congress created this foundation to lead a nationwide effort to remember America's fallen firefighters. Since 1992, the tax-exempt, nonprofit foundation has developed and expanded programs to honor our fallen fire heroes and assist their families and coworkers. http://www.firehero.org

National Fire Protection Association (NFPA). The goal of this nonprofit organization is to improve the quality of life worldwide by publishing and advocating consensus codes and standards, as well as promoting research, training, and education related to fire hazards. http://www.nfpa.org

National Incident Management System (NIMS). These guidelines were developed by the Department of Homeland Security and released in March 2004. The NIMS establishes a uniform set of processes and procedures that emergency responders at all levels of government use to conduct response operations to natural disasters and other emergencies, including acts of terrorism. http://www.fema.gov/emergency/nims/

National Safety Council (NSC). Founded in 1913 and chartered by the U.S. Congress in 1953, this nonprofit, nongovernmental, international public service organization is dedicated to protecting life and promoting health. Members include more than 48,000 businesses, labor organizations, schools, public agencies, private groups, and individuals. http://www.nsc.org

National Volunteer Fire Council (NVFC). This nonprofit membership association represents the interests of the volunteer fire, EMS, and rescue services. http://www.nvfc.org

Salvation Army. In addition to its other good works, this charitable social service organization provides relief for victims of disasters and other emergencies. http://www.salvationarmyusa.org

Underwriters Laboratories Inc. (UL). Founded in 1894, this independent, nonprofit, and highly respected organization conducts product safety and certification. http://www.ul.com

◆ FEDERAL ORGANIZATIONS

The fire service frequently cooperates with departments and agencies of the federal government, especially with those listed here. National policy and guidelines that originate in these organizations affect firefighters' work. In some cases, federal funds administered by these organizations can be used to fund fire service missions.

#10

Department of Defense (DOD). This civilian department of the U.S. government directs and controls the armed forces and assists the president in the direction of the nation's security. www.defenselink.mil

Department of Transportation (DOT). The Department of Transportation oversees the transportation network in the United States. www.dot.gov

Emergency Management Institute (EMI). Part of FEMA, this organization serves as the national focal point for the development and delivery of emergency management training to government officials at all levels, to volunteer organizations, and to the public and private sectors. http://training.fema.gov

Environmental Protection Agency (EPA). This federal agency, founded in 1970, is charged with protecting the natural environment of the United States—its air, water, and land. http://www.epa.gov

Federal Aviation Administration (FAA). This agency of the U.S. Department of Transportation oversees civil aviation. Its aim is to improve safety and efficiency while being responsive to the needs of the public. http://www.faa.gov

Federal Emergency Management Agency (FEMA). This agency is part of the Emergency Preparedness and Response Directorate of the Department of Homeland Security. The range of FEMA's activities is broad: advising on building codes and flood plain management; teaching people how to get through a disaster; helping equip local and state emergency preparedness units; coordinating the federal response to a disaster; making disaster assistance available to states, communities, businesses, and individuals; training emergency managers; and supporting the nation's fire service. (The U.S. Fire Administration and the Emergency Management Institute are part of FEMA.) http://www.fema.gov

Fire and Emergency Services Higher Education (FESHE) Conference. The U.S. Fire Administration sponsors this annual conference every June at the National Emergency Training Center in Emmitsburg, Maryland. Its aims are to produce national models for an integrated, competency-based system of fire and emergency services professional development and for emergency services higher education (associate through doctoral degrees). http://www.usfa.dhs.gov/nfa/higher_ed/index.shtm

National Emergency Training Center (NETC). The National Emergency Training Center is an advanced training facility for emergency service professionals, owned and administered by FEMA. It is located on a 107-acre campus in Emmitsburg, Maryland, that houses several agencies, including the U.S. Fire Administration (USFA), the National Fire Academy (NFA), and the Emergency Management Institute (EMI). Several research and development facilities are also located on the campus, including the Arson Burn Laboratory, Fire Prevention Laboratory, and Simulation/Exercise Laboratory. http://training.fema.gov/VCNew/firstVC.asp

National Fire Academy (NFA). This academy works to enhance the ability of fire and emergency services and allied professionals through education to deal more effectively with fire and related emergencies. http://www.usfa.dhs.gov

National Institute for Occupational Safety and Health (NIOSH). This federal agency, administered by the Centers for Disease Control, is responsible for conducting research and making recommendations for the prevention of work-related injury and illness. NIOSH conducts independent investigations of firefighter line-of-duty deaths. http://www.cdc.gov/niosh.

National Transportation Safety Board (NTSB). This independent federal agency is charged by the U.S. Congress with investigating every civil aviation accident in the United States and significant accidents in the other modes of transportation—railroad, highway, marine, and pipeline. The board also issues safety recommendations aimed at preventing future accidents. http://www.ntsb.gov

Occupational Safety and Health Administration (OSHA). The mission of this agency of the U.S. Department of Labor is to assure the safety and health of U.S. workers by setting and enforcing standards; providing training, outreach, and education; establishing partnerships; and encouraging continual improvement in workplace safety and health. http://www.osha.gov

U.S. Department of Homeland Security (DHS). This department of the federal government was created after the September 11, 2001, terrorist attacks. The DHS Emergency Preparedness and Response Directorate oversees domestic disaster preparedness training and coordinates government disaster response. The Federal

Emergency Management Agency (FEMA) was made a part of the Directorate, which also brings together agencies from other federal government departments: the Strategic National Stockpile and the National Disaster Medical System (Health and Human Services), the Nuclear Incident Response Team (Department of Energy), Domestic Emergency Support Teams (Justice Department), and the National Domestic Preparedness Office (FBI). http://www.dhs.gov

U.S. Fire Administration (USFA). A part of FEMA and the Department of Homeland Security, the mission of this administration is to reduce life and economic losses due to fire and related emergencies through leadership, advocacy, and support to the broader fire protection community. It is the parent organization of the National Fire Academy and the Emergency Management Institute. http://www.usfa.dhs.gov

U.S. Forest Service (USFS). Established in 1905, this agency of the U.S. Department of Agriculture manages public lands in national forests and grasslands. http://www.fs.fed.us

◆ STATE ORGANIZATIONS

In most cases, the mission of state agencies that support fire and emergency personnel is closely related to the mission of their national or federal counterparts, although their impact is largely limited to their individual states. The structure and function of these organizations (and sometimes the organization's name) can vary from state to state. In most states, however, you will find the following:

National Guard. These citizen soldiers may be called in during major incidents to help with disaster relief.

Office of the State Fire Marshal (OSFM). This office enforces state fire laws and promotes fire-related legislation.

Special Task Forces. These ad hoc teams of specialists are organized to carry out a particular mission.

State Association of Fire Educators. This association promotes education and provides resources and tools for ongoing training in the fire service industry.

State Fire Chiefs' Association. This professional membership organization promotes communication and cooperation among the state's fire chiefs and provides a forum in which they can voice their opinions and concerns.

State Fire Commission. This commission develops statewide training models or curriculums.

State Firefighters' Association. This association promotes the interests of firefighting professionals.

State Health Department. This department sets health policy for the state and may regulate other medical licensing and certifications.

State Office of Emergency Management (OEM). This department works to plan and prepare for emergencies, educate the public about preparedness, coordinate emergency response and recovery, and collect and disseminate critical emergency information.

State Police. This department is responsible for all state roadways. They can help with on-scene traffic control to secure a safe environment for other emergency personnel and can also be a resource during hazardous materials incidents.

◆ LOCAL ORGANIZATIONS

Local organizations can have a major impact on fire service operations, both in the short term on an emergency scene and in the long term for planning. The local police department is sometimes the first on the ground at an emergency and can help to secure the scene. The water department can help with vital information on water pressure problems and with water issues, of vital concern to fire departments. The planning and zoning commission can provide significant support for the fire department because its decisions impact population concentrations and building construction in the area the fire service protects. Other groups, such as the Lions Club, the Rotary Club, and the Chamber of Commerce, can also be useful by finding sources of revenue or donations or simply by helping the fire department with its public relations efforts.

◆ PERIODICAL PUBLICATIONS

Fire service periodicals provide the latest up-to-date information on firefighting methods, equipment, and other issues. Buyers guides, published annually by these periodicals, are particularly useful for departments. Fire service periodicals include the following:

Fire Chief. Examines issues that are of particular importance to the chief officers who lead and manage fire departments. http://www.firechief.com

Fire Engineering. Provides articles that are written by experts in the fire service and focus on lessons-learned technical issues. http://www.fireengineering.com

Firehouse. May be the most innovative, visually compelling, and factually accurate magazine in the firefighting industry. http://www.firehouse.com

FireRescue Magazine. Provides firefighters with a wide range of news, products, training opportunities, and job resources. http://www.firerescuemagazine.com

NFPA Journal. The world's leading publication on fire prevention, an important advocate for public safety and an authoritative information source. http://www.nfpajournal.org

◆ SUPPORT STAFF IN THE FIRE AND EMERGENCY SERVICES

Now that we have looked at the various organizations and agencies that assist the fire service, it is time to turn to the vital support role played by individuals. The range of their job descriptions is surprisingly wide and, in some cases, unexpected. These individuals may earn their paychecks from the fire service or from an entirely unrelated entity. In general, however, we classify their support as either managerial or technical. Managerial support includes job categories such as dispatcher, HazMat specialist, or arson investigator. Technical support personnel include information systems specialists and repair shop or radio shop technicians. We will discuss the roles played by higher-profile support personnel, but any firefighter will have his or her own list of unsung heroes.

DISPATCH

The primary purpose of the dispatcher is to support fire management operations from a remote location, performing a variety of staff and administrative duties. Dispatch centers receive calls for service, dispatch the calls, and track resources and their availability. The dispatcher may be a firefighter or a civilian. In most of the United States and Canada, the **enhanced 9-1-1** system enables a caller to dial 9-1-1-from any telephone and be linked to an emergency dispatch center operator who can provide assistance and also determine the caller's location. In the past, a red alarm pull-box was common in businesses, schools, hospitals, and on street corners. In some small towns, the system activated an air-horn or steam whistle, which would rally the volunteer fire company. If wired to the dispatch center, it caused an oscillator to beep the code of the box location to the dispatcher. The dispatcher identified the location and telephoned the firehouse to provide details. However, with the advance of the 9-1-1 phone system, these alarm boxes are slowly being phased out.

enhanced 9-1-1
An emergency phone system that automatically gives a dispatcher the caller's location.

HAZARDOUS MATERIALS (HAZMAT) TEAMS

A hazardous material incident is generally described as the intentional or accidental release of a radiological, toxic, corrosive, flammable, or dangerous chemical or a biological or nuclear agent into the environment. HazMat teams stand ready to assist the fire service when it responds at scenes involving these substances. The incident may involve either an intentional release, such as illegal drug lab operations, or an accidental release that happens in the course of processing or transporting hazardous materials. The HazMat team is also trained and equipped to handle the release of nuclear, chemical, or biological materials associated with terrorism.

ARSON UNITS

Arson is the act of deliberately setting fire to a building, car, or other property for fraudulent or malicious purposes. This is a crime in all states. Arson investigators are called to investigate fires of suspicious or unknown origin. The arson unit will typically report to the fire marshals, although in smaller fire departments, arson investigators may double as line firefighters. The unit must be available at any hour to work a fire scene.

GRAPHIC ARTS AND MAPS

New technology has created the need for individuals who develop that technology and keep it running smoothly. In the past, fire and emergency services used three-ring binders filled with paper maps with location information. These maps served their purpose, but the advent of the Global Positioning System (GPS) has revolutionized mapping. Originally developed by the Department of Defense, the GPS has come into civilian use in the last 12 years. GPS provides specially coded satellite signals that can be processed in a GPS receiver, enabling the receiver to compute position, velocity, and time in a matter of seconds and then generate real-time maps. Its use has measurably improved fire service response times. One fundamental limitation of GPS is that its receivers require an unobstructed view, so its use is limited in mapping downtown areas and heavily wooded locations.

INFORMATION SYSTEMS

Computers and the Internet have catapulted the fire service into a new age. A number of types of information systems are used within fire departments. These include many of the same programs used by other organizations and businesses to collect and store data, prepare budgets, keep personnel records, manage payroll, and so on. **Computer-aided dispatch systems (CAD)** are sometimes used in conjunction with the 9-1-1 telephone system to receive emergency calls for service and to coordinate the responses of emergency personnel and resources. Additionally, most fire departments utilize computer information systems to keep incident reports and to share them with other agencies. As the capabilities of technology-based systems continue to increase, so does the need to distribute them to firefighters in the field via hand held and on-board computing devices.

REPAIR GARAGES

A departmental maintenance garage or, at least, a mechanic specially trained or certified in fire apparatus repair can save the department time and ensure consistent maintenance standards.

RADIO SHOPS

Firefighters and emergency personnel put their lives on the line every day and need communication equipment that works everywhere, every time. New radio technology has produced lightweight portable radios that are able to withstand extreme temperatures and work reliably inside buildings. Today's complex radio networks utilize many different technologies. Choosing the best system for your department and your community means striking a balance between initial cost and long-term capabilities. Decisions must be made based on the affordability of the system and whether it has the potential to be updated as needs change. The radio shop should be familiar with the standard that guides emergency radio operations—**NFPA 1221**, *Standard for the Installation, Maintenance, and Use of Emergency Services Communications Systems.*

WEATHER FORECASTERS

Weather forecasters are also important information sources. Their daily input on local conditions supplies vital information on wind direction, humidity, and temperature—all key factors that influence how an incident will be handled, especially when it involves hazardous materials.

computer-aided dispatch systems (CAD)

Uses a wide range of software and hardware to help manipulate data for engineering purposes, coordinate the responses of emergency personnel and resources, and keep incident reports and share them with other agencies.

NFPA 1221

Standard for the installation, maintenance, and use of emergency services communications systems.

◆ SUMMARY

Many dedicated people and organizations are needed to keep our fire and emergency services up to date and running smoothly. They are the firefighters' partners, working with firefighters toward the common goal of protecting the public in the safest and most efficient ways possible. This introduction to the vast pool of help that is available is only a starting point in the search for additional resources and information. Knowing where to go for information and having the proper research skills is one of the most important tools for employees in any sector of the fire service.

On Scene

National, state, and local organizations can influence how public funding and other avenues of resources are allocated to fire and EMS services. Issues related to adequate training and equipment can also be brought to the attention of legislators and other decision makers by these organizations.

1. Can you name a few prominent organizations and associations that have directed fire and EMS issues to the forefront of the political arena?

2. What state organization regulates training for career and volunteer organizations?

Review Questions

1. Which organizations could be a source of information if your department is considering buying a fire apparatus?
2. List three reasons why a fire department requires a dispatch center.
3. Which organization could be a source of information about labor relations?
4. You are at the scene of a hazardous materials spill and need information on the substance involved. What free service could you contact for information?
5. Why is the information provided by a weather forecaster important during major incidents?
6. What national organization provides the following: shelter to a family displaced by a fire, clothing vouchers for hurricane victims, and food to emergency workers on extended duty?

Web Resources for Firefighters

American National Standards Institute. http://www.ansi.org/

American Red Cross. http://www.redcross .org/services/disaster/0,1082,0_319_,00 .html

American Rescue Dog Association. http://www.ardainc.org/

Board of Certified Safety Professionals. http://www.bcsp.org/

Building and Fire Research Laboratory. http://www.bfrl.nist.gov/

Chemtrec. http://www.chemtrec.org

Emergency Management Institute. http://www.training.fema.gov/emiweb/

Federal Aviation Administration. http://www.faa.gov/

Federal Emergency Management Agency. http://www.fema.gov

Fire Apparatus Manufacturers' Association. http://www.fama.org/

Fire Chief. http://firechief.com/

Fire Engineering. http://fe.pennnet.com/

FireRescue. http://www.nfrmag.com/

The Fire Sprinkler Network. http://www.sprinklernet.org

Firehouse.com. http://www.firehouse.com/

International Association of Arson Investigators. http://www.firearson.com/

International Association of Fire Chiefs. http://www.iafc.org/

International Association of Fire Fighters. http://www.iaff.org/

International Fire Service Training Association Headquarters. http://www.ifsta.org/

International Fire Service Accreditation Congress. http://www.ifsac.org/

National Association of State Fire Marshals. http://www.firemarshals.org/

National Fallen Firefighters Foundation. http://www.firehero.org/

National Fire Protection Association. http://www.nfpa.org/

National Incident Management System. http://www.fema.gov/emergency/nims/index.shtm

National Fire Academy. http://www.usfa.dhs.gov/training/nfa/

National Fire Academy Higher Education. http://www.usfa.fema.gov/training/nfa/higher_ed/

National Fire Protection Association. http://www.nfpa.org/

The National Institute for Occupational Safety and Health. http://www.cdc.gov/niosh/homepage.html

National Safety Council. http://www.nsc.org/

National Transportation Safety Board. http://www.ntsb.gov/

National Volunteer Fire Council. http://www.nvfc.org/

Occupational Safety and Health. http://www.osha.gov/

The Salvation Army. http://www.salvationarmyusa.org/

Texas Association of Fire Educators. http://www.tx-tafe.org/

Underwriters Laboratories Inc. http://www.ul.com/

U.S. Department of Defense. http://www.defenselink.mil/

U.S. Department of Transportation. http://www.dot.gov/

U.S. Fire Administration National Emergency Training Center. http://www.usfa.fema.gov/training/nfa/netc/

U.S. Department of Agriculture Forest Service. http://www.fs.fed.us/

U.S. Department of Homeland Security. http://www.dhs.gov/

U.S. Fire Administration. http://www.usfa.fema.gov/

U.S. Environmental Protection Agency. http://www.epa.gov/

On Scene Suggested Answers

CHAPTER 1

1. Organizations such as the International Association of Fire Chiefs (IAFC) have programs, grants, and scholarships that can help firefighters and departments move forward while embracing fire service history and traditions.
2. The United States Fire Administration (USFA) and other entities listed in the text. Facts and figures from these agencies can be utilized to formulate a plan to start to remedy these issues.

CHAPTER 2

1. Hurricanes Katrina and Rita of 2005 and other weather-related disasters. Volunteers have been trained to respond to these types of calls, and if they feel a sense of patriotic duty to respond to the call, they should not be penalized in their job. Job protection can enhance a person's loyalty to his or her employer and can encourage more volunteerism.

CHAPTER 3

1. The fire service has always attracted different types of individuals. Reasons range from the excitement of the job to the respect and honor they receive from the community they protect. Most firefighters will tell you that they feel a great deal of accomplishment and believe that they are making a difference in the community they serve.
2. Prepare before you ever submit the application of employment. You might not realize it but the interview process starts once the application is submitted when there's a position for hire. Research and practice can make all the difference between success and failure in the application and interview process.

CHAPTER 4

1. The National Professional Development Model. This was developed by the U.S. Fire Administration/National Fire Academy's Fire and Emergency Services Higher Education network.
2. The fire and emergency services field is taking on more responsibility in the industry. For this reason fire departments are setting training and education standards for initial hiring because of this occupation's complexity and increased interest in it. Employers need individuals who can read, write, fill out reports correctly, follow directions, and understand policies and procedures pertaining to the job.

CHAPTER 5

1. The chief should work to build a positive relationship with the volunteer organization. Future plans for building the paid department should include active participation of the volunteers.

2. The chief should approach the city administration to negotiate the purchase of the assets of the volunteer department. A combination department should be formed as a transition department. This would include allowing the volunteers who wish to become full-time paid firefighters to receive the state training required to achieve certification. These firefighters can be absorbed into the paid department as needed. This may reduce some of the opposition to change. As the paid fire department grows, the volunteer department can be reduced.

CHAPTER 6

1. The incident commander must prioritize the course of action. The first priority is the safety of the firefighters. A defensive action may be needed to prevent exposure to and contamination by the chemicals and pesticides.
2. a. Chemical exposure and contamination of personnel and equipment.
   b. Exposure of the residents in the surrounding area to toxic or hazardous airborne pollutants.
   c. Contaminated water runoff.
   d. Possible explosive hazard if fire expands to fertilizer storage area.
3. Water will be an effective extinguishing agent, but it must be used sparingly, and runoff will have to be controlled. Using water will cause the chemicals and pesticides to combine, forming a toxic and hazardous mixture that will contaminate personnel and equipment. Some fertilizers become explosive when wet and then exposed to heat.
4. In this situation, it would be better to let the fire burn to destroy the chemicals and pesticides. Limit firefighter exposure by implementing a defensive strategy and ensuring that all firefighters wear full personal protective equipment and SCBA. An evacuation of areas downwind of the fire would be advisable. The area should be diked to prevent runoff water from entering storm drains or waterways. The EPA should be notified. A hazardous materials response team would have the expertise and equipment to acquire air and water samples to determine the level of contamination. All equipment and personnel must be decontaminated before leaving the area. All contaminated personal protective equipment must be decontaminated and secured so that it will not be used by any personnel. No equipment will be placed back into service until thorough decontamination operations have been conducted. All firefighters who responded to the fire would be required to complete a medical exposure report for documentation in case of future complications.

CHAPTER 7

1. The implementation of a neighborhood fire inspection program may be a tool to allow firefighters an opportunity to get a visual of construction, deterioration, and other changes in the community. Although home inspections are only voluntary, just driving around the community will help firefighters get a better idea of possible problems that may arise during a fire.
2. The two-story addition to the residence might have been detected if a community preincident plan had been developed.

CHAPTER 8

1. The officer must be familiar with the state laws concerning warrentless entry. It would probably be wise to request law enforcement response to the scene.
2. This depends on the state law. Some states defend the rights of landowners if the fire poses no threat to other property.

Second On Scene

1. The U.S. Supreme Court has ruled that warrantless entry is allowed in the event of an emergency. It also has ruled that a fire is an emergency.

   Warrantless entry is allowed if an inspector observes a life-safety violation that can be viewed from a public way.

CHAPTER 9

1. The main issue is determining if the present water supply is adequate to supply this additional demand. If it isn't, a secondary water supply must be determined.
2. The chief should demand that storage facilities be constructed to ensure adequate water supply during drought. The size of the storage tanks will be based on flow requirements for the new development based on residential, industrial, and firefighting needs.
3. The chief should ask if there are plans to expand the development and what is the timeline to do so. He or she should ask the city administration if there are plans to develop industrial or business activities in the area. This will help determine the size of the infrastructure for initial construction.

CHAPTER 10

1. Command, Operations, Logistics, Planning, Finance and Administration, Information and Intelligence.
2. Yes.
3. The Environmental Protection Agency, the Department of Transportation, and the Occupational Safety and Health Agency.

CHAPTER 11

1. Request additional alarms and send apparatus downwind to protect exposures. Prepare for a strong defensive operation.
2. Life safety, evacuation of the buildings in close proximity to the fire and downwind from the fire, Crowd and traffic control, Water supply.
3. Ensure all buildings in the complex are evacuated, set up hose lines to protect exposures downwind, set up heavy stream devices for a defensive attack.

CHAPTER 12

1. Ever since the attacks on American soil, a spotlight has been directed at fire and EMS personnel, and many questions have been raised about the role and services that these professions provide. The IAFF has been instrumental in securing the safety and well-being of firefighters all over the world. FEMA and DHS also have been influential in bringing firefighting issues before political committees. Based on the text, there are several other organizations/associations that could also be mentioned in the answer to this question.
2. State fire marshal's office, state agency (e.g., Texas Commission on Fire Protection), or regulatory agency that has been established in your state. This could be different from state to state.

Answers to Review Questions

CHAPTER 1

1. Ancient Egypt.
2. 1648 at New Amsterdam, New York.
3. A group of men who carried wooden rattles that they would twirl to sound a fire alarm.
4. The apparatus Aldini was one of the earliest breathing apparatus. It was tested by the French in 1825.
5. Maltese cross.
6. The Fire and Emergency Services Higher Education organization was founded in 2000 by the National Fire Academy and the U.S. Fire Administration. Its purpose is to provide strategic approaches to professional development by moving the fire and emergency services from a technical occupation to a profession through higher education.

CHAPTER 2

1. The firefighter recruit is a position that has been created to help a person develop the skills to gain employment as a firefighter. The recruit trains as a firefighter, and a department or organization typically guarantees the recruit a job afterward, whereas a firefighter who attends an academy has to search for a job on graduation.
2. NFPA, FEMA, state and local agencies, the Forestry Service, police departments, medical directors, and so on.
3. The position requires advanced study and learning skills. The training required to become a paramedic is more extensive, and responsibilities at an emergency scene are dual. In many areas, paramedic training takes an additional six months to a year to complete.
4. Contract firefighters overseas, disaster assistance, hazardous materials specialists with a specialty in bioterrorism, and emergency medical specialty teams.
5. A fire prevention specialist teaches citizens about fire prevention and builds awareness to fire safety, whereas a training specialist is a person who provides firefighters or emergency personnel with programs to develop and maintain skilled firefighting performance.
6. Smokey Bear.
7. Dispatcher, fire prevention inspector, safety specialist, salesperson, insurance adjuster, and arson investigator.
8. Fire inspector, industrial fire brigade member, emergency medical technician, or paramedic.

CHAPTER 3

1. A valid driver's license and a GED or high school diploma.
2. Usually securing and filling out an application.
3. a. Subscribe to a notification service.
   b. Check newspapers.
   c. City human relations department.

4. Libraries and bookstores.
5. **a.** Engage in general physical training.
   b. Visit the fire station and use its equipment to prepare.
   c. Become a reserve/cadet or volunteer firefighter, which provides the training and practice at manipulating fire-related equipment.
6. **a.** Have firefighters give you practice with a "mock oral"
   b. Consider possible questions and practice your delivery on tape or in front of a mirror.
   c. Learn all that you can about the fire chief and his or her department before the interview.
7. To let the employer assess your performance, both mentally and physically.

CHAPTER 4

1. Primarily technical.
2. Firefighter I, Firefighter II, Hazardous Materials Operation, Fire Instructor I, Fire Instructor II, Fire Investigator, Public Fire and Life Safety Educator I, Marine Firefighting for Land Based Firefighters—Awareness Level, Fire Officer I, Fire Officer II, Inspector I, Hazardous Materials Technician, and Airport Firefighter.
3. Associate of Applied Science.
4. National and regional.
5. Cadet/reserve programs and volunteer firefighting.

CHAPTER 5

1. Administrative personnel are close to day-to-day operations in the fire station. This close relationship tends to reduce resistance to policy and procedure changes because station personnel are better informed about the reasons for the changes.
2. Pumpers are designed for structural firefighting or for support operations that require sustained pumping activity. Initial attack pumpers are designed to combat small grass and vehicle fires. They may be used on structural fires but cannot be used for sustained operations.
3. Aerial platforms are designed to operate with either a telescoping or articulating box beam or truss beam boom. They are different from aerial ladder platforms, which consist of a heavily reinforced telescoping aerial ladder with a platform at the tip. Personnel use the ladder for rescue and to move equipment and tools to and from the platform. The ladders on the aerial platforms are designed only for emergency evacuation of the platform personnel.
4. Ambulance, hazardous materials apparatus, heavy rescue apparatus, aircraft rescue, and firefighting apparatus
5. The Level B ensemble is similar to the Level A except that it does not give encapsulation protection from vapors. It offers splash protection only.
6. The PASS device is designed to emit a loud audible alarm when the wearer is motionless for more than 30 seconds. It may also be activated manually by the wearer if he or she becomes entrapped but not unconscious. Some PASS devices are also provided with a bright strobe light that serves as a location beacon.

CHAPTER 6

1. There are three components to an atom. The nucleus contains an equal number of protons and neutrons. The protons carry a positive electrical charge; the neutrons have no electrical charge. Negatively charged electrons circle the nucleus at different levels called shells.

2. When heat is transferred to matter, molecular movement increases. As the heat increases, molecules may collide with one another, and the matter may begin to decompose (pyrolyze) with some molecules escaping the surface of the matter as free radicals. Free radicals combine readily with oxygen in a chemical reaction known as oxidation, which is an exothermic or heat-releasing reaction. The newly created heat causes further pyrolysis in a chain reaction that keeps the fire burning.

3. Matter can exist in a solid, liquid, or gaseous state. When in a solid state, the molecules are packed closely together. They cannot move freely but only vibrate. The tighter the molecules are packed, the more dense the matter. The molecules in a liquid are more energetic and are able to move freely and slide past one another. This allows the liquid to flow and take the shape of the container in which it is placed. The molecules in a gas travel in a straight line at high speeds. Matter in the gaseous state occupies much more space than when it is in the solid or liquid state.

4. Heat energy can be classified as chemical, electrical, or mechanical. Chemical energy is released in oxidation reactions that produce heat. Chemical energy reactions can be in the form of spontaneous heating, heat of decomposition, heat of solution, or heat of reaction. Electrical heat energy is the passing of electrons from one atom to another. This movement is called current. There are six forms of heat energy: resistance, arcing, sparking, static, lightning, and induction. Mechanical energy comes in two forms: friction and compression. Friction heating occurs when two objects strike each other. Compression heating occurs when gaseous matter is compressed. As the gas is compressed, the molecules begin to collide with one another and with the walls of the container, and the temperature increases.

5. The four stages of fire are ignition, growth, fully developed, and decay. As soon as the ignition process is initiated, the growth stage begins. During the growth stage, heat from ignition is transferred to the molecules in the fuel, which begin to move more rapidly and collide with one another and escape from the surface of the fuel as vapor. Ignition of the vapor produces more heat, which radiates back to the fuel and initiates more molecular movement. If an adequate oxygen supply is available, this process will continue into the fully developed stage. The fully developed stage culminates with a phenomenon known as flashover. During this stage, burning fire gases travel laterally along the ceiling of the room and radiate heat down to the room's flammable contents. The contents begin to pyrolyze and give off vapors. The ignition of these vapors creates total area involvement. The decay stage occurs when adequate oxygen supply allows sufficient oxidation to deplete the fuel supply to the point that the fire will self-extinguish.

6. Fires are classified by the type of fuel that is burning. Class A fires are fueled by ordinary combustibles, such as wood, paper, and plastic. These products pyrolyze and can be extinguished by cooling with water. Class B fires involve flammable liquids and gases. These fires are extinguished by removing the oxygen from the fuel. Water-based foams create a film on the surface of the liquid, and the foam separates the vapors from the liquid. Class C fires are energized electrical fires. When the electricity is removed, a Class C fire becomes a Class A or B fire. Class D fires are combustible metal fires. Most of these fires require special extinguishing agents. Water may be used on some combustible metal fires but only if a very large quantity is available and used. Some metal fires burn at such high temperatures that water is decomposed into its constituent elements of hydrogen and oxygen and burns away, increasing instead of reducing the fire's intensity.

CHAPTER 7

1. a. NFPA 220
   Type I fire-resistive
   Type II noncombustible

Type III ordinary construction
Type IV heavy timber construction
Type V wood-frame construction

b. NFPA 5000®
Types I and II noncombustible
Types III, IV, and V combustible

2. **a.** First digit represents the fire-resistive rating in hours for the exterior load-bearing walls

   b. Second digit represents the fire-resistive rating in hours for the interior load-bearing walls, beams, girders, trusses, arches, and columns.

   c. Third digit represents the fire-resistive rating in hours for the floor and roof assemblies.

3. Type II construction is noncombustible but unprotected. The unprotected steel has no resistance to fire exposure and can fail when heated to 1,000°F.

4. Columns and foundations are designed to resist compression forces.

5. The floors of old-style Type III structures are designed to collapse without destabilizing the exterior walls. Modern tactics and strategy provide for aggressive interior fire attack. thus exposing today's firefighters to the hazard of structural collapse.

6. **a.** Prestressed concrete components are constructed at a factory and transported to the worksite. Tension is placed on the cable inside the form before the concrete is poured into the form. The concrete bonds with the cable.

   b. Poststressed concrete components are constructed on the worksite. Tension is placed on the cable after the concrete has cured. The cable is not bonded with the concrete.

CHAPTER 8

1. Building codes are designed to address fire safety through building construction and design. Fire prevention codes are enforced by the fire department and focus on fire safety as it relates to building maintenance and use, fire protection systems, and hazardous materials and processes.

2. In 1866, Iowa State Supreme Court Justice John Dillon ruled that local jurisdictions have only the legislative powers that are granted to them by the state because the Tenth Amendment to the Constitution of the United States gives power only to the states, not to local governments. The Cooley Doctrine claims that the Tenth Amendment gives power to the states "or to the people," interpreting "to the people" to include local jurisdictions. The Cooley Doctrine is based on the principle of home rule, which was embodied in the English Carta Civibus Londonarium (1100 AD) and in the Magna Carta (1215 AD). Dillon's Rule was upheld by the U.S. Supreme Court in 1903 and again in 1927.

3. The Fourth Amendment to the U.S. Constitution protects the rights of citizens from unlawful search and seizure. If a fire official is denied entry for the purpose of code compliance inspection, entry can be acquired only through the issuance of a warrant. Warrantless entry can occur only if a life-safety issue exists or if the area being inspected is visible from a public way.

4. Minimum codes are adopted by the state but allow local jurisdictions to implement more stringent code requirements as needed. Mini-maxi codes are statewide codes and prevent local jurisdictions from modifying or adopting any other regulations.

5. Model codes can be adopted by reference or by transcription. If they are adopted by reference, the adopting ordinance simply refers to the edition of the model code that is to be used. If they are adopted by transcription, the entire model code must be published in the adopting ordinance.

6. Prescriptive codes establish specific requirements for building and fire safety. The mandates are based on past fire incidents that resulted in a large loss of life and property. These codes are usually not changed unless there is another tragic incident. They do not identify the fire safety objectives. If the building is constructed using the mandated requirements, fire safety will be achieved. Performance codes specifically identify fire safety goals and objectives and mandate the use of building materials and assemblies that meet those objectives.

CHAPTER 9

1. A water source, a method to deliver the water to its final destination, a method to treat the water, and a method to distribute the water to the individual customers.
2. A gravity system, a direct pumping system, or a combination system.
3. Water distribution systems are categorized as either separate or dual systems.
4. The base-valve dry-barrel hydrant is constructed of cast iron and designed to be freestanding. A valve operating system is located on top of the hydrant for easy access. When the valve is closed, water drains from the hydrant barrel. This hydrant is the most common hydrant used in the fire service. It is found in areas where temperatures are susceptible to dropping below freezing. The barrel of the hydrant is dry because the base-valve is installed below the frost line. Water does not enter the barrel until the base-valve is opened.

   Wet-barrel hydrants do not require a base valve. Individual valves control each outlet on the hydrant. This simple design allows additional hose lines to be attached to the hydrant without disruption of the water flow. Wet-barrel hydrants have fewer moving parts and are easier to maintain than base-valve dry-barrel hydrants. They are located in areas that are not subject to freezing temperatures.

5. 500 GPM from the first standpipe and 250 GPM from each additional standpipe.
6. NFPA 13.
7. Outside screw and yoke (OS&Y) valve, post indicator valve (PIV), wall post indicator valve (WPIV), and post indicator valve assembly (PIVA)
8. The post indicator valve assembly.
9. Fixed-temperature, rate-of-rise, or a combination fixed-temperature/rate-of-rise.
10. Light-obscuration or light-scattering.
11. When Halon 1301 is decomposed by heat, the halogens react with hydrogen formed by the combustion process to create toxic by-products: hydrogen fluoride, hydrogen bromide, and bromine. Decomposition by-products of Halon 1211 include all of these compounds, plus hydrogen chloride and chlorine. Halon 1211 is the more toxic of the two agents. Total flooding CO_2 systems are designed to create a minimum carbon dioxide concentration of 34%. Life cannot be sustained in atmospheres with carbon dioxide concentrations above 9%. Asphyxiation is the primary danger posed by CO_2 systems. Dry chemical agents are nontoxic and nonconductive, but they can irritate the eyes, mucous membranes, and respiratory system. Some agents may cause chemical burns on moist skin. Personnel must always wear full protective equipment, including SCBA, when entering areas that have been extinguished with dry chemical agents.

CHAPTER 10

1. **a.** Unity of command means everyone answers to only one person.
   b. Chain of command means all communication and decision making travels in an orderly fashion up and down an interlinked chain.
   c. Span of control means the organization is set up so that people do not supervise more personnel than they can manage effectively, usually three to seven.

    d. Division of labor means the jobs to be done are divided into manageable workloads and allow for specialization where necessary.

    e. Delegation of authority means that when personnel are assigned a task, they are empowered with the authority as well as the responsibility to see that it is completed.

    f. Exception principle means the supervisor is advised of situations or occurrences that are out of the ordinary.

2. This chart should begin with the fire chief at the top and, by using various titles, proceed down through company officer to firefighter. The answer to this question can include any of the specialty positions of line and staff.

3. The incident management system was developed as a result of a disastrous wildfire in 1970 in California. There was no command and control, resources were not effectively used, and communication between agencies was nonexistent.

4. FIRESCOPE ICS was developed for large wildland fires and coordinating multiagency response. FGC was a simpler command system that focused on structure fires, hazardous materials, and mass casualty incidents.

5. ICS has five main components: command, planning, finance, logistics, and operations.

6. **a.** Command controls the incident at the strategic level and has authority to control resources, establish functional areas, and ensure the safety of all responding personnel.

    b. Operations is responsible for the implementation of the tactical assignments, fire, EMS, hazardous materials operations, and so on.

    c. Planning is responsible for evaluating incident information and developing the incident action plan. Its function is to provide the incident commander with up-to-date status reports on resource requirements and allocation, situation development, incident control, responder injuries, intelligence, and suggestions for stabilizing an incident.

    d. Logistics provides all of the support requirements for the incident operations, including facilities, family assistance centers, transportation, supplies, and equipment, and so on.

    e. Finance/Administration is responsible for documenting the cost of materials and personnel.

7. With the advent of the terrorist threat toward the United States, it is evident that emergency operations will include multiagency or multijurisdictional responses. The NIMS gives these organizations a common command structure, common terminology, and common communications. NIMS establishes who is in command and allows for a unified command on very large and complex incidents.

CHAPTER 11

1. Strategy is a broad-based plan for controlling an incident based on standard operating procedures and focuses on life safety, incident stabilization, and property conservation. Tactics involves the actual deployment of resources to resolve the situation and focuses on rescue, exposures, confinement of the hazard, extinguishment, overhaul, ventilation, and salvage.

2. The strategic plan identifies the task to be accomplished, identifies the resources needed, and specifies personnel requirements.

3. It is the most effective way to identify the hazard and determine the amount and availability of resources needed should an emergency occur.

4. Water supply, apparatus and equipment, available qualified personnel, weather conditions, topography, building construction, communications, and command structure.

5. OSHA 1919.134 and NFPA 1500.

CHAPTER 12

1. The Fire Apparatus Manufacturer's Association, the National Fire Protection Association, and fire service periodicals buyers' guides.
2. **a.** To receive calls for service.
   b. To dispatch calls.
   c. To track resources and their availability.
3. The International Association of Firefighters, state and local firefighters associations.
4. The Chemical Transportation Emergency Center (CHEMTREC).
5. Weather is one of the most changeable factors during any incident. On hazardous materials incidents, weather can be a major factor in how vapor clouds will act and can become a factor in determining if, when, and where evacuation is required.
6. The American Red Cross.

Stop and Think Suggested Answers

CHAPTER 1

Box 1.1

1. Implement the 16 Firefighter Life-Safety Initiatives.
2. Collect data and research to identify hazard areas that need to be improved.

CHAPTER 2

Box 2.1

1. Industrial fire brigades and contract firefighting in the United States and overseas.
2. Fire protection engineer or fire and emergency services consultants.

Box 2.2

1. Budget constraints.
2. Cost-effectiveness.
3. Human resource benefits are at a minimum. Specialized training is not available to local host nationals. The U.S. military is cutting back on its personnel, so it needs all members of its team to do what they are specifically trained to do—fight wars and protect our boundaries. Furthermore, overseas jobs offer incentives that conventional jobs do not, such as extremely high pay for hazard duty, travel opportunities, and tax-free incentives.

CHAPTER 3

Box 3.1

It is important for fire departments to have a baseline physical for each firefighter in case he or she ever becomes ill for unknown reasons. Unfortunately, there are still several fire departments that do not require a medical physical for an applicant when hired. Because firefighting is a hazardous occupation, firefighters must be monitored throughout their careers in case they are ever exposed to a hazardous material. This must be documented and kept on file in case the firefighter later develops symptoms from the exposure. Fire departments also need to develop and implement medical screenings and physical fitness standards that are equally applicable to all firefighters to make sure prior to hiring that the individual can perform the job in extreme working conditions. Firefighters must be in good health and in good physical condition to perform at a moment's notice.

Box 3.2

1. The definition given for diversity in this chapter is ethnic, gender, racial, and socioeconomic variety. However, diversity in America is much more than a simple list of differences. Diversity is at the core of what our country is, and it is our respect for diversity that draws so many people from around the world to our shores.

2. It is important to have diversity in the fire service because the service should reflect the entire composition of the community. Small towns to large cities often have residents from different ethnic backgrounds, and the fire service should reflect the workforce in these areas.

3. The benefits of having a diverse workforce in fire and emergency services is a better working relationship with the community. An individual who does not speak English will receive better patient care from a firefighter who speaks the same language. A female patient will feel more comfortable talking to a female firefighter/paramedic than to a male firefighter/paramedic. Diversity in the workforce will allow better customer service to the taxpayers who provide the funding for the service.

CHAPTER 4

Box 4.1

1. Rope rescue, surface water rescue, vehicle and machinery rescue, confined space rescue, structural collapse rescue, trench rescue, dive rescue, and wilderness rescue.

2. NFPA 1006, *Standard for Rescue Technician Professional Qualifications*.

3. Some NFPA standards require that an individual accrue a certain number of continuing education hours annually. Furthermore, most state fire regulatory agencies require that certified individuals must complete a certain number of continuing education hours annually to stay certified in a particular discipline.

Box 4.2

1. It is important to have a seamless transition between courses taken by fire and emergency services personnel to ensure that all necessary subjects are covered and to reduce duplication of subject matter between courses.

2. A national professional development model would assist local, state, and federal government agencies in deploying resources and fire and emergency services personnel in the event of a natural disaster or terrorist incident.

CHAPTER 5

Box 5.1

Rural communities may need specialized apparatus, such as wildland (brush truck) apparatus or mobile water supply apparatus, which are not needed in some urban areas.

Box 5.2

Entering a hazard zone with inadequate protective clothing is like entering a hazard zone with no protective equipment at all.

CHAPTER 6

Box 6.1

The rate of heat release and intensity of the combustion process depends on the form of the fuel. Solids, liquids, and gases react differently to various environmental conditions, depending on their chemical composition. The form of the fuel also determines the strategy and tactics used in mitigating the situation.

Box 6.2

Firefighters should familiarize themselves with the types of hazards that exist in their jurisdictions. This can be accomplished only by conducting thorough preincident inspections and developing a preincident plan.

Box 6.3

Firefighters should always be aware of potential electrical hazards. These hazards are not limited to downed power lines. Caution must be taken when using aerial apparatus around any power lines or during stormy weather. Polyester material, often part of the firefighter's uniform, has a tendency to develop a static charge as the wearer moves. Flammable atmospheres created by gas or vapor leaks can be ignited by static discharge.

CHAPTER 7

Box 7.1

When structural components are weakened by fire, they may shift from their normal position and direct loads in a way that the support structure was not designed to withstand. This may cause a domino effect in which structural components will fail in sequence as loads and forces shift, weakening structural integrity. This is also referred to as a *progressive collapse.*

Box 7.2

Even though Types I and II structures are noncombustible, they are limited in their fire resistance. Unprotected steel structures are susceptible to early collapse when exposed to intense heat. It is very important that firefighters obtain information about the contents of Types I and II structures in their respective areas of response. The types of contents determine the fire load and rate of heat release to which the structure will be exposed if the contents burn. This will help firefighters estimate the length of time that the building will maintain structural integrity under heavy fire conditions.

CHAPTER 8

Box 8.1

Unscheduled inspections may be performed in response to a written complaint concerning a code violation. In some jurisdictions, if the violation is witnessed from a public way, an unscheduled inspection may be requested.

Box 8.2

Failure to perform a follow-up inspection after documenting a code violation will result in the inspecting officer's being held legally liable for any damage, deaths, or injuries resulting from a fire that is related to the violation that was documented.

CHAPTER 9

Box 9.1

Many factors may cause a large drop in residual pressure. First, a review of the water main maps in the area is needed to determine the main size and age. If the main is equal to the size of the mains in areas tested and is determined to be adequate, a report of the pressure drop should be made to the water department because it may indicate a possible partial blockage of the main. This can greatly affect fire operations in the area. Low residual pressure will reduce the ability of an apparatus to provide an adequate fire flow. This will limit the handlines or master stream that the apparatus can supply.

Box 9.2

Attempting to use the extended hose line from a Class II system may prove to be a fatal error. The water pressure may be low or nonexistent. Even if the pressure is adequate, the GPM flow may not be, especially if this is a well-involved fire. Extending hose lines up a stairwell is a time-consuming, arduous task and may be the only solution. If the situation allows, an aerial ladder or platform can be raised to a window on the fire floor. If the apparatus is not equipped with handline outlets, the nozzle tip on the master stream device can be removed and replaced with a gated wye (a valve with reduced outlets to which handlines can be attached). This allows the master stream device on the aerial apparatus to be used as a Class I standpipe.

Box 9.3

Most newspaper printing rooms have total flooding or local application carbon dioxide extinguishing systems. When these systems activate, not only is the oxygen displaced in the room, but the condensed environment also forms a fog that limits visibility. It is imperative that all responding personnel wear full protective equipment, especially self-contained breathing apparatus.

CHAPTER 10

Box 10.1

1. The finance sector can be divided into smaller sectors with each sector having an assigned sector officer. The compensation and claims sector records all operational damage claims by citizens. The procurement sector ensures that no leased or rented equipment will be used unless there is a written contract of agreement. Records of all costs are documented by the cost sector. The time sector ensures that no personnel are allowed to work unless they are logged in and out and all hours worked are documented on a time sheet.
2. In order to receive aid from FEMA, an accurate accounting of all expenditures must be maintained. FEMA will reimburse 75% of the community's cost of a disaster. Small business loans are available to citizens who have suffered property loss.
3. For a community to qualify for reimbursement, FEMA requires that accurate spending records be maintained throughout the emergency. Written contracts must accompany all expenditures for leased or rented equipment, and records of operational costs for the equipment must be made available. Time sheets must be kept on all personnel working during the operation, including volunteers. If these records are not maintained properly, FEMA can deny any claim and leave the community to absorb all of the costs for response and recovery.

Box 10.2

1. Emergency response personnel should plan for a worst-case scenario and develop the NIMS structure that would be utilized should the event occur. After the plan is developed, emergency response personnel should conduct an exercise to determine the strengths and weaknesses of the plan. The exercise will also identify the responsibilities of all participants and their function in the National Incident Management System.
2. Conducting an exercise will help the incident commander identify the part of NIMS that needs to be activated. NIMS will identify which agencies need to be involved.

CHAPTER 11

Box 11.1

There is a great increase in occupancy load and live load forces. The increased amount of equipment, furniture, people, and the weight of the water for the aquarium adds live and dead loads that were not considered in the original design of the structure. The structure had to be reinforced to withstand these additional loads.

Box 11.2

1. Topography influences fire operations regardless of the geographical area. In some areas of the country, topography may be advantageous to emergency response and

operations. In other areas, the topography may be viewed as an obstacle to emergency response and operations. This can be determined only by establishing a preincident strategic plan and exercising that plan so that all participants understand their roles and responsibilities.

2. It is very difficult to gain permission to create a prefire plan for a residential structure. This can be done only by obtaining permission from the occupants of the residence or, under limited circumstances, by obtaining a warrant. The best way to obtain information on a private residence is during its construction. The building contractor must have a permit that allows for periodic inspections by the building inspectors. Although a fire company may not have legal authority to inspect the property, depending on the local fire and building codes, it may get permission from the building contractor to perform a walk-through to familiarize personnel with the type of construction and the basic floor plan.

3. Weather, in conjunction with topography, can create major problems for strategic and tactical planners. All areas of the country experience seasonal changes as well. Strategic planners should incorporate response and operational alternatives into the plan to handle possible extremes due to the time of year or to extreme weather conditions.

Box 11.3

1. Barn fires in a rural setting are very dangerous. These structures often contain such hazardous materials as flammable gases (acetylene and propane), flammable liquids (gasoline and diesel fuel), ammonium nitrate (fertilizer), and various types of pesticides and herbicides.

2. Many of the hazardous materials found in rural settings can also be found in urban areas. Flammable liquids and gases are common to both environments, as are pesticides and herbicides. Powdered chlorine is found more often in an urban setting, especially in residential areas that have private swimming pools. This can pose a severe problem to firefighters. Burning chlorine gives off vapors that cause hydrochloric acid to form wherever there is moisture. Standard structural firefighting protective clothing is not effective in this type of situation. If dry chlorine powder mixes with certain hydrocarbons, a violent exothermic reaction occurs, producing high heat and toxic fumes.

Box 11.4

The people in the involved structure should be rescued first. The five people in the exposed building on the north side of the fire should be rescued next, because they are the largest number immediately threatened by the fire. The 20 people on the south side of the fire should be rescued last, because they are the victims least exposed to danger. The fire is moving away from them.

CHAPTER 12

Box 12.1

1. The American Red Cross provides temporary shelter and food for fire victims.
2. The American Red Cross will provide a cantina that stocks food and beverages for first responders during an extended emergency operation.
3. The American Red Cross provides financial assistance to victims when it is deemed necessary.
4. The American Red Cross helps to establish shelters for displaced victims in the event of human-caused or natural disasters.

Expand Your Learning Activities

Newark Fire Department
Standard Operating Procedure

OCCUPATIONAL SAFETY AND HEALTH			
Subject:	***INCIDENT RESPONSE***	**SOP#**	**403.03**
		IAFF	
		Notice:	6/8/05
Written By:	CAPT. SPURGEON	**Initiated:**	7/1/05
Approved:		**Revised:**	
Fire Chief:	JACK STICKRADT	Date:	5/19/05
Safety Director:	KATHLEEN BARCH	Date:	6/6/05

I. Scope

   A. This standard applies to the driver of a vehicle owned or operated by the department while responding to an incident. It was developed to establish safety guidelines during *Priority One* and *Priority Two* responses.

   B. This standard was developed to ensure that the department's response to both *Priority One* and *Priority Two* incidents are safe, appropriate, and efficient.

II. Categories of Response

   A. *Priority One:* Those incidents that pose a significant risk to life or property. *Priority One* response requires the use of all audio (siren and air horns) and visual (lights) warning devices in accordance with ORC 4511.041. These devices must be in use during the entire duration of the response unless an incident commander downgrades the response to a *Priority Two.* Units responding on a *Priority One* status shall identify such by the radio transmission "*Priority One*" (i.e. "Rescue One is responding *Priority One* to 123 Your St.") The initial response to the following types of incidents shall be considered *Priority One* emergencies:

      1. *A reported fire in a structure.* This will receive a full first-alarm assignment.

         1. A full first-alarm assignment shall normally consist of all stations with a minimum of 2 engines, 1 ladder company, and 1 medic unit.

            a. A "company" is defined as a firefighting apparatus staffed by at least three personnel.

         2. If "nothing showing" is reported by the incident commander, all companies excluding the first due shall downgrade their response to *Priority Two.*

(continued)

Newark Fire Department
Standard Operating Procedure

Subject:	*INCIDENT RESPONSE*	SOP#	403.03

2. *A reported fire outside of a structure that involves the potential destruction of property or poses a risk to human or animal life* (e.g., dumpster, refuse, brush, etc.) This will receive 1 engine or ladder company assignment.

3. *A reported car fire.* This will receive 1 engine or ladder company assignment.

4. *Carbon monoxide checks with a reported illness.* This will receive 1 engine or ladder company and 1 medic unit assignment.

5. *Automatic fire alarms* shall receive a full assignment but are emergencies for the first due engine company only. All other units shall initially respond *Priority Two* and can be upgraded if needed.

6. *Inside gas leaks* shall receive the same initial assignment as an automatic fire alarm.

7. *Outside gas leaks* shall receive 1 engine or ladder company as the initial assignment.

8. *Mutual aid requests for fire support* shall receive 1 engine or ladder company as the initial assignment.

9. *Life-threatening Priority One medical incidents* (reports of unresponsiveness or unconsciousness) shall receive 1 medic unit and 1 engine or ladder company as the initial assignment.

10. *On all other Priority One medical incidents where an assist is required* the first due unit shall respond on a Priority One status and the additional unit shall respond Priority Two.

11. *On all Priority One medical incidents where an assist is not required* the response shall consist of 1 medic unit.

12. *Violent scenes* (stabbings, shootings, or fights in which a weapon is involved) shall receive the same assignment as a report of unresponsive or unconscious with the addition of 1 command unit. No units shall respond on a *Priority One* basis unless law enforcement personnel have secured the scene.

    1. In the event that a unit arrives on an incident scene that has not been secured, those unit(s) shall stage.

B. *Priority Two:* Those incidents that do not pose a significant risk to life or property. Audio and visual warning devices shall not be used during *Priority Two* responses unless ordered by an incident commander to upgrade the response to *Priority One* status. Units responding to such incidents shall identify themselves only as responding (i.e. "Rescue 1 is responding 123 Your St.") The initial response to the following types of incidents shall be considered *Priority Two:*

    1. Carbon monoxide checks without a report of illness.

    2. Public service calls to assist the public when there is no immediate threat to life or property.

III. Response Guidelines

A. Apparatus and vehicles engaged in *Priority Two* responses shall obey all applicable traffic safety rules and regulations and shall not exceed the posted speed limit.

B. Any and all traffic preemptive systems shall be utilized before exiting a division building prior to a *Priority One* response.

Newark Fire Department
Standard Operating Procedure

Subject:	*INCIDENT RESPONSE*	SOP#	403.03

C. Apparatus and vehicles engaged in a *Priority One* response shall at all times govern their response by the traffic, weather, and road conditions present at the time of response.

   1. The maximum speed of travel shall *not* exceed posted limits by more than 10 mph.

D. During a *Priority One* response, drivers shall bring their vehicles to a complete stop for any of the following:

   1. When directed by a law enforcement officer.
   2. Stop signs.
   3. Red traffic signals.
   4. Blind intersections.
   5. When the driver cannot account for all lanes of traffic in an intersection.
   6. When other intersection hazards are present.
   7. When encountering a stopped school bus with flashing warning lights.

E. Drivers shall proceed through an intersection only when the driver can account for all lanes of traffic in the intersection.

F. Drivers shall bring their vehicles to a complete stop at all railroad grade crossings and shall not cross the tracks until instructed do so.

IV.   Responsibilities

A. Drivers shall be directly responsible for the safe and prudent operation of their vehicles in all situations.

B. When a driver is under the direct supervision of an officer, the officer shall assume responsibility for the actions of the driver and shall be responsible for immediately correcting any unsafe condition.

Glossary

ability/agility test A demanding physical test that can include dry hose drag, charged hose drag, halyard raise, roof walk, attic crawl, ventilation exercise, victim removal, ladder removal and carry, stair climb with hose, ceiling breach and pull, crawling search, or stair climb with air bottles and hose hoist—each measuring physical strength and stamina.

absolute temperature Temperature that is measured from absolute zero.

absolute zero The temperature at which all molecular movement ceases. This is established at −459.67° Fahrenheit, 0° Rankine (based on the Fahrenheit scale), −273° Celsius, and 0 Kelvin (based on the Celsius scale).

advanced life support units (ALS) Medical units that provide advanced medical treatment and transport to the severely injured or ill.

aerial ladder A rotating, power-operated ladder mounted on a self-propelled automotive fire apparatus.

aerial ladder platform A power-operated aerial device that combines an aerial ladder with a personnel-carrying platform supported at the end of the ladder.

aerial platform apparatus A fire apparatus that carries a hydraulically operated elevating platform that can be utilized for rescue, fire, or observation operations.

andragogy The term relevant to lifelong learning for adult education methodology. Distance education is particularly suited for andragogical concepts.

apparatus Aldini One of the earliest recorded attempts to improve breathing difficulties in toxic environments was tested in 1825 in France. This was not the most desirable breathing apparatus for firefighters to wear, but scientific testing was conducted under actual fire conditions.

articulating boom A boom constructed of several sections that are connected by a hinged joint that allows the boom to function much like the human elbow.

aspect The slope of land that is exposed to the sun.

Associate of Applied Science Degree program usually two years in length.

atom The smallest particle that constitutes the matter of an element.

axial loads Loads that are placed perpendicular to the central axis of the support structure and utilize the full support capabilities of that structural component.

baccalaureate programs Degree programs usually four years in length. Students receive a BS or BA degree.

backdraft event Instantaneous explosion or rapid burning of superheated gases that occurs when oxygen is introduced into an oxygen-depleted space.

background check Verification that a candidate's experience or credentials are what they represent.

Bambi Bucket A brand of collapsible bucket.

barriers Natural or artificial firebreaks for controlling fire spread.

basic life support units (BLS) Medical units that provide medical and transportation needs for patients who do not have life threatening injuries or illness.

bioterrorism Relating to or involving the use of toxic biological or biochemical substances as weapons of war.

branch lines The smallest pipes of a sprinkler system on which the individual sprinkler heads are mounted.

British Thermal Unit (BTU) The amount of heat required to raise the temperature of one pound of water at 60°F by 1°F at constant pressure.

building codes Laws adopted by states, counties, or cities or other government sources to regulate all aspects of building construction.

burn building Training tower used so that new recruits can practice their skills.

caloric (or calorific) value The amount of heat necessary to raise the temperature of one gram of water by 1°C.

ceiling breach and pull This event is designed to simulate the critical task of breaching and pulling down a ceiling to check for fire extension. It uses a mechanized device that measures overhead push-and-pull forces and a pike pole. The pike pole is a commonly used piece of equipment that consists of a six-foot-long pole with a hook and point attached to one end.

certification programs- Shorter educational programs that provide a certificate verifying expertise in a specific trade or discipline.

chain reaction of self-sustained combustion When a fuel is heated to the point that it releases flammable vapors, which burn and generate more heat on the fuel, which releases more flammable vapors.

citation Same as a summons except it is issued by a law enforcement officer instead of a judge.

combination flame detectors Alarm devices that can detect abnormal conditions by several means.

combination system A water supply system that uses mechanical pumps and gravity to supply operating pressure for moving water.

compensation and claims unit Part of the finance branch of ICS that documents property loss by citizens during a disaster.

compressed air foam system (CAFS) A high-energy foam-generation system consisting of an air compressor (or other air source), a water pump, and foam solution that injects air into the foam solution before it enters a hoseline.

compressive force A load that causes structural elements to shorten.

compressive stress The stress applied to materials that results in their decrease in volume.

computer-aided dispatch systems (CAD) Use a wide range of software and hardware to help manipulate data for engineering purposes, coordinate the responses of emergency personnel and resources, and keep incident reports and share them with other agencies.

conflagration A fire that increases in size and spreads beyond human-made and natural barriers.

corbels Pieces of stone jutting out of a wall to carry weight.

cost unit Part of the finance unit that tracks the cost of equipment and supplies.

couvre feu A French name for a metal lid used to cover an open hearth. The English word *curfew* was derived from this name.

criminal conviction record check A search of an individual's history of any convictions under the law for felony, misdemeanor, and motor vehicle convictions.

crossmains Pipes that supply the branch lines from the main riser. They are slightly smaller than the riser but are larger than the branch lines.

dead load The structure and any equipment or building components that are permanently attached to the building.

Department of Homeland Security Created in 2002 primarily from a conglomeration of existing federal agencies in response to the terrorist attacks of September 11, 2001. Its purpose is to protect the nation against threats to the homeland. Its efforts affect firefighters and emergency personnel.

direct pumping system A water supply system that relies on mechanical pumps to move water.

discretionary authority Allows the fire official to use discretion when enforcing the law.

distribution grid   The system in which power travels from a power plant to individual customers.

distributor pipes   Piping located inside a building.

diversity   Ethnic, gender, racial, and socio-economic variety in a situation, institution, or group.

drafting pit   A poured-in-place concrete pit used to simulate drafting operations.

drill tower   A tower and training facility for firefighters.

eccentric loads   Loads that are placed perpendicular to the support structure but do not pass through the central axis, thus causing the support structure to bend.

electrons   Negatively charged particles that orbit the nucleus of an atom.

elevation   The height of the land above sea level.

Emergency Management Institute (EMI)   Serves as the national focal point for the development and delivery of emergency management training to enhance the capabilities of federal, state, local, and tribal government officials; volunteer organizations; and the public and private sectors to minimize the impact of disasters on the American public.

EMT   Emergency Medical Technician. A specified level of medical training that usually consists of around 100 hours of classroom and on-site training.

encrustation   The buildup of chemical deposits and biological growth on the inside walls of a water pipe, which reduces the inside diameter of the pipe.

endothermic reactions   Chemical reactions in which substances absorb heat energy.

energy   The capacity to do work.

engineering corrections   Corrections that enhance fire safety. These corrections may include the installation of fixed fire protection systems, such as automatic sprinklers.

enhanced 9-1-1   An emergency phone system that automatically gives a dispatcher the caller's location.

ergonomic   The scientific principle that bases design around human needs, and the profession that applies science and data to this concept.

exothermic   Chemical reaction between two or more materials that changes the materials and produces heat.

external exposures   External areas or persons in close proximity that can be threatened by a fire.

FAA   An agency in the U.S. Department of Transportation that oversees all aspects of civil aviation.

familia publica   "Servants of the commonwealth." These men were strategically positioned near the city gates to protect the city from fire.

Federal Emergency Management Agency (FEMA)   The federal agency responsible for supporting state and local municipalities during a disaster.

FESHE   Fire and Emergency Services Higher Education. Working with coordinators of two- and four-year academic fire and emergency medical services (EMS) degree programs, the U. S. Fire Administration's National Fire Academy (NFA) has established the FESHE network of emergency services–related education and training providers. The FESHE mission is to establish an organization of postsecondary institutions to promote higher education and to enhance the recognition of the fire and emergency services as a profession to reduce loss of life and property from fire and other hazards.

fire ecology   The study of the interrelationship of wildland fires and living and nonliving things in the environment.

fire flow   Quantity of water available for firefighting in a given area.

fire ground command system (FGC)   The incident management system developed for fire ground and emergency medical operations.

fire lines   Boundaries around a fire area to prevent access except for emergency vehicles and relevant professionals.

fire load   The amount of fuel in a compartment expressed in pounds per square foot obtained by dividing the amount of fuel present by the floor area.

fire prevention codes Body of laws designed to enforce fire prevention and safety.

fire science Study of the behavior, effects, and control of fire.

fire wardens Volunteers who patrolled the cities, inspected chimneys, and issued fines for noncompliance. Funds collected from fines were used to purchase new fire equipment.

firemark Metal marker that used to be produced by insurance companies for identifying their policyholders' properties.

fireplug A wooden plug inserted in logs that carried the town's water supply approximately every half block. These plugs could be removed to access the water supply for firefighting.

FIRESCOPE Firefighting Resources of California Organized for Potential Emergencies.

flash point Minimum temperature at which a liquid gives off enough vapors to form an ignitable mixture with air near the liquid's surface.

flashover (rapid fire progress) Stage of a fire at which all surfaces and objects within a space have been heated to their ignition temperature and flame breaks out almost at once over the surface of all objects in the space.

flow pressure The pressure created by the forward velocity of a fluid at the discharge orifice.

friction loss Pressure loss caused by turbulence as water flows through a conduit.

GIS Provide access not only to street maps but also to geographical data such as water sources, hazardous materials storage sites, residences with disabled persons, and municipal hydrants.

gravity system A water supply system that uses only gravity to move water.

The Great Fire of Boston Boston's largest urban fire and one of the largest fires in American history.

The Great Fire of London A devastating fire in central London that was considered one of the major events in the history of England.

groundwater Water located beneath the ground surface.

gusset plates Metal plates used to unite multiple structural members of a truss.

hand pumper A fire engine that has long bars (also known as brakes or pumping arms) running parallel to the body which operate the pump. Firefighters physically use the pumping arms to make a full up and down motion to build up pressure allowing water to spray out of a hose.

head pressure The height of the surface of the water supply above the discharge orifice is equal to 0.434 pound per square inch (psi) per foot of elevation.

heat of combustion Total amount of thermal energy that could be generated by the combustion reaction if a fuel were completely burned.

heavy rescue squad A team of firefighters who are highly skilled in various types of rescue operations that require the use of highly specialized equipment and training.

helitack Used during wildfires when the location of the fire(s) is inaccessible to firefighting crews.

higher education Obtaining knowledge for certifications, professional development, or college credit. The act of accumulating new facts to augment your personal and mental database.

high-pressure systems Carbon dioxide is stored at room temperature in standard Department of Transportation–approved cylinders at a pressure of approximately 850 pounds per square inch.

Homeland Security The department created after the events of 9/11 that is responsible for assessing the nation's vulnerabilities.

hotshot Highly trained firefighters used primarily in handline construction.

hydraulically operated Equipment that operates by utilizing pressure created by compressing oil in a cylinder.

hydrophoric materials Substances that react with water

IFSAC Focuses on ensuring that training and certification within its member jurisdictions meet strict National Fire

Protection Association Professional Qualification standards. IFSAC is divided into two Assemblies: the Certificate Assembly and the Degree Assembly.

incident command system (ICS) A blueprint for organizing the response to an emergency situation.

incident effectiveness The ability of the department's personnel to perform their functions safely and effectively during an emergency. This requires physical fitness, training, and equipment.

Incident Management System (IMS) A merger of FIRESCOPE'S organizational design and command structure with the tactical and procedural components of the Fire Ground Command system.

indicating valve A valve that can be visually inspected to determine whether it is open or closed.

infrared flame detectors Flame detectors that operate in extreme environment conditions.

infrared thermal imaging camera (TIC) Gives firefighters the ability to see heat generated by an object or person.

initial attack pumper A fire pumping apparatus that is utilized for fast response and rapid attack on a fire.

injunction A legal document from a judge that orders a person or entity to cease or perform certain activities.

internal exposures Areas or persons in close proximity that can be threatened by a fire within the involved structure.

ionization detectors Smoke detectors that use a finite amount of radioactive material to make the air within a sensing chamber conduct electricity.

latent heat The amount of heat energy absorbed or released by a substance when it changes physical states without a change in temperature.

latent heat of fusion The amount of heat absorbed by a substance when converting from a solid to a liquid at the same temperature.

latent heat of vaporization The amount of heat absorbed by a substance when converting from a liquid to a gas or vapor at the same temperature.

Level A Ensemble Highest level of skin, respiratory, and eye protection that can be afforded by personal protective equipment.

Level B Ensemble Personal protective equipment that provides the highest level of respiratory protection but a lesser level of skin protection.

Level C Ensemble Level B personal protective ensemble with filtered respirators for respiratory protection.

Level D Ensemble The standard work uniform worn at the fire station.

light-obscuration systems Smoke detection systems that are activated when smoke particles block a light beam from a photoelectric light-sensitive cell.

light-scattering systems Smoke detection systems that are activated when light beams are scattered by smoke particles and strike a light-sensitive photoelectric cell.

line operation Those activities that involve the direct physical tasks necessary for the completion of the mission.

load-bearing A component of a structure that is designed to bear the weight of the structure and loads and transmit this weight to the foundation.

local application systems Permanent suppression systems that are required to cover a protected area with 2 feet of foam depth within two minutes of system activation.

low-pressure systems Liquefied carbon dioxide at 0°F is contained in large refrigerated storage tanks pressurized to 300 pounds per square inch.

Maltese cross Symbol representing the traditions and ideals of the fire service. The arms and the tree of the Maltese cross are equal in length. The arms and the tree widen as they extend from a central point. In most cases, the edges are flat or curved slightly outward.

mandatory authority Makes the actions mandatory under the law.

manipulative training Hands-on learning of operations, equipment, and tools. In the fire service, this type of learning exhibits safety and proper techniques.

matter The substance or substances of which all physical things are composed. It occupies space, has weight, and can be measured. It exists as a solid, liquid, or gas.

mini-maxi code Establishes minimum and maximum building code requirements.

minimum codes Lowest acceptable standards.

mobile data terminal (MDT) Mobile computer that communicates with other computers on a radio system.

model codes Building codes developed and maintained by a standards organization separate from the governing body for building codes.

molecule Groups of two or more atoms.

multipurpose dry chemical systems Dry chemical systems that use extinguishing agents that are rated to be effective on more than one type of fuel.

National Board of Fire Underwriters (NBFU) Organized in 1866, the NBFU developed standards for fire safety in building construction and control of fire hazards.

National Fire Academy (NFA) Concerned with education—designing fire control courses and programs and making them available to firefighters across the nation.

National Fire Service Incident Management System Consortium (NFSIMSC) A nonprofit corporation with the mission of sorting out issues and ironing out difficulties that arise as the national incident management effort unfolds

National Incident Management System (NIMS) A reconstruction of NIIMS to accommodate additional needs arising from the nature of the new threat—terrorism—that the nation now faces

National Interagency Incident Management System (NIIMS) A total systems approach to risk management that integrates all levels of national emergency incident planning—federal, state, and local.

negligence The failure of a fire official to perform in the same manner a prudent person would under the same circumstances.

neutron A particle in the nucleus of an atom that has no electrical charge.

NFPA National Fire Protection Association.

NFPA 1001 Standard for firefighter professional qualifications.

NFPA 1041 A standard that requires instructors to be cognizant of the safety of their students to ensure that classroom and practical evolutions are conducted in a safe, controlled manner. Prerequisites and requirements are mandated.

NFPA 1221 Standard for the installation, maintenance, and use of emergency services communications systems.

NFPA 1403 A standard that provides the framework to improve safety during live fire evolutions and covers conditions such as safety, site preparation, water supply, training plan, fuel, and ventilation.

NFPA 1582 *Standard on Comprehensive Occupational Medical Program for Fire Departments.* Standard covers minimum medical requirements for firefighters, including full-time or part-time employees, and paid or unpaid volunteers.

NIMS National Incident Management System.

non-fire-suppression activities Any fire department activity that is not directly involved in fire suppression.

nonindicating valve A valve that has to be physically operated to determine if it is open or closed.

non-load-bearing A partition or wall that supports only itself or interior finish components.

nonpressurized dry hydrant system A hydrant system that is used to draft water from a static water source.

normal atmospheric pressure The pressure of the atmosphere at sea level is 14.7 pounds per square inch.

NREMT National Registry of Emergency Medical Technicians.

occupancy General term for how a building structure or home will be used and what it will contain.

occupation classification Classification given to a structure by a model code used in that jurisdiction and pursuant to the given use of that structure.

ordinance A law that is established by a local jurisdiction, such as a county, city, or town.

ordinary dry chemical systems Dry chemical systems designed primarily to extinguish flammable liquid fires (Class B fires).

OSHA Occupational Safety and Health Administration.

OSHA Code of Federal Regulations 29 CFR 1910.120 The federal code that mandates the level of training for anyone working with hazardous materials.

oxidation Chemical process that occurs when a substance is combined with oxygen.

paramilitary organization A group of civilians organized in a military fashion.

permit model Businesses that have a high potential for loss of life, such as nightclubs, restaurants, and hazardous materials processing and storage facilities, require a permit to operate.

personal alert safety system (PASS) Electronic lack-of-motion sensor that sounds a loud tone when a firefighter becomes motionless.

photoelectric detectors Smoke detectors that use small light source to detect smoke.

pictometry An enhancement of GIS that generates a 3-D perspective of a given area.

preincident planning Assessing risks and prioritizing potential hazards, then developing a strategic plan to mitigate the hazards.

prescribed fire A fire that is purposely ignited by fire personnel or agencies under controlled conditions for specific management objectives.

pressurized dry hydrant system Piping grid that connects several dry hydrants. A fire department connection is placed at the base of the grid near a water source. During firefighting operations, a pumping apparatus connects to the water supply and supplies water to the dry hydrant system.

primary feeder mains Large-diameter pipes that form the outside perimeter of the water distribution grid and receive water from the water source.

Pro Board National Board on Fire Service Professional Qualifications Accrediting Fire Service Training Organizations.

probationary period Introductory period of employment that allows the employee and agency to determine if the employee is suited for the job.

procurement unit Part of the finance branch that establishes contracts for equipment that may be needed during an emergency.

proton A particle in the nucleus of an atom that carries a positive electrical charge.

public safety department A department that employs a person who is triple-certified as a police officer, emergency medical technician (Basic or Advanced level), and firefighter.

pump-and-roll operations The capability of an apparatus to pump water while it is in motion.

pyrolysis The chemical decomposition of a substance by heat.

rate of heat release The speed at which a fuel releases its caloric value when burning.

regulations Mandated by law.

residual pressure The pressure that remains in the system when water is being discharged.

resistance The opposition of the flow of an electric current in a conductor component.

riser The largest pipe in a sprinkler system. It is connected directly to the main water supply line and contains the sprinkler alarm valve. It supplies water to the crossmain.

secondary feeder mains Piping that interconnects with the primary feeder main.

sedimentation The accumulation of mud, sand, and other debris inside a water pipe, reducing the inside diameter of the pipe.

self-contained breathing apparatus (SCBA) Wearable respirator supplying a breathable atmosphere that is carried by or generated by the apparatus and is independent of the atmosphere.

shape of the area Topographic shapes, such as canyons, ridges, mountains, and so on.

shear forces Loads that cause the structural element to deform or fracture in a direction parallel to the force, by sliding one section of the element along others.

siphona A large syringe used to deliver a stream of water.

slope The angle of the hillside.

smoke jumpers Firefighters who parachute from airplanes to suppress forest fires in remote locations.

Smokey Bear Created in 1944. The longest running public service campaign in U.S. history for fire prevention.

sole plate The bottom-most component of a wall assembly. The bottom of the wall studs are connected to this component, which in turn is connected to the foundation of the structure.

spalling The process in which parts of a concrete structure break or chip off when the structure is exposed to extreme heat.

span of control The number of units or personnel that a supervisor can effectively manage.

Sparky the Fire Dog Created in 1951. Used for NFPA's Risk Watch and Learn Not to Burn programs and has a special appeal to children's programs.

specific gravity Weight of a substance compared with the weight of an equal volume of water at a given temperature.

specific heat The amount of heat required to raise the temperature of a specified quantity of material by 1°C.

staff operation The administrative activities necessary to accomplish day-to-day operations.

stair climb Applicant must lift a prepared hose bundle from the floor onto the shoulder and climb a specific number of the stairs in a specified time frame.

Applicant must be wearing an air tank and harness (without valves, hose, or mask) during this evolution.

standard operating procedures (SOP) Rules by which an organization or fire department operates for its day-to-day functioning. Usually these procedures are documented in a handbook or related source.

standards A recommended course of action that is not mandated by law.

static electricity heating The heat generated by static electricity arcing.

static pressure Energy that is available to force water through pipes and fittings.

statute A law established by a state or federal legislative body.

steam engine A heat engine that performs mechanical work using steam as its working fluid.

stoichiometric reaction A reaction in which all reactants are present in fixed, definite proportions for the reaction to go to completion. The most violent of all combustion processes occurs when the fuel and oxygen proportions provide complete combustion, leaving no fuel or oxygen residue.

strategy Planning and directing the maneuvering of personnel and equipment into the most advantageous positions prior to actual operations.

subsidence A phenomenon in which the surface of the ground sinks as water is drawn from the underground aquifer faster that it can be replenished.

summons A written document from a judge or a law enforcement officer ordering a code violator to appear in court on a certain date.

surface water Water collected on the ground from a stream, lake, ocean, or wetland.

tactical training Comprehensive training, hands-on learning, and analysis for emergency incidents.

tactics The actual maneuvering of personnel and equipment in an operation to achieve short-term objectives.

target areas Areas that are of high value or where there is a potential for a large loss of life.

telescoping boom   Aerial device raised and extended through sections that slide within each other.

tensile force   A force that causes the structural element to lengthen.

tensile stress   Stress in a structure that tends to pull it apart.

time unit   Department that tracks all time sheets for every individual working on the emergency scene throughout the duration of the incident.

top plate   Horizontal part between where the roof and studs finish.

torsional loads   Loads that are placed only on the longitudinal axis of the structural component causing it to twist.

total flooding systems   Extinguishing systems that are designed to completely fill the protected area with extinguishing agent when activated.

training   Obtaining knowledge for the maintenance of skills required for actual tactical tasks performed during an emergency response.

training bureau   A division of an organization or department whose mission is to provide training to firefighters so that they have the knowledge, skills, and abilities to mitigate emergency incidents safely while minimizing the risks to themselves, civilians, and the environment.

tuberculation   The buildup of rust on the inside surface of a metal water pipe, which reduces the inside diameter of the pipe.

Type I combustion   Direct oxidation. Pyrolysis or decomposition of the matter is not required to occur in order to sustain combustion.

Type II combustion   Sequential oxidation. The molecular structure of the substance must be broken down to produce combustible gases, which will react with oxygen in the air.

ultraviolet flame detectors   Fire detection devices that are activated when ultraviolet light waves are detected.

unity of command   The concept that every member of the organization answers to just one supervisor.

vapor density   Weight of a given volume of pure vapor gas compared with the weight of an equal volume of dry air at the same temperature and pressure.

vapor pressure   The pressure exerted on the walls of a closed container by a liquid when the number of molecules escaping from the liquid reaches an equilibrium with the number of molecules returning to the liquid.

ventilation   The systematic removal of pressure, gases, and heat from structure or confined area.

warrants   Orders from a judge allowing fire officials the right to search or seize private property or arrest an individual.

watersheds   Geographic areas around a body of water that have a topography that allows all rainfall in that area to flow into and replenish that body of water.

wildland firefighting apparatus   A fire pumping apparatus with a high ground clearance and off-road capabilities.

Index

A

Ability/agility tests, 37
Absolute temperature, 100
Absolute zero, 99
Accelerators in sprinkler systems, 187–88
Accountability in NIMS, 219
Adjusters, 28
Administration
    IMS, 214
    model codes, 155–56
Administration buildings, 66–67
Adoption of model codes, 155
Advanced Life Support Units (ALS), 77
Advancement, 51–54
Aerial apparatus, 73–76
Aerial ladder platforms, 74
Aerial ladders, 68
Aerial platform apparatus, 75
Agility tests, 37
Aircraft in firefighting, 81–82
Aircraft rescue firefighters (ARFFs), 21–22, 80–81
Air pack, 10
Alarm-indicating device circuits, 190
Alarm systems, 190–92
Alarm valves in sprinkler systems, 181–82
Aldini, 10
Ambulances, 77–78
America Burning, 48
America Burning—Revisited, 12–13
American Fire Sprinkler Association (AFSA), 243
American Insurance Association (AIA), 149–50
American National Standards Institute (ANSI), 243
American Red Cross, 243
American Rescue Dog Association (ARDA), 244
American Water Works Association Standard C502–94, 169
    Standard C503–88, 170
    water distribution standards, 168
Andragogy, 57
Apparatus
    early, 6–10
    fire, 70–76
    in preincident planning, 226–27
    specialized, 76–82
Apparatus Aldini, 10
Apparatus bays, 68–69
Apparatus engineers, 22
Applications, job, 35
Aquifers, 161–62
Arched roofs, 129
Arcing heating, 106–7
Arson units, 24–25, 249
Articulating booms, 75–76
Aspect in topography, 230
Assistant chiefs, 210
Associate of Applied Science degrees, 56
Atkins v. Kansas, 152
Atmospheric pressure, 97
Atoms, 94–95
Augustus, Emperor, 2
Automatic detection systems, 192–96
Automatic sprinkler systems, 176–77
    building codes, 149
    components, 178–85
    types, 185–89
Automatic standpipe systems, 175
Automatic Vehicle Location (AVL), 29
Auxiliary appliances, 231
Auxiliary signaling systems, 191
Avillo, Anthony, 224
Axial loads, 120

B

Baccalaureate programs, 57
Backdraft events, 112
Backflow preventers, 178
Background checks, 40–41

Badges, 11
Balloon-frame construction, 140–41
Baltimore great fire, 149
Bambi Buckets, 82
Barriers, 230
Base-valve dry-barrel hydrants, 169–70
Basic Life Support Units (BLS), 77
Bathroom facilities, 69
Battalion chiefs, 210
Battalions, 209
Bays, apparatus, 68–69
Bilingual ability, 42
Bioterrorism, 30
Board for the Coordination of the Model
  Codes (BCMC), 129
Board of Certified Safety Professionals
  (BCSP), 244
Body language, 39–40
Booms, 75–76
Boots & Coots International Well Control,
  Inc., 29
Boston
  Cocoanut Grove nightclub fire, 148
  colonial era fire, 4
  fire engines, 7
  Great Fire, 8
Braithwaite, John, 7
Branch lines in sprinkler systems, 182–84
Breathing apparatus
  early, 10
  self-contained, 82, 87
British Thermal Units (BTUs), 99
Brunacini, Alan, 212
Bucket brigades, 3–4
Building and Fire Research Laboratory
  (BFRL), 244
Building codes, 148–49. *See also* Fire preven-
  tion codes and ordinances
Building construction, 118
  introduction, 118–19
  loads and forces, 119–22
  preincident planning, 231
  structural components, 122–29
  Type I, 129–34
  Type II, 134–35
  Type III, 134–39
  Type IV, 138–40
  Type V, 140–42
Building Construction and Safety Code, 151
Building Officials and Code Administrators
  International (BOCA), 150–51
Burn buildings, 47
Burning process, 110–12
Burns, Eddie W., Sr., 32, 62
Burns, Jackie, 32

C

Cadets, 18–19
California hydrants, 170
Caloric value, 102
Camara v. Municipal Court, 153
Cameras, thermal imaging, 88–89
Candidate Physical Ability Test (CPAT), 37
Captains, 210
Carbon dioxide extinguishing systems, 198–99
Career opportunities, 17. *See also* Selection
    process
  firefighting, 18–25
  future, 30
  introduction, 17–18
  non-firefighting, 25–28
  private sector, 28–29
  volunteer, 30
Carey, John, 176
Carta Civibus Londoniarum, 152
Ceiling breach and pull tests, 37
Ceiling joists, 127
Central station signaling systems, 192
Certificate Assembly, 60
Certifications
  programs, 56
  requirements, 42
  specialty, 51–53
Chain of command in NIMS, 218
Chain reaction of self-sustained
    combustion, 96
Champion Forest Volunteer Fire
    Department, 115
Charge, electrical, 105
Charles II, 3
Charlotte Fire Department, 54
Check valves, 178
Chemical energy, 101–5
Chemical Transportation Emergency Center
    (CHEMTREC), 244
Chicago
  GPS-based systems, 29
  great fire, 12
  Iroquois theater fire, 148
  World's Columbian Exposition, 149
Chicago Fire Department, 9
Chief's interviews, 38–40
Cincinnati fire engines, 8
Citations, 156
*City of Clinton v. Cedar Rapids and Missouri
    Railroad Company*, 152
Clappers in sprinkler systems, 186–87
Class A fires, 112–13
Class B fires, 113
Class C fires, 113

Class I standpipe systems, 173
Class II standpipe systems, 173–74
Class III standpipe systems, 174
Class K fires, 113
Clothing, 82–83
Coatings, fire-resistive, 132–33
Cocoanut Grove nightclub fire, 148
Codes. *See* Fire prevention codes and ordinances
Cole, Dana, 215
College degrees, 54
Combination fire departments, 208
Combination flame detectors, 196
Combination heat detectors, 194
Combination water transport systems, 163, 165
Combined-agent ARFF vehicles, 80–81
Combustion
   process, 110
   self-sustained, 96
Command sector in IMS, 212–13
Command structure in preincident planning, 235
Communications
   NIMS, 217
   preincident planning, 234–35
Communications operators, 25
Community Emergency Response Team (CERT), 30
Compensation and claims units, 214
Complete combustion reactions, 102
Compressed Air Foam (CAFS) extinguishing systems, 72, 202
Compression, heat of, 109
Compressive stress, 120, 122
Computer-aided dispatch (CAD) systems, 250
Concrete, 122, 132–34
Conduction, 100
Confinement issues, 237
Conflagrations, 2
Constitution of the United States, 152–53
Construction. *See* Building construction
Contingency plans, 230
Contract firefighter opportunities, 29
Control units in signaling systems, 190
Control valves, 166
Convection, 101
Cooking oils fires, 113
Cooley, Thomas, 152
Cooley Doctrine, 152
Cooling by extinguishing systems, 200
Corbels, 135
Corrective action, 156
Cost units in IMS, 214

Couvre feu, 3
Criminal conviction record checks, 41
Crossmains, 182–84
Ctesibius of Alexandria, 6
Customary system, 99

D

Dallas Fire-Rescue Department, 32
Dalmatians, 12
Davila, Martin D., 62–63
Dead-load weight, 126
Decay stage in burning process, 112
Decision making, tactical, 236–38
Decomposition, heat of, 103–4
Degree Assembly, 60
Degrees at a Distance Program (DDP), 57
Deluge sprinkler systems, 188–89
Density, vapor, 97–98
Department of Defense (DOD), 245
Department of Homeland Security (DHS)
   career opportunities, 30
   description, 246–47
Department of Transportation (DOT), 245
Deputy chiefs, 210
Detection systems, automatic, 192–96
Diesel pumps, 10
Dillon, John, 152
Dillon's Rule, 152
Direct pumping water transport systems, 163–64
Discretionary authority, 156
Dispatch centers, 249
Dispatcher careers, 25
Distance education, 57
Distribution grids, 166
Distributor pipes, 166
District chiefs, 210
District organizational units, 209
Diversity, 42
Documentation of inspections, 156
Dogs, 12
Doors
   bay, 68
   exit, 148
Dormitories, 69–70
Double-seated clapper valves, 186
Drafting pits, 47
Dress appropriateness, 39
Drill towers, 47
Drivers, 209–10
Driving records, 42
Dry-barrel hydrants, 169–70
Dry chemical extinguishing systems, 200–202
Dry hydrants, 170–71

Dry-pipe sprinkler systems, 186–87
Dual water distribution models, 168

E

Eccentric loads, 120
Education. *See* Training and higher education
Electrical energy, 105–9
Electrical fires, 113
Electrons, 95
Elevation in topography, 230
Eligibility lists, 41
Emergency incident management systems.
    See Incident Management System (IMS)
Emergency Management Institute (EMI), 245
Emergency medical services (EMS), 14
Emergency Medical Technicians (EMT), 14
Employment. *See* Career opportunities;
    Selection process
EMT Basic status, 41
Encasement, steel, 131–32
Encrustation in pipes, 167
Endothermic reactions, 103
Energized electrical fires, 113
Energy, 99
    chemical, 101–5
    electrical, 105–9
    heat, 99–101
    mechanical, 109–12
Engineering corrections, 149
Enhanced 9–1–1, 249
Environmental Protection Agency
    (EPA), 245
Equipment, 88–90
Ergonomics, 26
Exhausters in sprinkler systems, 187–88
Exit doors, 148
Exothermic reactions, 96
Exposure issues
    hazardous materials, 234
    tactical decision making, 237
External exposures to hazardous
    materials, 234
Extinguishers, 8–10
Extinguishment issues in tactical decision
    making, 237

F

Facilities
    fire department, 66–70
    history, 10–11
Factory Mutual Research Corporation
    (FMRC), 192
Fahrenheit scale, 99
Familia publica, 2

Federal Aviation Administration (FAA)
    ARFF regulations, 21, 81
    description, 246
Federal Emergency Management Agency
    (FEMA), 48, 51
    description, 246
    IMS reimbursement by, 214
Federal Fire Prevention and Control Act, 48
Federal firefighters, 20
Federal support organizations, 245–47
Feeder mains distribution grids, 166
Field exercises, 47
Fields of study, 2
Finance sector in IMS, 214
Fire academies, 18
Fire and Emergency Services Higher
    Education (FESHE)
    description, 246
    purpose, 14
    standards, 58–60
Fire apparatus, 70–76
Fire apparatus engineers, 22
Fire Apparatus Manufacturers' Association
    (FAMA), 244
Fire apparatus operators (FAOs), 209–10
Fire cadets, 18–19
Fire Chief magazine, 248
Fire companies, 209
Fire-cut design, 136
Fire Department Explorers, 48
Fire department training specialists, 28
Fire departments, 207
    facilities, 66–70
    organization, 209
    rank structure, 209–10
    sprinkler system connections, 181
    training by, 48–50
    types, 207–8
Fire dynamics, 93
    atoms and molecules, 94–95
    chemical energy, 101–5
    classification, 112–13
    combustion, 96
    electrical energy, 105–9
    heat energy, 99–101
    introduction, 94
    matter, 95–98
    mechanical energy, 109–12
Fire ecology, 21
Fire Engineering magazine, 248
Fire engines
    early, 7–10
    red, 11
Fire equipment salespersons, 26
Fire extinguisher service technicians, 26

Fire extinguishers, 8–10
Fire flow, 102
Fire ground command (FGC) systems, 211–12
Fire hydrants. *See* Hydrants
Fire inspectors, 28
Fire investigators, 28
Fire lines, 21
Fire load computation, 102
Fire marshals, 26–27
Fire prevention codes and ordinances, 147
    administration, 155–56
    and building codes, 148–49
    enforcement, 28
    introduction, 147–48
    legal authority history, 151–54
    model codes history, 149–51
    types and adoption, 154–55
Fire prevention specialists, 27
Fire Prevention Week, 27
Fire protection engineers (FPEs), 28–29
Fire protection systems, 160–61
    automatic detection systems, 192–96
    automatic sprinkler systems. *See* Automatic sprinkler systems
    fire hydrants, 169–72
    introduction, 161
    signaling systems, 190–92
    special extinguishing systems, 197–202
    standpipe systems, 172–76
    water supplies and distribution systems, 161–68
Fire pumps, steam-powered, 7
Fire-resistive coatings, 132–33
Fire-resistive ratings, 129
Fire science, 2
Fire science technology curriculum, 56
Fire services
    evolution, 2–6
    insignias and traditions, 11–12
Fire stations, 68
Fire strategy, 223–24. *See also* Preincident planning
Fire wardens, 4
Fireboats, 81
Firefighter I level, 42
Firefighter/paramedics, 20
Firefighters
    description, 209
    job overview, 19
    selection process. *See* Selection process
Firefighting aircraft, 81–82
Firefighting Resources of California Organized for Potential Emergencies (FIRESCOPE), 211–12

Firefighting services evolution, 2–6
Firehouse magazine, 248
Firehouses, 11
Firemarks, 4
Fireplugs, 5
FireRescue Magazine, 248
FIRESCOPE, 211–12
First impressions, 38
First Law of Thermodynamics, 99
Fixed foam extinguishing systems, 201
Fixed temperature heat detectors, 193–94
Flame detectors, 195–96
Flammable liquids fires, 113
Flammable metals fires, 113
Flash point, 97
Flashover, 112
Flat-roof construction, 127
Flexibility, steel, 131
Flow pressure in standpipe systems, 175–76
Flow rate
    calculations, 226–27
    hydrant testing, 172
    standpipe systems requirements, 175–76
Foam and water extinguishing systems, 202
Foam extinguishing systems, 201–2
Forces in building construction, 119–22
Ford, Henry, 11
Forest fires, 21
Forest rangers, 20
Foundations, 122–23
Fourth Amendment, 153
Frank v. Maryland, 153
Franklin, Benjamin, 4–5
Frequencies, communication, 235
Friction
    defined, 109
    in standpipe systems, 176
Full-time paid firefighters, 6
Fully developed stage in burning process, 112
Fully paid fire departments, 208
Functional sectors in IMS, 212–14
Fusion, latent heat of, 100

G

Gaseous matter, 97–98
General Services Administration Standard KKK-A-1822, 77
Geographical Information Systems (GIS), 29, 90
Gibbs apparatus, 10
Global Positioning Systems (GPS), 29, 90, 249
Goodfellow Air Force Base, 20
Graff, Frederick, 169
Gravity water transport systems, 163–64

Great Fire of Boston, 8
Great Fire of Chicago, 12
Great Fire of London, 3–4, 6
Greenburg, Morris, 170
Grinnell, Frederick, 176
Ground water, 161–62
Growth stage in burning process, 111–12
Guidelines, 19
Gusset plates, 125–26, 128

H

Halon extinguishing systems, 197–98
Hand-operated pumps, 6
Hand pumpers, 6
Hazard personnel, 14
Hazardous materials, 233–34
    exposures, 234
    protective ensembles, 84
    response and mitigation, 77
Hazardous materials apparatus, 80
Hazardous materials specialists, 24, 248–49
HAZWOPER (Hazardous Waste Operations
    and Emergency Response), 80
Head pressure, 163
Headquarter offices, 66–67
Heat detectors, 193–94
Heat energy, 99–101
    electrical, 106–7
    mechanical, 109–12
Heat of combustion, 102
Heat of compression, 109
Heat of decomposition, 103–4
Heat of reaction, 104–5
Heat of solution, 103–4
Heating
    arcing, 106–7
    sparking, 107–8
    spontaneous, 102–3
    static electricity, 108
Heavy rescue squads, 76
Heavy-timber construction, 138
Helicopters, 82
Helitacks, 21
High-expansion foam extinguishing
    systems, 202
High-pressure carbon dioxide extinguishing
    systems, 198
High-temperature cooking oils fires, 113
Higher education. *See* Training and higher
    education
Hiring conditions in selection process, 41
History of fire service, 1
    current situation, 13–14
    fire service insignias and traditions, 11–12

firefighting apparatus and fire vehicles,
    6–10
firefighting services evolution, 2–6
introduction, 1–2
personal protective equipment and fire
    facilities, 10–11
United States, 12–13
Home rule, 152
Homeland Security, 13
Homeland Security Presidential Directive
    5, 14
Horse-drawn fire wagons, 7
Hoses, early, 7
Hotshot crews, 21
Houston, subsidence in, 162
Humidity, 229
Hydrants, 169–72
    base-valve dry-barrel, 169–70
    dry, 170–71
    first, 5
    maintenance and flow testing, 172
    wet-barrel, 170
Hydraulic ventilation, 238
Hydraulically operated ladders, 74
Hydrophoric materials, 104

I

I-beam construction, 126
I-joists, 126–27
Ignition stage in burning process, 110
Immediately dangerous to life and health
    (IDLH) environments, 87
Incident action plans, 217
Incident command system (ICS), 209
    development, 210–12
    ratings, 215–16
Incident Command System paper, 215–16
Incident effectiveness, 207
Incident Management System (IMS), 212
    administration and finance sector, 214
    functional sectors, 212–14
    National Fire Service Incident Manage-
        ment System Consortium, 215
    National Incident Management System,
        216–19
    National Interagency Incident Manage-
        ment System, 215
    ratings, 215–16
Incident mobilization center locations and
    facilities, 217
Indicating valves in sprinkler systems, 179
Induction, 108–9
Industrial firefighters, 24
Information management in NIMS, 219

Information systems, 250
Information transfer procedures, 235
Infrared flame detectors, 196
Infrared radiation, 101
Infrared thermal imaging cameras, 88–89
Initial attack pumpers, 72
Initiating device circuits in signaling
  systems, 190
Injunctions, 156
Insignias, fire service, 11–12
Inspection model, 154
Inspections
  documentation, 156
  warrantless entry, 152–53
Instructors, 53–54
Insurance adjusters, 28
Insurance companies, 28
Insurance Office, 4
Integrated communications in NIMS, 217
Intelligence management in NIMS, 219
Internal combustion fire engines, 9
Internal exposures to hazardous materials, 234
International Association of Arson
  Investigators (IAAI), 244
International Association of Fire Chiefs
  (IAFC), 244
International Association of Firefighters
  (IAFF), 244
International Building, Fire, and Property
  Maintenance Codes, 150
International Code Council (ICC), 150–51
International Conference of Building Offi-
  cials (ICBO), 150–51
International Fire Marshals Association
  (IFMA), 244
International Fire Service Accreditation
  Congress (IFSAC), 57, 59–60
  description, 244
  training and certification, 60–61
International Fire Service Training Associa-
  tion (IFSTA), 244
Internet courses, 57–58
Interviewing skills, 38–40
Inventors, 29
Ionization detectors, 194
Iroquois theater fire, 148

J

Job performance requirements (JPRs), 54
Job searches, 35
Jobs
  growth estimates, 17–18
  selection process. *See* Selection process
Joists, ceiling, 127

K

K factor coefficient, 184
Katrina, hurricane, 22–23
Kellogg Brown & Root (KBR), 29
Kitchens, 69
Knights of St. John, 12

L

Laboratory testing, 155
Ladders, 68, 74
Language skills, 42
Latent heat, 100
Latent heat of fusion, 100
Latent heat of vaporization, 100
Latter, Alexander, 8
Law of Conservation of Mass and Energy, 99
Learn Not to Burn program, 27
Legal authority for code enforcement,
  151–54
Level A ensembles, 85
Level B ensembles, 85–86
Level C ensembles, 85–86
Level D ensembles, 86
Liaison officers, 213
Lieutenants, 210
Life-Safety Code, 151
Life-safety conditions, 153
Life-safety hazards in preincident planning,
  231–32
Light detectors, 195–96
Light-obscuration smoke detector systems,
  194–95
Light rescue squads, 77
Light-scattering smoke detector
  systems, 196
Lightning, 108–9
Line operation, 210
Line-type rate-of-rise detectors, 194
Liquid matter, 96–97
Liquids fires, 113
Living areas, 69
Load-bearing walls, 123
Loads in building construction, 119–22
Local alarm systems, 190–91
Local application carbon dioxide extinguish-
  ing systems, 198
Local government for code enforcement, 151
Local support organizations, 248
Lockheed Martin, 29
Logistics sector in IMS, 214
London great fire, 3–4, 6
Lote, Thomas, 7
Louis F. Garland Department of Defense
  Fire Academy, 20

Low-pressure carbon dioxide extinguishing systems, 198
Lubricants, hydrants, 172

M

Magna Carta, 152
Maintenance garages, 250
Maintenance of hydrants, 172
Major firefighting ARFF vehicles, 80
Malta, 12
Maltese Cross, 11–12
Management by objectives in NIMS, 217
Manby, George, 8
Mandatory authority, 156
Manipulative training, 47
Manual dry standpipe systems, 175
Manual wet standpipe systems, 175
Mapp v. Ohio, 153
Maps, 249
MAST trousers, 76
Master-coded alarm systems, 191
Mathews, Samuel R. C., 169
Matter, 95–98
Mechanical energy, 109–12
Mechanical ventilation, 238
Medical exams, 40
Membership organizations, 243–45
Memory, steel, 131
Mentoring and peer networking, 48
Merrill, William Henry, 149
Metals fires, 113
Military organizations, 207
Mill construction, 138
Mini-maxi code, 154–55
Minimum codes, 154–55
Minorities, 42
Mobile data terminals (MDTs), 89–90
Mobile water supply apparatus, 73
Mock interviews, 38
Model codes. *See also* Fire prevention codes and ordinances
history, 149–51
types, 155
Molecules, 94–95
Montreal Protocol, 197
Multipurpose dry chemical systems, 200

N

National Aeronautics and Space Administration (NASA), 11
National and international membership organizations, 243–45
National Association of State Fire Marshals (NASFM), 244

National Board of Fire Underwriters (NBFU), 149
National Building Code, 149–50
National Electric Code, 151
National Emergency Training Center (NETC), 246
National Fallen Firefighters Foundation, 245
National Fire Academy (NFA), 14
courses, 51
description, 246
National Fire Academy Higher Education website, 57
National Fire Protection Association (NFPA)
codes, 150
description, 245
exam standards, 40
hydrant testing, 172
manufacturing standards, 66
pipe size recommendations, 167
sprinkler standards, 177
water distribution standards, 168
National Fire Service Incident Management System Consortium (NFSIMS), 215
National Guard, 247
National Incident Management System (NIMS), 14, 207, 211, 216–19
National Institute for Occupational Safety and Health (NIOSH), 246
National Interagency Incident Management System (NIIMS), 211, 215, 245
National Registry Emergency Medical Technicians (NREMT), 14
National Safety Council (NSC), 245
National Transportation Safety Board (NTSB), 246
National Volunteer Fire Council (NVFC), 245
Natural Resource Conservation Service (NRCS), 171
Natural ventilation, 238
Neally's Smoke Excluding Mask, 10
Negligence, 156
Nelson, Bill, 226–27
Nero, Emperor, 2–3
Neutrons, 94
New Amsterdam fire wardens, 4
New York City
fire engines, 7
fire wardens, 4
plug uglies, 5
NFPA 10, 113
NFPA 12, 197–98
NFPA 13, 177, 187
NFPA 14, 173–75
NFPA 17, 200
NFPA 25, 172

NFPA 70, 151, 190
NFPA 71, 192
NFPA 72, 190, 192
NFPA 101, 151, 177
NFPA 220, 129, 138–40, 143
NFPA 251, 129
NFPA 291, 172
NFPA 414, 81
NFPA 1001, 66
NFPA 1035, 27
NFPA 1041, 53
NFPA 1221, 250
NFPA 1402, 67
NFPA 1403, 47–48, 67
NFPA 1404, 87
NFPA 1500, 82, 87, 236
NFPA 1582, 40
NFPA 1901, 72, 77
NFPA 1902, 72
NFPA 1903, 73
NFPA 1904, 76
NFPA 1906, 72
NFPA 1971, 82
NFPA 1975, 82, 86
NFPA 1976, 84
NFPA 1977, 84
NFPA 1981, 87
NFPA 1991, 85
NFPA 1992, 85
NFPA 1994, 85
NFPA 5000, 129, 134, 138, 140, 143, 151
NFPA Journal, 248
Non-firefighting career opportunities, 25–28
Noncoded alarm systems, 190–91
Nonfire-suppression activities, 67
Nonindicating control valves, 166
Nonload-bearing walls, 123
Nonpressurized dry hydrant systems, 170
Normal atmospheric pressure, 97

O

Occupancy
    defined, 119, 224
    preincident planning, 228
Occupation classification, 154
Occupational health and safety specialists, 26
Occupational Safety and Health Administration (OSHA)
    description, 246
    Hazmat training requirements, 80
Office of Emergency Management (OEM), 247
Office of the State Fire Marshal (OSFM), 247
Offices, 70

Oklahoma State University, 57, 60
On-site testing, 155
Online courses, 57–58
Operations sector in IMS, 213
Ordinances, 148
Ordinary combustibles fires, 112–13
Ordinary dry chemical extinguishing systems, 200
Organizational structure, 206
    fire departments, 207–10
    incident command system, 210–12
    introduction, 207
OSHA Code of Federal Regulations 29 CFR 1910.120, 80
OSHA Code of Federal Regulations 29 CFR 1910.134, 87
Outside Stem and Yoke (OS&Y) valves, 179
Overhaul operations, 237–38
Overseas firefighter opportunities, 29
Oxidation, 95

P

Paid firefighting departments, 208
Paid firefighting service, 4, 6
Palace of Electricity, 149
Panic hardware, 148
Paramilitary organizations, 207
Parmelee, Henry, 176
Parmelee Sprinklers, 176–77
Passenger airliner protection, 13
Patent Office fire, 169
Peer networking, 48
Performance-based models, 155
Performance codes, 155
Periodicals, 248
Permit model, 153–54
Personal alert safety systems (PASS), 82, 87–88
Personal history in selection process, 40–41
Personal protective equipment (PPE), 82
    clothing, 82–87
    history, 10–11
    PASS, 87–88
    SCBA, 87
Personnel
    in NIMS, 219
    in preincident planning, 226–27
Philadelphia volunteer fire company, 4
Photoelectric detectors, 194
Physical fitness, 42
Physical requirements, 37
Pictometry, 29
Pier and beam foundations, 123
Pike poles, 37

Pipe size requirements, 167–68
Pitched roofs, 128–29
Plain concrete, 132
Planning, preincident. *See* Preincident
    planning
Planning responsibilities in IMS, 213
Platform framing, 142
Platforms, 68
Plug uglies, 5
Point of strain for steel, 131
Portable fire extinguishers, 8
Portland fire, 149
Post-and-beam construction, 140
Post indicator valves, 179–81
Post-tensioned concrete, 133
Power supplies in signaling systems, 190
Preaction automatic sprinkler systems,
    187–88
Preemployment medical exams, 40
Preincident planning, 223
    apparatus and personnel, 226–27
    auxiliary appliances, 231
    building construction, 231
    command structure, 235
    communication systems, 234–35
    hazardous materials, 233–34
    introduction, 223–24
    life-safety hazards, 231–32
    location and extent of fire, 234
    occupancy, 228
    overview, 224–26
    street conditions, 232
    time of day, 232
    topography, 229–30
    water supply, 226–27
    weather, 228–29
Prerequisites for firefighter position, 35
Prescribed fires, 21
Prescription-based models, 155
Prescriptive codes, 155
Pressure
    automatic sprinkler systems, 186
    head, 163
    pipes, 167
    standpipe systems, 175–76
    vapor, 97
Pressure regulators
    SCBA, 11
    standpipe systems, 176
Pressure surges, 186
Pressurized dry hydrant systems, 170
Prestressed concrete, 133
Primary feeder mains distribution grids, 166
Primary power supplies in signaling systems,
    190

Private sector career opportunities,
    28–29
Pro Board, 62
Probationary periods, 19, 41
Procurement units, 214
Promotions, 51–54
Proprietary signaling systems, 192
Protected construction, 134–35, 138–39
Protective equipment, 82
    clothing, 82–87
    history, 10–11
    PASS, 87–88
    SCBA, 87
Protons, 94
Proximity ensembles, 84
Public information officers (PIOs), 27,
    212–13
Public safety departments, 208
Public safety officers (PSOs), 208
Public water systems, 5
Pump-and-roll operations, 72
Pumpers, 71–72
Pumps
    diesel, 10
    hand-operated, 6
Pyrolysis process, 96, 139

R

Radiation, 101
Radiation shielding extinguishing systems,
    200
Radio shops, 250
Rank structure, 209–10
Rankin scale, 99
Rapid fire progress events, 112
Rapid intervention team (RITs), 236
Rapid intervention vehicles (RIVs), 80
Rate of heat release, 102
Rate-of-rise detectors, 193–94
Rattle Watch, 4
Reactions, 102–5
Reading comprehension tests, 36
Red Cross, 243
Red fire engines, 11
Regulations, 81
Relative humidity, 229
Remote station signaling systems, 191–92
Repair garages, 250
Rescue apparatus, 79
Rescue issues in tactical decision making,
    236–37
Rescue squads, 76–77
Residual pressure in pipes, 167
Resistance heating, 106

Resources, 65
   fire apparatus, 70–76
   fire department facilities, 66–70
   introduction, 66
   NIMS, 217, 219
   personal protective equipment, 82–88
   specialized apparatus, 76–82
   tools and equipment, 88–90
Retard chambers in sprinkler systems, 186–87
Ridge, Tom, 216
Right of entry, 153–54
Risers
   sprinkler systems, 182–84
   standpipe systems, 173
Risk Watch program, 27
Rome, 2–3
Roof configurations, 127–29
Ross, Uncle Joe, 8
Royer, Keith, 226–27
Rural areas in preincident planning, 227

S

Safety officers, 213
Salvage operations, 238
Salvation Army, 245
San Francisco
   earthquake, 12
   residential fire, 231
Sand lime mortar, 137
Search and rescue (SAR) teams, 22–23
Seasonal changes, 231
Secondary backup power supplies in signaling systems, 190
Secondary feeder mains distribution grids, 166
Sedimentation in pipes, 167
See v. The City of Seattle, 153
Selection process, 34
   hiring conditions, 41
   interviewing skills, 38–40
   introduction, 34
   personal history, 40–41
   preparation, 35–38
   self-marketing, 41–42
Self-contained apparatus foam extinguishing systems, 201
Self-contained breathing apparatus (SCBA), 87
   description, 82
   early, 10–11
Self-marketing, 41–42
Self-sustained combustion, 96
Semi-fixed type A foam extinguishing systems, 201

Semi-fixed type B foam extinguishing systems, 202
Semiautomatic dry standpipe systems, 175
Separate water distribution models, 168
Shank, Able, 8
Shape of fire area, 230
Shear forces in concrete, 132
Shut-off valves, 179
SI (Systeme International) system, 99
Signaling systems, 190–92
Single clapper valves, 187
Siphona, 6
Sleeping accommodations, 69–70
Slope, 230
Smoke detectors, 194–95
Smoke jumpers, 21
Smokey Bear, 27
Smothering dry chemical extinguishing systems, 200
Soda-acid extinguishers, 9
Sole plates, 142
Solid matter, 96
Solution, heat of, 103–4
Southern Building Code Congress International (SBCCI), 150–51
Spalling, 133–34
Span of control in NIMS, 217, 219
Sparking heating, 107–8
Sparky the Fire Dog, 27
Speaking trumpets, 6
Special extinguishing systems, 197–202
Special Operations Bureaus, 48–50
Special Task Forces, 247
Specialized apparatus, 76–82
Specialized fire brigades, 208
Specialty Certifications, 51–52
Specific gravity, 98–99
Specific heat, 100
Spontaneous heating, 102–3
Spot-type rate-of-rise detectors, 194
Sprinkler heads, 183–85
Sprinkler systems, 176–77
   building codes, 149
   components, 178–85
   types, 185–89
Squirts, 6
Staff operation, 210
Stair climb tests, 37
Standard Grading Schedule, 149
Standard operating procedures (SOP), 19
Standards
   base-valve dry-barrel hydrants, 169–70
   defined, 66
   exam, 40

higher education, 58–60
  manufacturing, 66
  sprinkler, 177
  water distribution, 168
Standpipe systems, 172–76
State Association of Fire Educators, 247
State Fire Chiefs' Association, 247
State Fire Commission, 247
State Firefighters' Association, 247
State Health Department, 247
State Police, 247
States
  code enforcement, 151
  support organizations, 247
Static electricity heating, 108
Static pressure in pipes, 167
Station work uniforms, 82
Statutes, 154
Steam engines, 7
Steam-powered fire pumps, 7
Steel construction, 130–32
Steel-reinforced concrete, 133–34
Stockton fire department, 143–44
Stoichiometric reactions, 102
Straight-chassis apparatus, 73–74
Strategy, 223–24. *See also* Preincident
    planning
Street conditions, 232
Stresses in building construction, 119–22
Structural components in building construc-
    tion, 122–29
Structural firefighting ensembles,
    82–83
Subsidence, 161–62
Summons, 156
Support for fire emergency services, 242
  federal organizations, 245–47
  individuals, 248–50
  introduction, 242–43
  local organizations, 248
  national and international organizations,
    243–45
  periodicals, 248
  state organizations, 247
Surface water, 161–62

T

Tactical training, 46–47
Tactics, 236
  decision making, 236–38
  introduction, 223–24
Target areas for hydrants, 172
Task Force 1, 22–23
Taylor, John, 177

Technology curriculum, 56
Telescoping booms, 75–76
Temperature, 229
Tensile stress, 120, 130
Tenth Amendment, 152
Tests
  employment, 35–37
  hydrant, 172
  laboratory, 155
Texas Engineering Extension Service
    (TEEX), 22–24
Texas Task Force 1, 22–23
Thermal imaging cameras (TICs), 88–89
Tillermen, 74
Time of day factors, 232
Time units in IMS, 214
Tools, 88–90
Top plates, 142
Topography, 229–30
Torsional loads, 121–22
Total flooding carbon dioxide extinguish-
    ing systems, 198–99
Traditions in fire service, 11–12
Training and higher education, 45
  advancement and promotion, 51–54
  facilities, 67–68
  higher education, 46, 54–58
  IFSAC, 60–61
  introduction, 45–46
  Pro Board accreditation, 62
  requirements, 19
  standards, 58–60
  training, 46–51
Training bureaus, 54
Training instructors, 53–54
Training officers, 28, 54
Transfer of command in NIMS, 217
Transportation Security Administration
    (TSA), 10
Triple certification, 13–14
Trouble signal power supplies in signaling
    systems, 190
Trusses, 123–26
Tuberculation in pipes, 167
Turnover rate, 14
Two-in/two-out rule, 236
Type I ambulances, 77
Type I combustion, 110
Type I construction, 129–34
Type I pumpers, 71
Type II ambulances, 77
Type II combustion, 110
Type II construction, 134–35
Type II pumpers, 72
Type III ambulances, 77

Type III construction, 134–39
Type III pumpers, 72
Type IV construction, 138–40
Type V construction, 140–42

U

Ultraviolet flame detectors, 196
Undergraduate college degrees, 54
Underwriters Electrical Bureau, 149
Underwriters Laboratories (UL), 149
   central station signaling systems, 192
   description, 245
   fire hydrants, 170
   water distribution standards, 168
Unified command in NIMS, 218–19
Uniform Building Code (UBC), 150
Union Volunteer Fire Company of Philadel-
   phia, 4
United States, fire danger views in, 12–13
United States Fire Administration (USFA)
   description, 247
   education programs, 48, 51
   fire danger report, 12
   fire department count, 207
United States Forest Service (USFS), 247
United States Supreme Court decisions,
   152–53
Unity of command in NIMS, 218–19
Unprotected construction, 134–35, 138–39
"U.S. Experience with Sprinklers", 177

V

Valves
   distribution grids, 166
   sprinkler systems, 178–82, 186–87
van der Heiden, Jan, 7
Vapor density, 97–98
Vapor pressure, 97
Vaporization, latent heat of, 100
Vehicles, early, 6–10
Ventilation issues, 238
Verification of code compliance, 155
Vigiles, 2
Violations documentation, 156

Volunteer career opportunities, 30
Volunteer experience, 42
Volunteer fire departments
   first, 4–5
   overview, 207–8

W

Wall post indicator valves, 179–80
Walls, 123
Warrantless entry, 152–53
Warrants, 153
Water flow control valves, 179
Water gongs, 181–83
Water supplies and distribution systems
   automatic sprinkler systems, 178
   components, 162–66
   pipe size requirements, 167–68
   preincident planning, 226–27
   transport systems, 163
   treatment systems, 163–65
   water sources, 161–62
Watersheds, 162
Weather considerations, 228–29
Weather forecasters, 250
Wet-barrel hydrants, 170
Wet-pipe sprinkler systems, 186
Wildland areas in preincident planning, 227
Wildland firefighting
   apparatus, 72–73
   ensembles, 83–84
   factors, 230
   firefighters, 20–21
William the Conqueror, 3
Wind, 229
Women firefighters, 42
Wood trusses, 123–26
World Trade Center, 12
World's Columbian Exposition, 149
Written notice, 156
Written tests, 35–36

Z

Zone-coded alarm systems, 191
Zone-noncoded alarm systems, 191